Society of Earth Scientists Series

Series Editor

Satish C. Tripathi, Lucknow, India

AF904184

The Society of Earth Scientists Series aims to publish selected conference proceedings, monographs, edited topical books/text books by leading scientists and experts in the field of geophysics, geology, atmospheric and environmental science, meteorology and oceanography as Special Publications of The Society of Earth Scientists. The objective is to highlight recent multidisciplinary scientific research and to strengthen the scientific literature related to Earth Sciences. Quality scientific contributions from all across the Globe are invited for publication under this series. Series Editor: Dr. Satish C. Tripathi.

More information about this series at http://www.springer.com/series/8785

Surya Kant Mani Tripathi

Catalogue of Selected Angiosperm Pollen Grains from Palaeogene and Neogene Sediments of India

The Society of Earth Scientists
2008

Springer

Surya Kant Mani Tripathi
Birbal Sahni Institute of Palaeosciences
Lucknow, India

ISSN 2194-9204 ISSN 2194-9212 (electronic)
Society of Earth Scientists Series
ISBN 978-3-030-42434-3 ISBN 978-3-030-42435-0 (eBook)
https://doi.org/10.1007/978-3-030-42435-0

© The Editor(s) (if applicable) and The Author(s), under exclusive license to Springer Nature Switzerland AG 2020
This work is subject to copyright. All rights are solely and exclusively licensed by the Publisher, whether the whole or part of the material is concerned, specifically the rights of translation, reprinting, reuse of illustrations, recitation, broadcasting, reproduction on microfilms or in any other physical way, and transmission or information storage and retrieval, electronic adaptation, computer software, or by similar or dissimilar methodology now known or hereafter developed.
The use of general descriptive names, registered names, trademarks, service marks, etc. in this publication does not imply, even in the absence of a specific statement, that such names are exempt from the relevant protective laws and regulations and therefore free for general use.
The publisher, the authors and the editors are safe to assume that the advice and information in this book are believed to be true and accurate at the date of publication. Neither the publisher nor the authors or the editors give a warranty, expressed or implied, with respect to the material contained herein or for any errors or omissions that may have been made. The publisher remains neutral with regard to jurisdictional claims in published maps and institutional affiliations.

This Springer imprint is published by the registered company Springer Nature Switzerland AG
The registered company address is: Gewerbestrasse 11, 6330 Cham, Switzerland

Series Editor Foreword

The palynological studies of Indian Palaeogene and Neogene sediments assumed significant importance owing to association with oil reserves. As such extensive studies have been done in the past but the literature is scattered and there was an urgent need to prepare a Catalogue of angiosperm pollen taxa recorded so far from Indian Palaeogene and Neogene sediments. Dr. Surya Kant Mani Tripathi, Senior palynologist has extensively worked in this field and compiled details about 130 genera and 402 species in this catalogue. I am also thankful to Dr. R. K. Saxena, former senior scientist, Birbal Sahni Institute of Paleosciences, Lucknow for critically reviewing the manuscript. Hope this catalouge will be very useful in further research.

The mega-event of 36th International Geological Congress-2020 in India opened new chapter on the Geology of India. On such an occasion Society of Earth Scientist Series by Springer decided to bring out 36th IGC Commemorative Volumes on various recent geological and geophysical studies of India. As such veteran geoscientists were requested to prepare comprehensive account as monographs or edited volumes. I am personally thankful to all the editors and authors for timely submission of high quality manuscripts for inviting interest of global community of geoscientists.

Lucknow, India

Satish C. Tripathi

Preface

Importance of palynological studies is now well established. Vital role of these studies in oil and coal explorations, biostratigraphy, correlation and dating of sedimentary sequences, palaeoenvironmental interpretations, reconstruction of past vegetation, evolution of morphographic characters, etc. is well understood. Because of rich oil reserves, Indian Palaeogene and Neogene sediments received attention of palynologists at various research organizations and universities, which resulted into rapid accumulation of publications. Progressively, it has become very difficult to keep the track of genera and species proposed from time to time. Due to non-availability and poor dissemination of literature, comparison of newly recorded palynofossils is generally hampered and often remains incomplete. Such inherent problem can only be solved if original descriptions of genera and species are available at one place. This catalogue is an attempt towards this direction.

The publication incorporates description, distribution, remarks and text-figures of selected angiosperm pollen taxa proposed from time to time from Palaeogene and Neogene sediments of India. Taxa that have been included in this publication are widely and frequently recorded from Indian Palaeogene and Neogene sediments. A brief discussion on migration of some pollen taxa has also been elucidated. Geological ages are taken from original publications. Geographical names have been changed as per their present usage, e.g. Varkalla and Warkalli are changed to Varkala, Cutch and Kachchh to Kutch, Cannanore to Kannur, Trichinopoly to Tiruchirapalli, Madras State to Tamil Nadu, Dharamsala to Dharmsala, Bombay to Mumbai, Kundara to Kundra, Sirmur to Sirmaur, Simla to Shimla, Bhaunagar to Bhavnagar, part of Punjab to Himachal Pradesh and part of Assam to Meghalaya.

Lucknow, India

Surya Kant Mani Tripathi

Acknowledgements

I am grateful to the authorities at Department of Science and Technology, New Delhi for granting me the project under USERS Scheme, which enabled me to write this catalogue. Author is thankful to the Director, Birbal Sahni Institute of Palaeosciences, Lucknow for providing me the infrastructural facilities at the Institute. I am also thankful to Mr. Pawan Kumar, Technical Assistant 'D' at BSIP, Lucknow for making the text-figures of pollen. Sincere thanks are due to Ms. Saumyata Tripathi for making the text-figures of pollen, arranging the figures in the form of Plates and for making computer-generated maps used in the catalogue. Author is thankful to Dr. R. K. Saxena, Ex-Scientist 'F', Birbal Sahni Institute of Palaeosciences, Lucknow for the discussions during writing of the catalogue. I am also grateful to him for reviewing the manuscript and constructive suggestions.

Contents

About the Author

Dr. Surya Kant Mani Tripathi obtained Master's Degree in Botany from Allahabad University in 1974 and joined Birbal Sahni Institute of Palaeosciences, Lucknow in November, 1974 as Research Scholar. Very shortly he was selected as one of the members of Scientific Staff of the Institute. He was bestowed with Ph. D. degree in 1983 for his biostratigraphic work on Tertiary pollen from Meghalaya, India. He also completed a Professional Course in Geology from the Department of Geology, Lucknow University, Lucknow in the year 1996.

He carried out research on Microfossils from Indian late Cretaceous and Tertiary sediments from Meghalaya, Arunachal Pradesh, Rajasthan, Gujarat and Andaman and Nicobar Islands. He also carried out collaborative works with scientists of many National/International Institutes and Universities. He visited many European countries and worked at different departments and museums. He published many books, original research papers, monographs, catalogues and popular articles in National and International journals. He guided two Ph.D. students and many students working for post-graduate dissertations. He worked for up-gradation and renovation of Museum at Birbal Sahni Institute of Palaeosciences, Lucknow. Dr. Tripathi carried out studies under two projects sponsored by Department of Science and Technology, New Delhi.

Dr. Tripathi undertook joint research programmes with a scientist of Plant Cell Biology Unit, Szeged University, Szeged, Hungary. These research programmes were conducted under the Exchange of

Scientists Programme between Indian National Science Academy, New Delhi and Hungarian Academy of Sciences, Budapest. Dr. Tripathi received a Diploma, a Commemorative Medal and a Millennium Medal in the years 1996, 1999 and 2001 from Botany Department, J. A. University, Szeged, Hungary.

Chapter 1
Introduction

Palynology has emerged as a significant tool for various applied aspects in the field of Earth sciences in the last few decades. Pollen grains, forming an integral part of palynofossil assemblages, are male gametophytes of plants. After maturity, these are separated from the parent plants and are widely spread in the atmosphere. Due to presence of highly resistant wall, these can easily be preserved in sediments. The wall of spore/pollen is made up of *sporopollenin*, a polymer of carotene and carotenoid esters. Pollen grains possess morphological characters like, shape and size, number, type and position of apertures and exine stratification and ornamentation, which are very important in plant systematics. Fossil pollen and spores are commonly given artificial generic and specific names and are generally classified based on their morphological characters only. If comparable to the modern plants, these are assigned botanical affinity. This system, the *Artificial System of Classification*, is convenient and is easy to apply in stratigraphic studies. Like extant plants, nomenclature of fossil spore/pollen is also governed by the *International Code of Nomenclature for algae, fungi and plants* (Turland et al. 2018).

Last few decades witnessed enormous growth in palynological studies. It resulted into stupendous increase in the number of form genera and species established to describe the extremely varied pre Quaternary record of dispersed spores and pollen grains. Potonié (1956, 1958, 1960, 1966, 1970) published '*Synopsis der Gattungen der Sporae dispersae*' in five parts and provided diagnosis of each genus and its type species. After these publications further compilation of any such work could not eventuate due to continuous publication of new genera or species names, quite often in journals having limited distribution. In 1976, J. Jansonius and L. V. Hills initiated a big project and they published *Genera File of Fossil Spores*. In this publication, the description and taxonomic details of pollen/spore genera are given and these are arranged in alphabetical order on separate cards. It is a very valuable publication and its practical use in research can easily be estimated.

Indian workers also made efforts to collate and describe Indian angiosperm genera (Thanikaimoni et al. 1984; Venkatachala et al. 1989, 1996). Thanikaimoni et al. (1984) restudied and re-photographed some angiosperm pollen and attempted

© The Editor(s) (if applicable) and The Author(s), under exclusive license to Springer Nature Switzerland AG 2020

S. K. M. Tripathi, *Catalogue of Selected Angiosperm Pollen Grains from Palaeogene and Neogene Sediments of India*, Society of Earth Scientists Series,
https://doi.org/10.1007/978-3-030-42435-0_1

to establish relationship between Indian and African pollen. These authors also provided affinity of these pollen grains with angiosperm genera or family. Venkatachala et al. (1989) also worked on similar lines but restricted their studies to Palaeocene-Eocene marker pollen from India. Venkatachala et al. (1996) made a critical morphological assessment of Indian Palaeogene and Neogene angiosperm pollen whose type specimens were available at Birbal Sahni Institute of Palaeosciences, Lucknow. These studies were undertaken with a view to make the discussed genera more useful in biostratigraphy and age determination. All these works are up to generic level only. No such effort has been made so far where all Indian angiosperm pollen species are restudied, figured and described along with all taxonomic details and occurrences. The present project was undertaken to provide description and occurrences of selected Tertiary angiosperm pollen. In all 130 genera and 402 species have been chosen for inclusion. It is expected that the publication will help the palynologists.

Chief objective of this book is to provide taxonomic details of fossil angiosperm pollen having wide range of geological as well as geographical records. Genera having unique morphological characters have also been included. The catalogue also provides a summarized account of palynofloras along with their biostratigraphic and palaeoclimatic applications in various regions of India. The palaeogeographical implications of some genera are also discussed. Common morphologic terms used to describe angiosperm pollen have also been given. A brief discussion on some groups of pollen genera having affinity with the family Arecaceae and Dipterocarpaceae and polycolpate pollen is added.

1.1 Organization of the Book

The publication is broadly divided into three chapters: Introduction, Taxonomic details and Discussion. The introductory part provides brief information about pollen, general pollen morphology, historical background of Palaeogene-Neogene palynology in India and an overview of these sediments. In the second chapter a complete list of genera (according to the nature of apertures they possess) described in the catalogue is given. In this part, a key is given which will be helpful in identification of Indian angiosperm pollen genera. The second chapter deals with description of stratigraphically and ecologically significant angiosperm pollen genera and species from India, their illustration and records in India along with horizon, locality and age as per given by the original author(s). Descriptions of genera and species under them are in alphabetical order. Along with each generic name, original/validating author and year of publication is mentioned. In further lines the information available are: direct bibliographical reference to the page on which the name is first proposed or validated, name of the type species, relevant bibliographical information on the type species, generic description and remarks on the nomenclatural status of the taxa, if required. Text-figures of different species are grouped together in the form of Plates at the end of text. To facilitate comparison of different species, placement of taxa in Plates is done according to the aperture(s) they possess. The Discussion part mainly deals with summarized accounts of palynofloras from different Palaeogene and Neogene sediments of India, their biostratigraphic application in various regions and their palaeogeographical and palaeoclimatic implications.

1.2 Historical Background of Palaeogene and Neogene Palynology in India

Professor Birbal Sahni realized the importance of plant microfossils in the study of various stratigraphical problems and initiated palaeopalynological researches in India as early as 1939. This work was taken up in Assam Basin under the oil exploration project sponsored by Burmah Oil Company. A short, yet significant, report containing the results of this project was published (Sahni et al. 1947). Ghosh (1941) published first palynological report on Palaeogene-Neogene sediments from Assam Basin of northeastern India and recorded single bisaccate pollen. Thereafter, Sen (1948) and Meyer (1958) published stray reports of spore/pollen assemblages. Later, detailed biostratigraphical studies were carried out on Palaeogene-Neogene stratigraphic sequences of northeastern India (Biswas 1962; Baksi 1962; Sah and Dutta 1966, 1968; Dutta and Sah 1970). Early palynological reports were confined to morphotaxonomic enumerations only, but in later years, especially after 1960, serious efforts were made to utilize the palynological data in biostratigraphy and correlation of strata in various sedimentary basins.

At Birbal Sahni Institute of Palaeobotany, Lucknow studies in palaeopalynology were taken up in the right earnest in early fifties and since then researches on various surface/subsurface sedimentary rocks covering all parts of India are being carried out. Other centres where palynological studies have been pursued vigorously are different laboratories of Oil and Natural Gas Corporation Limited, Geological Survey of India, and some other organizations including the universities.

1.3 Pollen Morphology

Angiosperm pollen grains arise from Pollen Mother Cell (PMC) inside the anther. Generally, each PMC produces four pollen grains in the form of a tetrad. After separation, each pollen grain gets free and is ready to perform the reproductive functions. Unable to reach the suitable destination, cytoplasm of pollen perish but the outer layer, the exine, being very resistant to destruction can remain intact for a very long time. The exine is constituted by *sporopollenin*, an oxidative polymer of carotenes and carotenoid esters. The exine is provided with characteristic features important for identification and tracing its botanical affinity. Various workers proposed different terminologies from time to time to describe pollen grains (Wodehouse 1935; Erdtman 1943, 1947, 1958, 1972; Kremp 1965; Faegeri and Iverson 1989). An attempt has been made to discuss these terms in a simplified manner. Pollen characters are described under the categories *Polarity, Symmetry, Shape, Size, Aperture* and *Sporoderm* (spore/pollen wall).

Polarity In a tetrad, part of pollen close to the centre (attachment area) is known as the *proximal* side and the part away from it is the *distal* side. The line between proximal and distal poles is the axis of symmetry and the middle plane perpendicular to it is the *equatorial plane*. Surface features perpendicular to equatorial plane are known as *meridional.* Pollen with similar proximal and distal faces are *isopolar* and those with dissimilar faces are *heteropolar* or *anisopolar.*

Symmetry and Shape Pollen grains divisible into two equal halves are *symmetrical* (radial or bilateral) but those with no plane of symmetry are a*symmetrical.* The shape of radiosymmetric isopolar pollen is determined by the ratio between polar axis (P) and equatorial axis (E). Based on P/E ratios different shapes have been identified (peroblate, oblate, sub oblate, spheroidal, subprolate, prolate and perprolate).

Aperture Pollen may be *aperturate* or *inaperturate*. Most of the pollen grains possess apertures hence, are *aperturate* but some do not have apparent apertures and are *inaperturate*. Aperture is an opening or a thin area in the exine. Pollen tube comes out from the aperture at the time of germination. Another function of the aperture is to regulate the increase or decrease of size of pollen due to desiccation or retention of moisture. The process is known as *harmomegathy. Inaperturate* pollen grains do not exhibit apparent area for the exit of pollen tube. Such pollen grains are called *omniaperurate* and the tube may emerge from any place of the pollen wall. *Monoaperturate* pollen has single aperture and may be *monosulcate* or *monoporate.* Monoaperturate pollen are generally produced by monocotyledons. Pollen having more than two apertures are produced by dicotyledons. Depending upon number of apertures, pollen grains may be di- to polyaperturate. Position of apertures may be equatorial, meridional or global. Apertures may be *simple* or *composite*. In simple aperture, there is no differentiation between outer and inner units. In composite aperture, outer unit is termed as *ectoaperture* and is located in the outer wall of exine, the *sexine*. The inner aperture is called as *endoaperture* and is situated in the inner wall of exine, the nexine.

Nature, number and position of apertures are very significant in identification and description of dispersed pollen. Distally placed simple aperture having length/breadth ratio less than two (more or less isodiametric) is named as *ulcus* but that having a length/breadth ratio more than two (furrow-like) is called *sulcus.* Meridionally placed furrow-like aperture is known as *colpus*. A *pore* is more or less rounded, simple and equatorially placed. Globally distributed furrow-like apertures are called as *rugae* but those having more or less rounded shape are known as *foramina*. Pollen having rugae are called as *rugate* or *pericolpate* or *pantocolpate*.

Pollen with globally distributed pores are known as *foraminate*, *periporate*, or *pantoporate*. Pollen grains having meridionally placed furrow-like aperture with a rounded pore in the middle is *colporate* while those having rounded composite aperture are known as *pororate*.

Transversely elongated apertures are called *lalongate* whereas, the longitudinally elongated ones are known as *lolongate*. Pollen with short colpi are known as *brevicolpate* and those with long colpi are called as *longicolpate*. Pollen having more or less rounded ectoaperture as well as the endoaperture are known as *brevissimicolpate*. Pollen grains with apertures at the angles are called as *angulaperturate* while those having these at the middle of angles are known as *planaperturate*. If contours between angles are concave, pollen are *sinuaperturate* and if it is lobed the pollen are known as *fossaperturate*. Pollen are termed as *syncolpate* if colpi anastomose near the poles but are known as *parasyncolpate* if the anastomosing colpi leave a triangular area at the apocolpial region. Pollen having elongated spirally arranged apertures are known as *spiraperturate*.

Features Associated with Apertures Apertures are usually covered with a membrane, the *aperture membrane*. It may be smooth or provided with small granules. An *operculum* is formed if the membrane is thick at the central part. Often thickenings are present around the aperture. If present around the pore, it is called as *annulus* and if present around the colpus, it is known as a *margo*. Pollen with apertures on small dome-like protrusions are called *aspidate*. A space between ectoaperture and endoaperture is called *atrium* and a lateral space between two apertures is termed as *vestibulum*. A band-like thickening in the exine running from one pore apparatus to the other one is known as *arcus*.

Pollen Wall Failing to reach the proper destination, the cytoplasm and intine of pollen are destroyed but exine, being resistant to destruction, remains intact for a very long time. It can be fossilized easily. Exine is divided into two layers: outer *sexine* and inner, *nexine*. Sexine is generally sculptured and nexine is amorphous exhibiting no visible features. Beneath exine, a layer is present which is known as *intine*. Intine is composed of cellulose and pectic substances and can be destroyed easily. The sexine is divisible into two layers: *ectosexine* and *endosexine*. The ectosexine is separated from nexine by endosexine. This layer may be columellate, granulate or alveolar. Endosexinous rod-like structures standing over nexine are called *columellae*. The columellae can support a complete layer known as *tectum* and pollen having tectum are called *tectate*. Pollen can be *intectate* or *atectate*. In intectate pollen there is no tectum and the columellae are distally free whereas, in atectate pollen exine lacks columellae hence also lacks tectum. Presence of columellae is a salient feature of angiosperm pollen exine but in gymnosperm pollen this layer is generally alveolate.

Exine Ornamentation Structures present on outer surface of pollen wall are termed as *sculpture*. These are positive structures and are present over the tectum. Surface without any sculpture are called as *psilate*. Sculptural elements are named based on their shape and size and exine possessing these may be *scabrate*, *papillate, verrucate, gemmate, spinulate, echinate, baculate, clavate, striate* etc. Tectum is often provided with perforations and is incomplete. Based upon shape, size and arrangement of these perforations different patterns have been identified and exine possessing these features is called *punctate, foveolate, scrobiculate* or *reticulate*. Mesh-like structures form reticulum which is made up of *lumen* (a depressed structure) surrounded by *muri* or ridges (slightly raised portion). The mesh of a reticulum consisting of a lumen and half of the muri is known as *brochus*. The top of a brochus is made up of tectum supported by infratectal rods (columellae). A reticulum with brochi more or less uniform size is called as *homobrochate* whereas that having brochi of different size is *heterobrochate*. Muri supported by single row of columellae are *simplicolumellate* but it is *dupli-* or *triplicolumellate* if supported by two or three rows of columellae. A reticuloid pattern formed by columellae or pila instead of muri is known as retibaculate/retipilate.

Some prefixes are commonly used to describe pollen. These are: *Crassi*—thick, *tenui*—thin, *ecto*—outer, *endo*—inner, *zoni*—used for aperture(s) arranged in equatorial zone, *syn*—fused or united, *peri*-used for apertures distributed all over the body of pollen, *ana*—distal position, *cata*—proximal position, *iso/homo*—similar, *hetero*—dissimilar.

1.4 Palynology and Biostratigraphy

Stratigraphy deals with description, classification and arrangement of rocks in chronological order. Parameters considered for identification of the original sequence of rocks mainly depends upon determination of order of superposition (the order in which the rocks were laid down), lithology and fossil contents. The order of superposition can often be disturbed due to various past geological activities. Three main categories of stratigraphic units (lithostratigraphic, biostratigraphic and chronostratigraphic) have been identified. *Lithostratigraphic units*, completely based on lithological characters identifiable in the field, are amongst the most practical units employed in systematic geological works. *Biostratigraphic units* (zone, subzone, zonule) are body of strata characterized by the qualitative and quantitative representation of its fossil contents contemporaneous with the deposition of strata (both in situ and transported). A zone is the basic unit in biostratigraphic classification and it is independent of lithostratigraphic unit. A biostratigraphic unit is a recognizable body of strata based on presence or abundance of fossil(s). These are generally named after characteristic fossil(s) found in that particular unit. Chiefly three types of biostratigraphic zones have been identified. These are Assemblage-zone, Range-zone and Acme-zone.

Assemblage-Zone It is a body of strata characterized by specific group of taxa constituting an assemblage. It should have a specifically designated and delimited type section or reference section. Taxa characteristic of a particular assemblage zone should have continued vertical as well as lateral occurrence to a reasonable extent. Characteristic fossil(s) of an assemblage-zone are known as *Guide fossils* or *Index fossils*.

Range-Zone or Acrozone It is a body of strata representing the horizontal and vertical range of occurrence of a particular taxon. Such zone is named after the taxon whose range of occurrence limits a zone. A range-zone should have a type section.

Acme-Zone (synonyms—Peak-zone, Flood-zone, Epibole, Abundance-zone) It is a body of strata representing abundance of some taxa or a taxon but not its total range. It is named after the taxon whose maximum occurrence is identified within the zone.

1.5 Overview of Indian Palaeogene and Neogene Sediments

These sediments are reported from northern (Lesser as well as Tethys Himalayas), eastern, northeastern, southern and western parts of India (Fig. 1.1). Early Palaeogene rocks of Lesser Himalaya are marine in nature but those of the later part and Neogene are estuarine or fresh-water deposits. Collectively, these sediments are known as Subathu-Murree-Siwalik succession and range in age from Eocene to Pliocene. In eastern India, these sediments are found in Andaman and Nicobar Islands and as subsurface sediments in Bengal Basin. These sediments range in age from Palaeocene to Pliocene. Extensive Palaeogene-Neogene sediments are exposed in northeastern part of India covering Assam, Meghalaya, Arunachal Pradesh and adjoining states. In south, these rocks occur along the eastern coastal margin of Cauvery Basin and Malabar and Coromandal coasts on the western side. In western India, sediments of these ages are very well developed in Rajasthan and Gujarat states and are known for having many workable lignite seams. Complete sequences of Palaeocene to Pliocene sediments are developed in western and southern parts of the Kutch Basin, Gujarat. In Rajasthan and Cambay basins, these sediments are largely covered with alluvium. In Rajasthan, these rocks range in age from Palaeocene to Early Eocene whereas; those of Cambay Basin are of Palaeocene to Pliocene age. A brief account of Palaeogene-Neogene exposures in different parts of India from where palynological information is available is given below.

Northern India Extensive Palaeogene-Neogene deposits are present throughout the Lesser Himalayan belt covering the states of Jammu and Kashmir, Punjab, Haryana, Himachal Pradesh, Uttarakhand and Uttar Pradesh. Subathu Formation represents oldest Palaeogene succession in Lesser Himalaya of Himachal Pradesh. These are shallow marine to brackish-water deposits. These sediments have yielded rich palynological assemblages including dinoflagellate cysts. Subathu Formation is

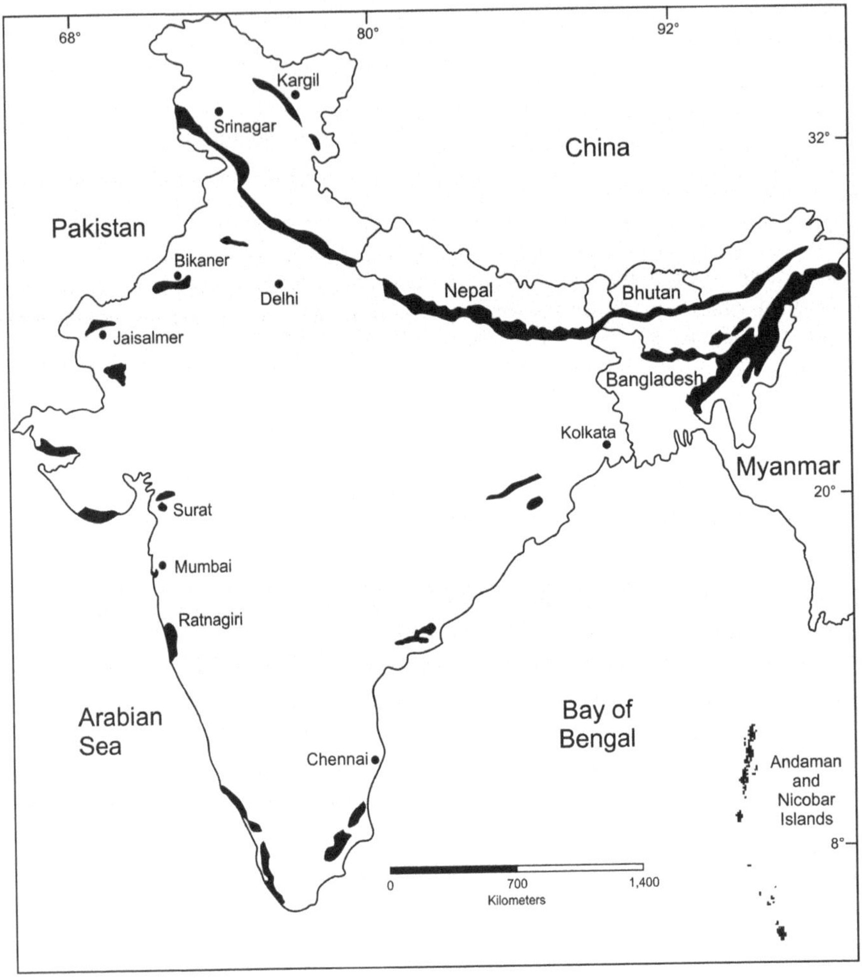

Fig. 1.1 Map of India showing distribution of Tertiary exposures. Modified after Saxena and Tripathi (2012)

overlain by Dagshai and Kasauli formations, which are Oligocene to Early Miocene in age. Thick Neogene successions, named as Siwalik Group, are exposed in southern part of Himalayan ranges. Palynological assemblages from this group are poor although, many reports have been published.

Northeastern India Palaeocene to Pliocene deposits are extensively developed in Assam, Meghalaya and adjoining states. In these areas, sediments were deposited over Precambrian basement and Sylhet Traps in different facies of shallow marine regime. Due to facies variations, these sediments are differently named in various parts of northeastern region. Palaeogene-Neogene sediments are well developed at the southern fringes of Shillong Plateau. Here Palaeocene-Eocene rocks are divided into Therria, Lakadong, Umlatodoh, Prang and Kopili formations. Oligocene sediments, named as Barail Group, are divided into Laisong, Jenam and Renji formations. Rocks of this group are overlain by Neogene sediments and are divided into Surma and Tipam groups. The Surma Group is divisible into Bhuban and Bokabil formations whereas, Tipam Group is divided into Tipam Sandstone and Girujan Clay formations. Tertiary sedimentary sequences of Assam and Meghalaya have been extensively worked out palynologically resulting into numerous publications.

In the extreme northeastern part of India, Palaeogene-Neogene sediments are exposed in Arunachal Pradesh. These are represented by Yinkiang Group and Siwalik Supergroup. The Yinkiang Group, ranging in age from Late Palaeocene to Early Eocene, is divided into Geku and Dalbuing formations. Neogene sediments in Arunachal Pradesh are represented by Tipam, Girujan and Namsang formations. Few palynological reports are available from Arunachal Himalayas.

Southern India (Kerala, Karnataka and Tamil Nadu) Based on palynological studies, subsurface Palaeocene-Eocene sediments have been identified in Ambalapuzha, Alappuzha District, Kerala. Miocene rocks in Kollam and Varkala are divided into Quilon beds and Warkalli Beds. Rich palynological assemblages are described from these beds. Palynofossils indicating Miocene age have been described from sediments of the west coast of Karnataka and are exposed along Kullur-Kavur Road near Mangaluru. Miocene sediments are extensively spread over in Tamil Nadu. These sediments are named as Cuddalore Formation. Neyveli, located in Cuddalore, Tamil Nadu bears huge deposits of lignite. Lignite samples have been thoroughly investigated palynologically and many papers have been published.

Western India The Western part of India shows presence of Palaeogene-Neogene rocks in Gujarat, Rajasthan and Konkan Coast (southern Maharashtra and Goa). Palaeocene sediments in Gujarat are represented by Vagadkhol Formation, which is devoid of fossils. It unconformably overlies the Deccan Traps. Sediments equivalent to this formation in Kutch, Gujarat are named as Matanomadh Formation (= Madh Series). Plant fossil records and spore/pollen assemblages show that these sediments are of Palaeocene age. This is overlain by Naredi, Harudi and Fulra Limestone formations (= Berwali Series), which range in age from Early- Middle Eocene. In the Saurashtra region of Gujarat, Oligocene rocks are found in subsurface and are named as Tarkeshwar Formation. Sediments of this formation have yielded rich palynological assemblages. In Kutch Oligocene and Miocene rocks are represented by Maniyara Fort Formation (= Bermoti Series) and Khari Nadi, Chhasra and Sandhan formations (= Khari Series). Numerous publications dealing with palynological studies on Tertiary sediments of Gujarat are available. Along the Konkan Coast, Sindhudurg Formation, containing thin lignite seams, is developed in Ratnagiri and Sindhudurg districts of Maharashtra. This formation is Miocene in age and is palynologically productive.

Palaeogene sequences in Rajasthan are reported from the basins of Jaisalmer, Barmer and Bikaner. The Jaisalmer Basin, represented by Sanu, Khuiala and Bandah formations, is palynologically less productive. In Barmer Basin, Early Palaeogene rocks are named as Akli Formation, which is succeeded by Mata-Ji-Ka Dungar and Kapurdi formations. These sediments range in age from Late Palaeocene to Eocene. Rich palynological assemblages are recorded from the Akli Formation of Barmer Basin.

Deccan Volcanic Province Deccan Traps, extensively spread over an area of 500,000 km^2 in western and central India, cover Precambrian to Cretaceous rocks. Studies suggest that Deccan volcanism was initiated during Maastrichtian and continued till Early Palaeocene. At different levels in Deccan Traps many fossiliferous horizons are found. These are identified as infra-, inter- or supra-trappean beds. In many localities, these rocks have been studied for their faunal and floral contents, especially from Anjar (Kutch, Gujarat), Mohgaon Kalan and Jabalpur (Madhya Pradesh), Rajamahendraverman and Narsapur (Andhra Pradesh) and Mumbai.

Andaman and Nicobar Islands Andaman and Nicobar Islands are part of Sunda Arc System. It extends from Myanmar (Burma) to Sumatra. The main chain of islands is Cretaceous to Oligocene in age and the peripheral chain of islands is Neogene in age. The sediments of main Andaman Islands are deep sea flysch deposits and are classified into Baratang and Port Blair formations. Sediments of Baratang Formation are mainly argillaceous whereas the overlying rocks of Port Blair Formation are arenaceous.

Offshore Basins Palaeogene-Neogene sequences in the offshore basins are developed at the eastern coasts, western coasts and the Andaman and Nicobar Islands. At the eastern coast, the Krishna-Godavari Basin shows the presence of complete sequence of Cenozoic. From the same coast, Cauvery Basin sediments have also been palynologically worked out. These sediments range in age from Palaeocene to Miocene. At the west coast, almost continuous sequences of sediments occur in Konkan-Kerala Shelf. Bombay Offshore Basin has been thoroughly investigated because of the Oil Exploration Programmes. Almost continuous sequence from Palaeocene to Pleistocene has been noticed in this basin. Palaeogene-Neogene sequences in offshore basins are well developed in Kutch and Saurashtra.

References

Baksi SK (1962) Palynological investigation of Simsang River Tertiaries, South Shillong Front, Assam. Bull Geol Mining Metall Soc India 26:1–22

Banerjee D (1964) A note on the microflora from Surma (Miocene) of Garo Hills, Assam. Bull Geol Mining Metal Soc India 29:1–8

Biswas B (1962) Stratigraphy of the Mahadeo, Langpar, Cherra and Tura formations, Assam, India. Bull Geol Mining Metall Soc India 25:1–48

Dutta SK, Sah SCD (1970) Palynostratigraphy of the Tertiary sedimentary formations of Assam-5. Stratigraphy and palynology of South Shillong Plateau. Palaeontographica. Abteilung B 131 (1–4):1–72

Erdtman G (1943) An introduction to pollen analysis. Chronica Botanica, Waltham, USA

Erdtman G (1947) Suggestions for the classification of Fossil and recent pollen grains and spores. Svensk Botanisk Tidskrift Utgifven af Svenska Botaniska Foreningen. Stockholm 41(1):104–114

Erdtman G (1958) Towards terminological unification in pollen and spore morphology. Grana Palynologica (N.S.) 3:3–5

Erdtman G (1972). Pollen morphology and plant taxonomy—Angiosperms: an introduction to palynology, vol 1. Hafner Publishing Company, New York

Faegeri K, Iverson J (1989) Text book of pollen analysis, 3rd edn. Hafner Publishing Company, New York

Ghosh AK (1941) Fossil pollen in the Tertiary rocks of Assam. Sci Culture 6:674

Kremp GOW (1965) Morphologic encyclopedia of palynology. The University of Arizona Press, Tucson

Meyer BL (1958) Palynological investigation of some samples from Nahorkatiya, Assam. J Palaeontol Soc India 3:156–157

Potonié R (1956) Synopsis der Gattungen der Sporae dispersae I. Teil: Sporites. Beihefte zum Geologischen Jahrbuch 23:1–103

Potonié R (1958) Synopsis der Gattungen der Sporae dispersae II. Teil. Sporites (Nachtrage), Saccites, Aletes, Praecolapates, Polyplicates, Monocolpates. Beihefte zum Geologischen Jahrbuch 31:1–114

Potonié R (1960) Synopsis der Gattungen der sporae dispersae III. Teil: Nachtrage sporites, Fortsetzung Pollenites mit generalregister zu Teil 1-III. Beihefte zum Geologischen Jahrbuch 39:1–189

Potonié R (1966) Synopsis der Gattungen der sporae dispersae IV Teil: Nachtrage zu allen Gruppen (Turmae). Beihefte zum Geologischen Jahrbuch 72:1–244

Potonié R (1970) Synopsis der Gattungen der sporae dispersae. V Teil. Nachtrage zu allen gruppen (Turmae). Beihefte zum Geologischen Jahrbuch 87:1–172

Sah SCD, Dutta SK (1966) Palynostratigraphy of the sedimentary formations of Assam-1. Stratigraphical position of the Cherra Formation. Palaeobotanist 15(1–2):72–86

Sah SCD, Dutta SK (1968) Palynostratigraphy of the Tertiary sedimentary formations of Assam-2. Stratigraphic significance of spores and pollen in the Tertiary succession of Assam. The Palaeobotanist 16:177–195

Sahni B, Sitholey RV, Puri GS (1947) Palaeobotany in India—6. Correlation of Tertiary succession in Assam by means of microfossils. J Indian Bot Soc 26:261–263

Saxena RK, Tripathi SKM (2012) Seven decades of Indian spore-pollen floras: a compendium. AASP Contrib Ser 45:1–182. AASP Foundation, Dallas, TX, USA

Sen J (1948) Microfossils of Assam coal-fields-1. The coal seam at Laitryngew and the age of the Cherra Sandstone. Bull Bot Soc Bengal 2(2): 1–11

Thanikaimoni G, Caratini C, Venkatachala BS, Ramanujam CGK, Kar RK (1984) Selected Tertiary angiospermous pollen from India and their relationship with African Tertiary pollen, vol 19. Institut Francais de Pondichéry, Travaux de la Section Scientfique et Technique 92p

Turland NJ, Wiersema JH, Barrie FR, Greuter W, Hawksworth DL, Herendeen PS, Knapp S, Kusber W-H, Li D-Z, Marhold K, May TW, McNeill J, Monro AM, Prado J, Price MJ, Smith GF (eds) (2018). International code of nomenclature for algae, fungi, and plants (Shenzhen code). In: Adopted by the nineteenth international botanical congress Shenzhen, China, July 2017. Regnum Vegetabile, vol 159. Koeltz Botanical Books, Glashütten. https://doi.org/10.12705/Code.2018

Venkatachala BS, Caratini C, Tissot C, Kar RK (1989) Palaeocene-Eocene marker pollen from India and tropical Africa. Palaeobotanist 37(1):1–25

Venkatachala BS, Saxena RK, Singh HP, Kar RK, Tripathi SKM, Kumar M, Sarkar S, Mandal J, Rao MR, Singh RS, Mandaokar BD, Ambwani K (1996) Indian Tertiary angiosperm pollen: a critical assessment. Palaeobotanist 43(2):106–138

Wodehouse RP (1935) Pollen grains—Their structure, identification and significance in science and medicine. McGraw Hills, New York

Chapter 2
Groups of Pollen Genera Described

Inaperturate

Assamiapollenites Singh 1975
Incrotonipollis Jansonius & Hills 1981
Retipilonapites Ramanujam 1966 ex Potonié 1970

Monocolpate/Monosulcate

Arecipites Wodehouse 1933
Arengapollenites Kar 1985
Clavapalmaedites Rao & Ramanujam 1978
Crotonisulcites Rao & Ramanujam 1978
Kapurdipollenites Tripathi 1994
Liliacidites Couper 1953
Longapertites von Hoeken-Klinkenberg 1964
Matanomadhiasulcites Kar 1985
Neocouperipollis (Venkatachala and Kar 1969) Kar & Kumar 1987
Nymphaeoipollis Venkatachala & Kar 1969
Monocolpopollenites Pflug & Thomson in Thomson & Pflug 1953
Palmidites Couper 1953
Paravuripollis Rao & Ramanujam 1978
Pinjoriapollis Saxena & Singh 1981
Proxapertites van der Hammen 1956
Quilonipollenites Rao & Ramanujam 1978
Retipollenites Gonzalez Guzman 1967
Retiverrumonosulcites Tripathi 1994
Spinizonocolpites Muller 1968
Spinomonosulcites Singh & Misra 1991c
Tuberculozonisulcites Kar & Sharma 2001

© The Editor(s) (if applicable) and The Author(s), under exclusive license to Springer Nature Switzerland AG 2020

S. K. M. Tripathi, *Catalogue of Selected Angiosperm Pollen Grains from Palaeogene and Neogene Sediments of India*, Society of Earth Scientists Series, https://doi.org/10.1007/978-3-030-42435-0_2

Dicolpate/Docolporate

Dicolpopollis Pflanzl 1956 emend. Potonié 1966
Umbelliferoipollenites Venkatachala & Kar 1969

Tricolpate

Albertipollenites Srivastava 1969
Bacubrevitricolpites Rao & Ramanujam 1982
Bithnokiapollis Kar & Sharma 2001
Cruciferoipollenites Navale & Misra 1979
Dipterocarpuspollenites Kar 1992
Foveotricolpites Pierce 1961
Icacinoipollenites Navale & Misra 1979
Intrareticulites Kar 1985
Ladakhipollenites Mathur & Jain 1980
Loranthipites Rao & Ramanujam 1982
Plumbaginacipites Navale & Misra 1979
Polygonaceaepites Baksi 1962
Rousea Srivastava 1969
Tricolpites Cookson 1947 ex Couper 1953
Warkallipollenites Ramanujam & Rao in Thanikaimoni et al. 1984

Trisyncolpate/Trichotomosulcate

Barringtoniapollenites Kar & Sharma 2001
Clavasyncolpites Rao & Ramanujam 1982
Marginipollis Clarke & Frederiksen 1968
Neotrichotomosulcites Samant & Phadtare 1997
Pilatrisyncolpites Kar et al. 1994 emend. Mandal & Rao 2001
Retitrisyncolpites Mandal et al. 1994
Trisyncolpites Kar 1979

Polycolpate

Ctenolophonidites von Hoeken-Klinkenberg 1966
Ocimumpollenites Kar 1996

Tricolporate/Tricolporoidate

Alangiopollis Krutzsch 1962
Angulocolporites Kar 1985
Araliaceoipollenites Potonié 1951 ex Potonié 1960
Asperitricolporites Venkatachala & Rawat 1972
Bombacacidites Couper 1960
Calophyllumpollenites Sah & Kar 1974
Caprifoliipites Wodehouse 1933
Chiranthodendronpollenites Kar & Sharma 2001
Compositoipollenites Potonié 1951 ex Potonié 1960

Coneopollis Venkatachala & Rawat 1973
Coramandalipollis Venkatachala & Rawat 1973
Cupanieidites Cookson & Pike 1954
Dakshinipollenites Navale & Misra 1979
Densiverrupollenites Tripathi & Singh 1984
Dermatobrevicolporites Kar 1985
Favitricolporites Sah 1967
Hippocrateaceaedites Ramanujam 1966
Intectocolporites Mandal & Guleria 2006
Lacrimapollis Venkatachala & Rawat 1973
Lakiapollis Venkatachala & Kar 1969
Lanagiopollis Morley 1982
Margocolporites Ramanujam 1966 ex Srivastava 1969
Meyeripollis Baksi & Venkatachala 1970
Nyssapollenites Thiergart 1937 ex Potonié 1960
Palaeocoprosmadites Ramanujam 1966
Paleoaraliaceaepites Biswas 1962
Paleocaesalpiniaceaepites Biswas 1962
Paleosantalaceaepites Biswas 1962 ex Dutta & Sah 1970
Pellicieroipollis Sah & Kar 1970
Ratariacolporites Kar 1985
Retitrescolpites Sah 1967
Rhoipites Wodehouse 1933
Sastriipollenites Venkatachala & Kar 1969
Spinulotricolporites Saxena & Misra 1987
Striacolporites Sah & Kar 1970
Symplocoipollenites Potonié 1951 ex Potonié 1960
Talisiipites Wodehouse 1933
Tinalipollenites Mandal & Kumar 2000
Tricolporocolumellites Kar 1985
Tricolporopilites (Kar 1985)
Tricolporopollis Dutta & Sah 1970
Verrucolporites Sah & Kar 1970

Tetra-/Penta- to Polycolporate

Iugopollis Venkatachala & Rawat 1972
Meliapollis Sah & Kar 1970
Padappakkarapollis Rao & Ramanujam 1982
Polybrevicolporites Venkatachala & Kar 1969
Polycolporites Mehrotra 1983
Polygalacidites Sah & Dutta 1966
Prectenolophonidites Samant & Phadtare 1997

Monoporate/Monoaperturate

Dracaenoipollis Sah & Kar 1970
Echimonoporopollis Saxena et al. 1991
Graminidites Cookson 1947 ex Potonié 1960
Jacobipollenites Ramanujam 1966
Nymphaeacidites Sah 1967

Diporate/Diporocolpate

Clavadiporopollenites Ambwani & Singh 1996
Echidiporites Mandal 1997
Piladiporocolpites Kar 1995
Psiladiporocolpites Kar 1995
Psilodiporites Varma & Rawat 1963
Retidiporocolpites Kar 1995

Triporate

Acanthotricolpites Kar 1985
Dorreenipites Biswas 1962
Gemmatriporopollis Saxena & Khare 1996
Grevilloideaepites Biswas 1962
Myricipites Wodehouse 1933
Ornatriporites Rao & Ramanujam 1982
Proteacidites Cookson 1950 ex Couper 1953
Sonneratioipollis Venkatachala & Kar 1969
Triorites Erdtman 1947 ex Cookson 1950
Triporopollenites Pflug & Thomson in Thomson & Pflug 1953
Varispinitriporites Kar 1985

Panporate/Polyporate

Clavaperiporites Ramanujam 1966
Cryptopolyporites Venkatachala & Kar 1969
Malvacearumpollis Nagy 1962
Polygonacidites Sah & Dutta 1968
Polypornia Naumova 1939 ex Potonié 1960
Pseudonothofagidites Venkatachala & Kar 1969
Thymelaepollis Sah & Kar 1970

Pollen in Tetrads/Polyads

Kielmeyerapollenites Sah & Kar 1974
Ornatetradites Rao & Ramanujam 1982

2.1 Key to the Identification of Indian Angiosperm Pollen

The key presented here is a simplified guide that can help to identify Indian angiosperm pollen at generic level. Characters making the base of the key are shape of pollen, features associated with aperture(s) and exine characters. The key contains generic names that have been selected for inclusion in this catalogue. It is recommended that initial characters given in the key must be followed by detailed examination of morphologic features of each genus given by original author(s) along with the figures.

Pollen Without Apparent Germinal (Inaperturate)

Spheroidal to subspheroidal

exine pilate, baculate, granulate, exinal projections forming reticulum

Assamiapollenites Singh 1975

exine moderately thick, retipilate, coarsely reticulate

Retipilonapites Ramanujam 1966 ex Potonié 1970

exine intectate, thick, provided with warts, which are arranged in crotonoid pattern

Incrotonipollis Jansonius & Hills 1981

exine heavily reticulate, reticulum not placed on basal layer

Retipollenites Gonzalez Guzman 1967

Pollen with Single Germinal (Monosulcate)

Body oval to elliptical, sulcus relatively long

exine foveolate, punctate

Arecipites Wodehouse 1933

exine flecked to psilate

Palmidites Couper 1953

Sulcus relatively short

exine psilate, granulate or finely structured

Palmaepollenites Potonié 1951 ex Potonié 1958

Sulcus moderately long and wide with thick margin

exine thick, microreticulate, covered with verrucae, clavae of variable size

Retiverrumonosulcites Tripathi 1994

Sulcus long, wide, sulcus margin thin

exine retipilate to retibaculate

Matanomadhiasulcites Kar 1985

exine foveolate, reticulate, reticulum variable, muri composed of baculae or clavae

Liliacidites Couper 1953

Sulcus long, moderately wide

exine echinate

Neocouperipollis (Venkatachala and Kar 1969) Kar & Kumar 1987

exine spinulate, spinose, baculate, verrucate

Spinomonosulcites Singh & Misra 1991c

exine echinate, baculate, spines/baculae arranged in alternate fashion at the sulcal margin

Arengapollenites Kar 1985

Lanceolate to elliptical, colpus long

exine laevigate, faintly structured

Pinjoriapollis Saxena & Singh 1981

Sulcus very long, extending up to more than half of the pollen circumference

exine reticulate

Longapertites von Hoeken-Klinkenberg 1964

Sulcus very long, zonisulcate, generally splitting into two halves

exine psilate, pitted, foveolate, reticulate

Proxapertites van der Hammen 1956

exine scrobiculate, sometimes microreticulate

Nymphaeoipollis Venkatachala & Kar 1969

exine finely reticulate, covered with baculae or spines

Spinizonocolpites Muller 1968

Sulcus very long, zonisulcate

exine densely pilate, clavate

Paravuripollis Rao & Ramanujam 1978

exine foveolate, imperfectly tectate, covered with densely or sparsely placed verrucae

Kapurdipollenites Tripathi 1994

exine tuberculate

Tuberculozonisulcites Kar & Sharma 2001

Heteropolar, sulcus long, broad with thick margin

exine tectate, heavily ornamented with pila/clava with triangular heads aligned in reticulate fashion

Crotonisulcites Rao & Ramanujam 1978

Heteropolar, ovoid, sulcus long to very long with thin margin

exine densely and coarsely ornamented with free clava/pila having triangular to rounded heads locally aligned to give reticuloid pattern

Clavapalmaedites Rao & Ramanujam 1978

exine reticulate, retipilate, clavate

Quilonipollenites Rao & Ramanujam 1978

Pollen with Two Germinals (Dicolpate/Dicolporate)

Oval, bisulcate

exine microreticulate to micro-verrucate

Dicolpopollis Pflanzl 1956 emend. Potonié 1966

exine psilate to punctate, body constricted in the middle

Umbelliferoipollenites Venkatachala & Kar 1969

Pollen with Three Simple Germinals (Tricolpate)

Subprolate equatorially, colpi short, narrow with thick or thin margin

exine intectate, baculate with rounded bacular heads

Bacubrevitricolpites Rao & Ramanujam 1982

exine thick, tegillate, spinulate

Icacinoipollenites Navale & Misra 1979

Subtriangular to spheroidal, angulaperturate, colpi long or short

exine reticulate, reticula of uniform size

Albertipollenites Srivastava 1969

exine reticulate, lumina smaller at apocolpia and colpi margins

Rousea Srivastava 1969

Subcircular to subtriangular, may be colporoidate also, colpi long

exine thin retipilate, pila heads arranged in stria-reticulate pattern

Bithnokiapollis Kar & Sharma 2001

exine moderately thick, microreticulate, lumina variable in shape

Cruciferoipollenites Navale & Misra 1979

Subcircular, colpi long

exine pilate, pila closely placed

Dipterocarpuspollenites Kar 1992

exine, tectate, foveolate

Foveotricolpites Pierce 1961

Subtriangular to subcircular, colpi long

exine intrareticulate:

Intrareticulites Kar 1985

Oval to subcircular, colpi long,

exine laevigate to faintly structured

Ladakhipollenites Mathur & Jain 1980

exine microfoveolate with long, papillate rods or clavae

Plumbaginacipites Navale & Misra 1979

Spherical to ovoid, mesocolpia bulging

exine finely reticulate

Tricolpites Cookson 1947 ex Couper 1953

Subcircular to three-lobed

exine intectate, baculate-clavate with pentagonal heads

Warkallipollenites Rao & Ramanujam in Thanikaimoni et al. 1984

Triangular with three arms, angles flared, sides concave, colpi long

exine tectate, smooth or finely patterned

Loranthipites Rao & Ramanujam 1982

Body zoned, outer part like a flange with radiating striae, colpi long

exine of flange granulate-reticulate

Polygonaceaepites Baksi 1962

Pollen with Three Simple United Germinals (Trisyncolpate/Trichotomosulcate)

Triangular to rounded triangular, colpi wide

exine intectate, densely covered with clavae

Clavasyncolpites Rao & Ramanujam 1982

Prolate spheroidal to prolate, colpi margins thick, thicker at poles

exine psilate or foveolate

Marginipollis Clarke & Frederiksen 1968

Triangular in polar view, colpi long

exine thin, foveolate to reticulate

Neotrichotomosulcites Samant & Phadtare 1997

exine tectate, tectum perforate, surface reticulate

Retitrisyncolpites Mandal et al. 1994

exine moderately thick, psilate

Pilatrisyncolpites Kar et al. 1994

Oval in equatorial view, subcircular in polar view, colpi long

exine thin, pilate to baculate

Barringtoniapollenites Kar & Sharma 2001

Subcircular in polar view, elliptical in equatorial view, colpi margin thick, appearing like a triradiate ridge in equatorial view

exine pilate to baculate

Trisyncolpites Kar 1979

Pollen with More Than Six Simple Germinals (Polycolpate)

Rounded in polar view, stephanocolpate

exine thin, smooth to finely granular provided with endexinous thickenings fusing at poles forming an irregular ring-like pattern

Ctenolophonidites von Hoeken-Klinkenberg 1966

exine thick, tectate, columellate, some columella fused at top but free at base, reticulate, muri pluricolumellate, columella also present in lumina

Ocimumpollenites Kar 1996

Pollen with Three Compound Germinals (Tricolporate)

Spheroidal to oval, colpi wide with thickened margin

exine reticulate to striate

Alangiopollis Krutzsch 1962

Triangular to subtriangular in polar view, angulaperturate, colpi long, pores distinct to indistinct

exine microreticulate

Angulocolporites Kar 1985

Oval to bipyramidal

Exine tectate, baculate, surface psilate to finely reticulate

Araliaceoipollenites Potonié 1951

Roundly oval, apertures at three sides, colpi short, pore distinct to indistinct

exine reticulate, reticulation finer at corners

Bombacacidites Couper 1960

Subcircular to circular in polar view with three constrictions due to apertures, colpi long, pore costate

exine thick, laevigate or finely reticulate

Calophyllumpollenites Sah & Kar 1974

Small, ellipsoidal to roundly triangular in polar view, colpi long or short, ora lalongate

exine tectate, reticulate

Caprifoliipites Wodehouse 1933

Subcircular in polar view, oval in equatorial view, colpi long, funnel-shaped in polar view, pores distinct

exine moderately thick, columellate, scrobiculate-reticulate

Chiranthodendronpollenites Kar & Sharma 2001

Spherical in polar view, colpi short, pores big, costate

exine thick, tegillate, coarsely reticulate

Coneopollis Venkatachala & Rawat 1973

Spherical to roundly triangular in polar view, colpi short, ora distinct, costate

exine vermiculate to pseudoreticulate

Coramandalipollis Venkatachala & Rawat 1973

Triangular to subcircular, angulaperturate, colpi non- to syncolpate or parasyncolpate, pores indistinct

exine scabrate to reticulate

Cupanieidites Cookson & Pike 1954

Triangular to subtriangular in polar view, angulaperturate, colpi long, narrow, apocolpi small, ora indistinct

exine of mesocolpi bulged, thin but thick at colpal margin, tectate

Dakshinipollenites Navale & Misra 1979

Subcircular to triangular, colpi short or long, pores distinct

exine well-stratified, retipilate

Favitricolporites Sah 1967

Triangular to subtriangular, colpi long, colpi margins thick, ora prominent, costate exine finely reticulate, sexine forming knob-like processes on either side of colpus

Hippocrateaceaedites Ramanujam 1966

Spheroidal, colpi margin thick, short, ora distinct, large, costate

exine thick, tectum perforate, columella prominent, finely reticulate

Lacrimapollis Venkatachala & Rawat 1973

exine eureticulate to eurugulate, muri composed of distinct capita, each supported by a group of columellae

Lanagiopollis Morley 1982

Oblate to suboblate, colpi long, margocolpate, ora faint, lolongate

exine reticulate or retipilate

Margocolporites Ramanujam 1966 ex Srivastava 1966

Rounded or oval, meridian often rhombic, equator roundly triangular, colpi moderately long with thickened border, pores indistinct

exine infrapunctate to finely reticulate

Nyssapollenites Thiergart 1937 ex Potonié 1960

Oval, colpi long, ora costate, forming equatorial girdle

exine semi-tectate, reticulate to microreticulate

Paleoaraliaceaepites Biswas 1962

Three-lobed in equatorial view, lobes thickened, colpi long, broad, ora distinct

exine thick, retipilate

Paleocaesalpiniaceaepites Biswas 1962

Rounded triangular to subspherical in polar view, colpi small to long, ora distinct, large, generally zonorate

exine variably thick, smooth, granulate or finely reticulate

Paleosantalaceaepites Biswas 1962 ex Dutta & Sah 1970

Triangular to subtriangular in polar view, rarely di- or tetracolporate, colpi short, ora lalongate, costate

exine thick, reticulate, muri simple to duplibaculate

Pellicieroipollis Sah & Kar 1970

Triangular to subtriangular in polar view, colpi small, pores distinct, costate

exine, foveolate, folded in apertural region

Ratariacolporites Kar 1985

Rounded triangular to spheroidal, colpi long, ora indistinct

exine retipilate to reticulate

Retitrescolpites Sah 1967

Oval in equatorial view, colpi long, ora distinct, lalongate

exine finely reticulate-pitted

Rhoipites Wodehouse 1933

Oval in equatorial view, subcircular in polar view, apertures ill-developed to developed, ora indistinct to distinct

exine, striate-reticulate

Striacolporites Sah & Kar 1970

Triangular to subtriangular in polar view, colpi short to long, ora prominent

exine punctate to microreticulate

Symplocoipollenites Potonié 1951 ex Potonié 1960

Subspheroidal in polar view, apertures subequatorial, colpi short, pore circular

exine reticulate, pitted reticulate or foveolate

Tricolporopollis Dutta & Sah 1970

Triangular to subtriangular, colpi small, pore large, costate

exine thick, laevigate to finely intrastructured

Dermatobrevicolporites Kar 1985

Subcircular to circular in polar view, colpi small, pores well-developed, oval-elliptical, broader than colpus

exine thin, laevigate to infrastructured

Lakiapollis Venkatachala & Kar 1969

exine spinulose, spine short, delicate, sparsely placed

Spinulotricolporites Saxena & Misra 1987

Subtriangular in polar view, colpi short, ora prominent, large lalongate

exine thin, smooth

Palaeocoprosmadites Ramanujam 1966

Triangular to subcircular, planaperturate

Exine uneven, tectum semiperforate, surface scabrate

Tinalipollenites Mandal & Kumar 2000

Oblate, triangular in equatorial view, colpi long, narrow, pores aspidate, annulate exine granulate

Talisiipites Wodehouse 1933

Ovoidal in equatorial view, colpi long, narrow, ora circular to lalongate

exine tectate, suprategillar ornamentation spinulate

Asperitricolporites Venkatachala & Rawat 1972

Spheroidal to subspheroidal in polar view, apertures subequatorial, colpi short, ora circular to slightly lalongate exine thin, verrucate to gemmate

Densiverrupollenites Tripathi & Singh 1984

Spherical, colpi short, pores distinct to indistinct exine spinose, spines with bulbous base and curved tips

Compositoipollenites Potonié 1951 ex Potonié 1960

Triangular with rounded angles, sides convex, colpi nearly united at poles, pores indistinct exine thick, studded with gemmate-tuberculate processes which are bigger at equatorial angles

Meyeripollis Baksi & Venkatachala 1970

Three-lobed, colpi long, funnel-shaped, margocolporate, ora distinct to indistinct

exine granulose to intrabaculate

Sastriipollenites Venkatachala & Kar 1969

Circular to subcircular in polar view, colpi short, narrow, pore circular, costate

exine incompletely tectate, tectal elements forming pila, clava, gemmae or verrucae, which are supported by groups of columellae

Tricolporopilites Kar

Subcircular to oval, colpi short, pore distinct, lalongate, costate

exine columellate or pilate, pila forming negative reticulum in surface view

Tricolporocolumellites Kar 1985

Oval-elliptical in equatorial view, subcircular in polar view, colpi long, pores indistinct

exine thick, verrucate

Verrucolporites Sah & Kar 1970

exine intectate, columella free, baculate, baculae closely placed

Intectocolporites Mandal & Guleria 2006

Pollen with Four or More Compound Germinals (Tetra-/Penta-/Polycolporate)

Subcircular to pentagonal in polar view, suboblate to subprolate in equatorial view, tetracolporate (rarely tri- tetra- and pentacolporate), colpi long to short, tenuimarginate, ora well-defined, lolongate, costate

exine thin to very thick, psilate to weakly structured

Meliapollis Sah & Kar 1970

Broadly oval to rectangular, tetracolporate, a dark equatorial band surrounding apertures present

exine baculate, tectate

Iugopollis Venkatachala & Rawat 1972

Subcircular to circular, polycolporate, colpi short, pores distinct

exine thick, intrabaculate

Polybrevicolporites Venkatachala & Kar 1969

Circular to subcircular, polycolporate, colpi long, pores distinct

exine thin, scabrate

Polycolporites Mehrotra 1983

Spheroidal to subspheroidal, polycolporate, colpi long or short, generally synorate

exine smooth to finely structured

Polygalacidites Sah & Dutta 1966

Subspheroidal to stellate spheroidal, 6–8 colporate, brevicolpate, ora lalongate, costate

exine differentially sculptured, nexine with undulating or vermiform thickenings, sexine finely pitted

Padappakkarapollis Rao & Ramanujam 1982

Circular in polar view, polycolporate, colpi long to short, ora distinct

exine thick, perforate, apocolpial area with primitive to distinct nexinal thickenings

Prectenolophonidites Samant & Phadtare 1997

Pollen with Single Germinal (Monoporate)

Circular to subcircular, aperture circular to subcircular

exine weakly intrastructured

Dracaenoipollis Sah & Kar 1970

Spherical to subspherical or ovoidal

exine spinose, interspinal area psilate, punctate or microreticulate

Echimonoporopollis Saxena et al. 1991

Spherical, pore annulate

exine thin, psilate to finely granulate

Graminidites Cookson 1947 ex Potonié 1960 emend. Krutzsch 1970

Spherical, pore large, rounded or slightly oval, thickened to annulate, appearing as a rugged cavity in structured wall

exine pilate, pila interspersed with spines, tectum perforate

Jacobipollenites Ramanujam 1966

Spheroidal to prolate spheroidal, pore may be operculate

exine thick, finely structured

Nymphaeacidites Sah 1967

Pollen with Two Germinals (Diporate/Diporocolpate)

Tyre-shaped, circular in equatorial plane, concave at poles, diporate, pores circular, juxtaposed, sunken, annulate, each pore surrounded by a whorl of clavae

exine microreticulate to foveolate, muri fused at places to form verrucae, gemmae or clavae

Clavadiporopollenites Ambwani & Singh 1996

Barrel-shaped, diporate, pores simple, one on each end of pollen

exine thin, psilate to finely structured

Psilodiporites Varma & Rawat 1963 emend. Venkatachala & Rawat 1972

exine thin, tectate, spinose, spines mixed with bacula and coni

Echidiporites Mandal 1997

Oval to elliptical; diporocolpate, pores equatorial, distinct, big, pore margin spinose, spines delicate, colpus distinct, connecting the two pores, colpus margin spinose

exine thick, laevigate

Psiladiporocolpites Kar 1995

exine thick, pilate, pila moderately long, sparsely placed, interpilar exine laevigate

Piladiporocolpites Kar 1995

exine thin, tectate, columellate, reticulate

Retidiporocolpites Kar 1995

Pollen with Three Germinals (Triporate)

Oval to spindle-shaped, middle part not constricted; triporate, zoniporate, pores oval, lalongate

exine psilate

Sonneratioipollis Venkatachala & Kar 1969

Subcircular, pores distinct, large with thickened margins

exine thin, spinose, spines of two types – bigger and smaller

Varispinitriporites Kar 1985

Circular, pores circular, medium to large

exine thin, columellate, spinose, spines variable in size and shape, occasionally gemmae interspersed with spines, interspinal area finely reticulate

Acanthotricolpites Kar 1985

Subtriangular to subcircular, pores large, subcircular, equatorially placed

exine gemmate, area between gemmae granulate or finely baculate, conate or microreticulate

Gemmatriporopollis Saxena & Khare 1996

Subtriangular to roundly triangular, pore margin incrassate, ornate, studded with gemmae or tubercles

exine intectate, gemmate to tuberculate

Ornatriporites Rao & Ramanujam 1982

Heteropolar, rounded triangular to subcircular in polar view, trilatiporate, pores circular, subequatorial, costate, located on well-developed projections

exine thick, finely reticulate, reticulation weak at poles

Dorreenipites Biswas 1962

Triangular to subtriangular in polar view, pores circular, costate, trilatiporate, almost equatorial, borne on feeble projections

exine, semitectate, granulate to microreticulate

Grevilloideaepites Biswas 1962

Triangular to subtriangular

exine baculate, clavate, baculate or tuberculate forming pitted or reticulate appearance in surface view

Proteacidites Cookson 1950 ex Couper 1953

Approximately spheroidal, pores protruding

exine psilate to finely structured

Myricipites Wodehouse 1933

Triangular, sides straight to slightly concave

exine thick, baculate, reticuloid in surface view

Triorites Erdtman 1947 ex Cookson 1950

Triangular to subcircular

exine faintly granulate to faintly reticulate

Triporopollenites Pflug & Thomson in Thomson & Pflug 1953

Pollen with Multiple Germinals (Panporate)

Spheroidal, apertures distinct or indistinct, usually not well-define

exine thick, reticulate

Polygonacidites Sah & Dutta 1968

exine tegillate, suprategillar ornamentation granulate, conate or finely echinate

Polypornia Naumova 1939 ex Potonié 1960

Circular with regular lobed margin, pores rimmed, invaginated

exine finely granulose

Pseudonothofagidites Venkatachala & Kar 1969

Spheroidal to subspheroidal, pores distinct to indistinct

exine moderately thick, baculate to spinose

Thymelaepollis Sah & Kar 1970

exine thick, densely clavate, clavae heads polygonal forming loose reticulum

Clavaperiporites Ramanujam 1966

exine baculate, bacula closely placed

Cryptopolyporites Venkatachala & Kar 1969

exine pilate, covered with conical and acuminate spines

Malvacearumpollis Nagy 1962

Pollen Found in Tetrads or Polyads (Tetrads/Polyads)

Tetrahedral tetrads, individual grain tricolporate, colpi distinct, pore margin thickened

exine thick, tegillate, retipilate

Kielmeyerapollenites Sah & Kar 1974

Tetrahedral tetrads, individual grain rounded to suboblate, porate, pores confined to proximal face

exine tectate, clavate, gemmate or verrucate, surface between sculptural elements punctate

Ornatetradites Rao & Ramanujam 1982

2.2 Description of Angiosperm Pollen

A

Genus *Acanthotricolpites* Kar 1985

Type Species *Acanthotricolpites bulbospinosus* Kar 1985

Generic Description (Kar 1985, p. 72): Pollen grains subcircular in polar view. Tricolpate, colpi long. Exine spinose, spines with bulbous base and pointed tip, inter-spinal space densely covered with grana and bacula.

Emended Description (Singh and Misra 1991c, p. 224): Pollen circular, isopolar, usually folded; tri- to tetraporate; pores medium to large, more or less circular, unthickened, difficult to recognize due to heavy ornamentation and multiple exinal folds rendering sulcate (false sulcus) nature to the grains; exine thin, spinose; spines broad-based, long and variable in shape and size and shed off easily leaving scars, gemmae occasionally present interspersed with spines; exine columellate, nexine usually indistinct; interspinal area microreticulate appearing to be microgranulate in different foci.

Botanical Affinity Arecaceae.

Remarks Initially described as tricolpate (Kar 1985), this genus is triporate to tetraporate (Singh and Misra 1991c). Different species of this genus possess variety of spines over the exine surface. Sometimes spines are so dense that apertures are obscured. Due to presence of similar kind of spines and obscured apertures, it may be difficult to distinguish between *Acanthotricolpites* and *Spinomonosulcites* in a casual observation. *Spinomonosulcites* is monosulcate pollen with spinose exine and exhibit great deal of variation with regard to the shape, size and density of spines. Morphological features of pollen assignable to both of these genera strongly suggest their affinity with the family Arecaceae.

Species Recorded from India

Acanthotricolpites brevispinosus Saxena & Khare 2004 (Plate 23, Fig. 11)

Description (Saxena and Khare 2004, p. 85): Pollen grains circular to subcircular. Size range 44–49 × 42–47 μm (excluding spines). Triporate, pores subequatorial, 15–17 μm in diameter, pore margin unthickened. Exine up to 1.5 μm thick, spinose, spines 3–4 μm long with pointed tips, uniformly distributed, interspinal area microreticulate.

Occurrence Late Palaeocene-Middle Eocene subsurface sediments, Tiruchirapalli, Tamil Nadu (Saxena and Khare 2004).

Acanthotricolpites bulbospinosus Kar 1985 (Plate 23, Fig. 12)

Description (Kar 1985, p. 73): pollen grains subcircular in polar view, 42–55 × 40–53 μm. Tricolpate, colpi generally indistinct, broad. Exine 1–2 μm thick, sexine as thick as nexine, spinose, spines strongly built, 5–8 μm long, 1–2 μm broad at base, spines gradually taper to form pointed tip, spines sparsely placed, generally 4–8 μm apart. Interspinal space densely granulate-baculate.

Description (Singh and Misra 1991c, p. 224): Pollen grains isopolar, 40–65 μm, usually folded; 3–4 porate; pores 10–16 μm with unthickened margins; exine 1–2 μm thick (excluding spines), spinose, spines (1–3 μm broad; 3–9 μm long) gradually tapering from base to apex, straight or curved, shed off leaving scars; exine columellate (columellae up to 1 μm long), nexine usually undistinguishable; interspinal area microreticulate.

Occurrence Early Eocene, Naredi Formation, Kutch, Gujarat (Kar 1985); Palaeocene, Lakadong Sandstone, Jaintia Hills, Meghalaya (Mandal 1987); Miocene, Cuddalore Formation, Neyveli Lignitefield, Tamil Nadu (Singh and Misra 1991c); Late Palaeocene-Early Eocene, Palana Formation, Rajasthan (Kar and Sharma 2001); Late Palaeocene, Akli Formation, Rajasthan, Barmer Basin (Tripathi et al. 2003).

Acanthotricolpites complexus Singh & Misra 1991c (Plate 23, Fig. 13)

Description (Singh and Misra 1991c, p. 226): Pollen grains isopolar, 48–60 μm, more or less circular; 3–4 porate; pores 10–12 μm in diameter; easily traceable only in polar view; exine 1–2 μm thick (excluding spines), spinose; spines (5–8 μm long) vary in shape, attached flatly to the exine surface; exine columellate (columellae less than 1 μm long), sexine and nexine indistinguishable; interspinal area microreticulate.

Occurrence Miocene, Cuddalore Formation, Neyveli Lignitefield, Tamil Nadu (Singh and Misra 1991c).

Acanthotricolpites intermedius Singh & Misra 1991c (Plate 23, Fig. 14)

Description (Singh and Misra 1991c, p. 226): Pollen grains isopolar, circular, 38–85 μm in size (including spines), usually folded; 3–4 porate; pores more or less circular, margin unthickened, sometimes indiscernible due to exinal folds, pore diameter 8–16 μm; exine spinose, 1–2 μm thick (excluding spines), spines with bulbous base and gradually tapering ends, elongated, vary slightly in shape, 2–3 μm broad × 4–10 μm long, exine columellate (columellae less than 1 μm high), nexine indistinguishable; surface ornamentation in interspinal area faint to distinct, microreticulate.

Occurrence Miocene, Cuddalore Formation, Neyveli Lignitefield, Tamil Nadu (Singh and Misra 1991c).

Acanthotricolpites karii Saxena & Khare 2004 (Plate 23, Fig. 15)

Description (Saxena and Khare 2004, p. 86): Pollen grains circular to subtriangular. Size range 50–52 × 49–51 μm (excluding spines). Triporate, pores distinct, 14–18 μm in diameter, pore margin unthickened. Exine up to 2 μm thick, spinose, spines 7–10 μm long with thick bases and pointed tips, sparsely distributed, interspinal area punctate to scabrate.

Occurrence Late Palaeocene-Middle Eocene subsurface sediments, Tiruchirapalli, Tamil Nadu (Saxena and Khare 2004).

Acanthotricolpites kutchensis (Venkatachala and Kar 1969) Singh & Misra 1991c (Plate 24, Fig. 1)

Basionym: 1969 *Couperipollis kutchensis* Venkatachala & Kar, *The Palaeobotanist* 17(2), p. 161, Plate 1, Figs. 15 and 16.

Synonym: 1987 *Neocouperipollis kutchensis* (Venkatachala and Kar 1969) Kar and Kumar, *The Palaeobotanist* 35(2), p. 172, Fig. 2.

Description (Singh and Misra 1991c, p. 226): Pollen grain circular, 30–90 μm, usually folded, isopolar, tri- to tetraporate; pores large, 15–20 μm, more or less circular, unthickened, sometimes indistinct due to thin nature of exine and heavy ornamentation; exine 1–2 μm thick (excluding spines), spinose; spines (3–4 μm broad and 3–10 μm long) with distinct bulbous base and pointed tips, exine columellate, interspinal area microreticulate.

Occurrence Naredi Formation, Matanomadh, Kutch, Gujarat (Venkatachala and Kar 1969); Miocene, Cuddalore Formation, Neyveli Lignitefield, Tamil Nadu (Singh and Misra 1991c); Late Palaeocene-Early Eocene, Palana Formation, Bikaner, Rajasthan (Kar and Sharma 2001); Late Palaeocene, Akli Formation, Barmer, Rajasthan (Tripathi et al. 2003).

Acanthotricolpites microreticulatus Saxena & Khare 2004 (Plate 24, Fig. 2)

Description (Saxena and Khare 2004, pp. 85, 86): Pollen grains circular-subcircular to triangular, size range 60–69 × 55–67 µm (excluding spines). Triporate, pores distinct, 29–32 µm in diameter, pore margin unthickened. Exine 1.5–2 µm thick, tectate, spinose, spines with bulbous base and pointed tip, 6.5–9 µm long, uniformly placed, interspinal area microreticulate.

Occurrence Late Palaeocene-Middle Eocene subsurface sediments, Tiruchirapalli, Tamil Nadu (Saxena and Khare 2004).

Acanthotricolpites multitypicus Singh & Misra 1991c (Plate 24, Fig. 3)

Description (Singh and Misra 1991c, p. 226): Pollen grains isopolar, circular, 50–75 µm (including spines), usually folded, 3–4 porate; pores large (15–16 µm), unthickened, more or less circular, distinct in polar view; exine (1–2 µm thick excluding spines), spinose, spines (2.5–4 µm broad × 5–8 µm long), vary in shape and size, gemmae (0.5–4 × 1–7 µm) interspersed with spines; exine columellate (columellae about 1 µm long); interspinal area microreticulate.

Occurrence Miocene, Cuddalore Formation, Neyveli Lignitefield, Tamil Nadu (Singh and Misra 1991c).

Acanthotricolpites robustus Saxena & Khare 2004 (Plate 24, Fig. 4)

Description (Saxena and Khare 2004, p. 86): Pollen grains circular to subtriangular. Size range 38–42 × 36–39 µm (excluding spines). Triporate, pores distinct, 11–15 µm in diameter, pore margin unthickened. Exine up to 2 µm thick, spinose, spines robustly built with blunt tips, 4.5–5.5 µm long, densely placed, interspinal area faintly microreticulate.

Occurrence Late Palaeocene-Middle Eocene subsurface sediments, Tiruchirapalli, Tamil Nadu (Saxena and Khare 2004).

Acanthotricolpites tiruchirapalliensis Saxena & Khare 2004 (Plate 24, Fig. 5)

Description (Saxena and Khare 2004, p. 85): Pollen grains circular to subtriangular. Size range 32–44 × 32–42 μm (excluding spines). Triporate, pores 6–11 μm in diameter, pore margin unthickened. Exine 1–1.5 μm thick, spinose, spines 4.5–6 μm long with pointed tips, uniformly distributed, interspinal area punctate to laevigate.

Occurrence Late Palaeocene-Middle Eocene subsurface sediments, Tiruchirapalli, Tamil Nadu (Saxena and Khare 2004).

Genus *Alangiopollis* Krutzsch 1962

Type Species *Alangiopollis barghoornianus* (Traverse 1955) Krutzsch 1962

Generic Description (Krutzsch 1962, p. 279): Large tricolporate pollen with relatively short polar axis; figure spherical or oval to lenticular; amb approximately circular; exocolpi usually widely gapping, not very long, narrowing towards the poles to a V-shape, hardly depressed in the equatorial region; endopores large (up to over 10 μm diameter) round to somewhat oval, cavernae variously thickened; structure and sculpture variable, in part very strongly reticulate-striate; structured part of the wall (especially along the side) thicker than the smooth inner wall layer (s).

Botanical Affinity Alangiaceae.

Species Recorded from India

Alangiopollis arcotense Navale & Misra 1979 (Plate 13, Fig. 2)

Description (Navale and Misra 1979, p. 230): Pollen grains isopolar, rounded triangular, subcircular in polar view, 70–120 × 70–120 μm in size. Tri- to tetra-colporate, colpi distinct, length varying from medium to short, 20–25 μm deep, bordered by thickened margo, colpi membrane psilate, ends acute to rounded. Pore distinct, lalongate, circular or rarely lolongate, pore margin thickened. Exine 2.5–5 μm thick, sexine thicker than or equal to nexine, surface tegillate, reticulate, retipilate, simplicolumellate, capita more or less rounded, columellae short and thin, lumina small, of varying shapes, smaller near colpi margins.

Occurrence Miocene, Neyveli lignite, Tamil Nadu (Navale & Misra 1979).

Alangiopollis gemmatus Navale & Misra 1979 (Plate 13, Fig. 1)

Description (Navale and Misra 1979, p. 230): Pollen grain isopolar, subtriangular, size 63 × 60 μm. Tricolporate, angulaperturate, colpi short, faintly visible, pore simple, distinct, pore diameter 3–3.5 μm. Exine 3–4.7 μm thick (including surface ornamentation). Surface provided with wart-like projections—gemmae, gemmae of various sizes, 1.5–3 μm, irregularly distributed, exine area between gemmae microgranulate.

Occurrence Miocene, Neyveli lignite, Tamil Nadu (Navale and Misra 1979).

Genus *Albertipollenites* Srivastava 1969

Type Species *Albertipollenites rosalindiae* Srivastava 1969

Generic Description (Srivastava 1969, p. 54): Tricolpate, angulaperturate, oblate to spheroidal, colpi short or long, meridional; amb triangular to circular, sides straight to convex; sexine thick, tectate; ornamentation reticulate, reticula of uniform size all over the pollen body; lumina size 1 μm or larger.

Botanical Affinity Dipterocarpaceae.

Species Recorded from India

Albertipollenites aquifoliaceaeformis (Biswas 1962) Mandal & Rao 2001 (Plate 8, Fig. 9)

Basionym: *Tricolpites aquifoliaceaeformis* Biswas 1962, *Bulletin of the Geological, Mining and Metallurgical Society of India,* 25: 1–48, Plate 5, Fig. 18.

Description (Biswas 1962, p. 38): Equatorial compression; exine moderately thick, with minute rods; tricolpate; 21 μm.

Occurrence Early- Middle Eocene, Tura Formation, Meghalaya (Biswas 1962).

Albertipollenites baculatus (Jain et al. 1973) Mandal & Rao 2001 (Plate 8, Fig. 10)

Basionym: *Tricolpites baculatus* Jain et al. 1973, *Geophytology* 3: 150–165, Plate 1, Figs. 23–26, 34, 36.

Description (Jain et al. 1973, p. 155): Pollen grains tricolpate, spherical, 30–35 μm, colpi long. Exine moderately thick, tegillate, surface baculate-spinose. Sculptural elements up to 3 μm high.

Occurrence Palaeocene, Barmer clay bed, Rajasthan (Jain et al. 1973); Palaeocene, Matanomadh Formation, Kutch, Gujarat (Saxena 1979).

Albertipollenites crassireticulatus (Dutta and Sah 1970) Mandal & Rao 2001 (Plate 8, Fig. 11)

Basionym: *Tricolpites crassireticulatus* Dutta & Sah 1970, *Palaeontographica,* 131: 1–72, Plate 6, Figs. 9–12.

Synonym: 1973- *Tricolpites crassisexinous* Venkatachala & Rawat, *The Palaeobotanist* 20: 224, Plate 4, Fig. 9.

Description (Dutta and Sah 1970, p. 32): Size range 25–37 μm, amb roundly triangular to subspheroidal, tricolpate, colpi with bulging mesocolpia; exine rather thick, sexine as thick as nexine, pilate, tegillate, surface sculpture coarsely reticulate, crassimurate.

Occurrence Palaeocene, Cherra Formation, Meghalaya (Dutta and Sah 1970); subsurface Oligocene-Miocene sediments, Cauvery Basin (Venkatachala and Rawat 1973).

Albertipollenites karii Mandal & Rao 2001 (Plate 8, Fig. 13)

Description (Mandal and Rao 2001, p. 344): Pollen grain subprolate, 58 × 70–66 × 73 μm in equatorial view. Tricolpate, colpi 52 μm long, margin smooth. Exine 5–6 μm at equator, generally decreases to 3 μm at poles, nexine almost uniform, 1–1.5 μm, columellae stout, 3.5 μm long, 1 μm broad at base and gradually widens upward, apex 1.5–2 μm apart at equator, closer (1–1.5 μm) at poles; tectum smooth, nearly uniform, 1.5 μm. Surface reticulate, reticulum nearly squarish all over the surface, uniform in size; muri simplicolumellate, 1–1.5 μm thick; lumina 3 × 4 μm, free columellae rarely present within lumina.

Occurrence Late Palaeocene, Lakadong Sandstone Member, Meghalaya (Kar and Kumar 1986).

Remarks *Tricolpites* sp. A, earlier described by Kar and Kumar (1986, p. 196, Plate 9, Fig. 1) was proposed as a new species as *Albertipollenites karii* by Mandal and Rao (2001).

Albertipollenites kumarii Mandal & Rao 2001 (Plate 8, Fig. 14)

Description (Mandal and Rao 2001, p. 346): Pollen grain 100 μm (polar axis) and 61 μm (folded equatorial axis) in equatorial view. Tricolpate, colpi 93 μm long, nearing the poles, margins smooth. Exine 5.5 μm thick at equator, 4.5 μm at poles;

nexine uniformly thick, ca. 1 μm, tectum smooth, 1–1.5 μm thick; columellae of two types; one set supports tectum, 2.5–3 μm long, 1–1.5 μm broad, 2–3.5 μm apart, other set mainly present in equatorial areas between two long columellae, smaller, 2 μm long, 1 μm broad and does not touch tectum (free). Surface reticulate, reticula vary in size; muri 1.5–2 μm broad, simplicolumellate; lumina vary in shape (oval-elongated) and size, largest lumen 4 × 3 μm, 1–3 columellae within each lumen.

Occurrence Late Palaeocene, Lakadong Sandstone Member, Meghalaya (Kar and Kumar 1986).

Remarks *Tricolpites* sp. B, described by Kar and Kumar (1986, p. 198, Plate 9, Fig. 8) was proposed as a new species as *Albertipollenites kumarii* by Mandal and Rao (2001).

Albertipollenites kutchensis Mandal & Rao 2001 (Plate 8, Fig. 15)

Description (Mandal and Rao 2001, p. 346): Pollen grain spherical in polar view, 85 × 88 μm. Tricolpate, longicolpate, colpi 33 μm deep, wide at equator. Exine thick, 7 μm at mesocolpium and gradually thins out towards colpi, 4 μm at colpi margins, nexine as thick as or thicker than sexine, tectate, columellae 3.5 μm long, mixed with pila, capita 1.5 μm broad, closely placed. Surface uniformly reticulate, lumina elongated, 1.5–2.5 μm, muri simplicolumellate.

Occurrence Middle-Late Palaeocene, southern Kutch, Gujarat (Kar and Saxena 1981).

Remarks *Tricolpites crassireticulatus* Dutta and Sah (1970) in Kar and Saxena (1981, p. 106, Plate 3, Fig. 56; non Dutta and Sah 1970) was proposed as *Albertipollenites kutchensis* by Mandal and Rao (2001).

Albertipollenites proboscideus (Biswas 1962) Mandal & Rao 2001 (Plate 9, Fig. 1)

Description (Biswas 1962, p. 42): Polar compression, exine thick, uniformly reticulate; tricolpate.

Occurrence Palaeocene, Tura Formation, Meghalaya (Biswas 1962).

Remarks *Tricolpites proboscideus* Biswas (1962, p. 42, Plate 8, Fig. 44) was transferred to

Albertipollenites proboscideus by Mandal and Rao (2001).

Albertipollenites retibaculatus (Saxena 1979) Mandal & Rao 2001 (Plate 9, Fig. 2)

Description (Saxena 1979, p. 133): Pollen grains subcircular-subtriangular in polar view, 58–114 µm in size. Tricolpate, colpi long, well-developed, mesocolpia wide. Exine 2.5–7 µm thick, sexine thicker than nexine, retibaculate to retipilate, bacula/pila thick and strongly built, tegillate, sculptural elements provide pseudoreticulum on surface view.

Restated Description (Mandal and Rao 2001, p. 348): Pollen grains subtriangular in polar view, 88–90 µm. Tricolpate, colpi wide at equator, 38 µm deep. Exine 3 µm thick, sexine thicker than nexine (1 µm), uniformly thick, tectate, columellae slender, 2–3 µm apart, 1.5 µm long, capita 1.5 µm in diameter, mostly fused. Surface retipilate, reticulum almost uniform, lumina irregular, 1–3 µm wide.

Occurrence Palaeocene, Matanomadh Formation, Kutch, Gujarat (Kar and Saxena 1981).

Remarks *Tricolpites retibaculatus* Saxena (1979, p. 133, Plate 2, Figs. 27, 28) was transferred to *Albertipollenites retibaculatus* by Mandal and Rao (2001).

Albertipollenites robustus (Sah and Kar 1970) Mandal & Rao 2001 (Plate 9, Fig. 3)

Description (Sah and Kar 1970, p. 131): Pollen grains subcircular, 40–54 × 38–52 µm. Tricolpate, colpi well-developed. Exine thick, reticulate, meshes duplibaculate.

Restated Description (Mandal and Rao 2001, p. 348): Pollen grain oblate, 44 × 60 µm in equatorial view. Tricolpate, colpi long, 30 µm, ends pointed. Exine 4 µm thick, sexine uniformly thick, about 3 µm; nexine layer not very distinct at all places, about 1 µm at mesocolpia, tectate, tectum 2.5 µm thick, smooth, columellae slender, 1 µm long and 1.5 µm apart. Surface reticulate, muri mostly duplibaculate, 2 µm wide, lumina oval to elongate, 2.5–4.5 µm, smaller on apocolpial areas.

Occurrence Early Eocene, Naredi Formation, Kutch, Gujarat (Sah and Kar 1970).

Remarks *Retitricolpites robustus* Sah and Kar (1970, p. 131, Plate 1, Fig. 30) was transferesd to *Albertipollenites robustus* by Mandal and Rao (2001).

Genus *Angulocolporites* Kar 1985

Type Species *Angulocolporites microreticulatus* Kar 1985

Generic Description (Kar 1985, p. 76): Pollen grains triangular-subtriangular in polar view. Tricolporate, angulaperturate, colpi long, pores distinct or indistinct. Exine microreticulate.

Species Recorded from India

Angulocolporites microreticulatus Kar 1985 (Plate 13, Fig. 12)

Description (Kar 1985, p. 76): Pollen grains triangular-subtriangular in polar view, 30–45 × 29–44 μm. Tricolporate, angulaperturate, colpi long, funnel-shaped, pores generally indistinct, margin not thickened. Exine 1.5–2.5 μm thick, sexine thicker than nexine, laevigate to microreticulate.

Occurrence Early Eocene, Naredi Formation, Kutch, Gujarat (Kar 1985).

Genus *Araliaceoipollenites* Potonié 1951 ex Potonié 1960

Type Species *Araliaceoipollenites euphorii* Potonié (1951) Potonié 1960

Generic Description (Potonié 1960, p. 97): Shape oval to bipyramidal, poles more or less flattened; equator more or less triangular, meridian more or less rhombic; deeply tricolporate, at poles exoexine thicker than ectexine, at equator equally thick, outline more or less crenate especially at the poles; optical section of exoexine more or less radially striate (columellae) in polar regions; germinals internally as narrow equatorial rugae (lalongate ora), more externally more or less circular.

Botanical Affinity Araliaceae.

Species Recorded from India

Araliaceoipollenites baculatus Jain et al. 1973 (Plate 13, Fig. 3)

Description (Jain et al. 1973, p. 156): Pollen grains tricolporate, 37–43 μm in size, amb ovoid-rhomboidal, colpi long, pore circular-lalongate. Exine thick, sexine as thick as nexine, surface baculate, bacula up to 5 μm high.

Occurrence Palaeocene, Barmer Sandstone, Barmer, Rajasthan (Jain et al. 1973).

Araliaceoipollenites descretus Venkatachala & Rawat 1973 (Plate 13, Fig. 4)

Description (Venkatachala and Rawat 1973, p. 245): Pollen grains more or less oval in equatorial view, 36.4–39.5 × 28.5 μm. Tricolporate; colpi long, extending almost the whole length of the grain, more or less uniformly broad with tapering ends; ora conspicuously lalongate, 7.2 × 4.0 μm. Exine 2–2.5 μm thick, tectate, sexine thicker than nexine, reticulate.

Occurrence Oligocene-Miocene, Cauvery Basin, Tamil Nadu (Venkatachala and Rawat 1973).

Araliaceoipollenites potoniaei Ramanujam 1966 (Plate 13, Fig. 5)

Description (Ramanujam 1966, p. 168): Pollen grains isopolar, prolate to sub-prolate, 3-zonicolporate, 28–34 × 21–27 μm. Equatorial zone protruding or pouting laterally. Colpi thick, longicolpate, ends rather blunt or pointed, sharply bent at equatorial zone, ora conspicuously lalongate, as thick as colpi. Exine 2.5 μm thick, slightly thicker at poles, columella distinct only at poles, surface smooth.

Occurrence Miocene-Pliocene, Cuddalore Formation, Neyveli Lignitefield, Tamil Nadu (Ramanujam 1966).

Araliaceoipollenites psilatus Dutta & Sah 1970 (Plate 13, Fig. 6)

Description (Dutta and Sah 1970, p. 38): Size range 25–34 μm; amb roundly triangular to sometimes sub-prolate; 3-colporate, angulaperturate, colpi distinct, long, tenuimarginate; ora narrow, probably lalongate; exine fairly thick, endoexinal thickenings prominent in both polar and equatorial views, surface sculpture psilate.

Occurrence Palaeocene, Cherra Formation, Meghalaya (Dutta and Sah 1970).

Araliaceoipollenites quilonensis Rao & Ramanujam 1982 (Plate 13, Fig. 7)

Description (Rao and Ramanujam 1982, p. 76): Pollen grains isopolar, prolate and somewhat rhomboidal equatorially, poles smoothly arched, 18–22.5 × 15–17 μm; zonaperturate, tricolporate, longicolpate, colpae almost reaching poles, wide and often bent at equator, gradually tapering towards poles, ora prominently lalongate, up to 8 μm across. Exine 2.5 μm thick, thicker at poles than elsewhere, sexine thicker than nexine, columella distinct, particularly at polar regions, surface psilate to locally finely punctate.

Occurrence Miocene, Quilon beds, Kerala (Rao and Ramanujam 1982).

Araliaceoipollenites reticulatus Dutta & Sah 1970 (Plate 13, Fig. 8)

Description (Dutta and Sah 1970, p. 38): Size range 25–34 μm; subprolate in equatorial view, more or less oblate-spheroidal in polar view; 3-zonaperturate, colporate, colpi distinct, fairly deep; pores faint; exine thick, sexine as thick as nexine, surface sculpture coarsely pitted-reticulate, meshes distinct at poles, becoming faint or even indistinct towards the equator.

Occurrence Palaeocene, Cherra Formation, Meghalaya (Dutta and Sah 1970).

Araliaceoipollenites rugulatus Venkatachala & Rawat 1972 (Plate 13, Fig. 9)

Description (Venkatachala and Rawat 1972, p. 312): Pollen grains roundly triangular in polar view, 27.8 × 27.8 μm; tricolporate. Colpi long, up to 10 μm deep in polar view, wider in equatorial region and tapering towards poles; ora more or less circular. Exine up to 2.5 μm thick, tectate, sexine thicker than nexine, rugulate, rugulae thin, less than 1 μm.

Occurrence Palaeocene-Eocene, Cauvery Basin, Tamil Nadu (Venkatachala and Rawat 1972).

Genus *Arecipites* Wodehouse 1933

Type Species *Arecipites punctatus* Wodehouse 1933 (designated by Potonié 1958, p. 97)

Generic Description (Wodehouse 1933, p. 497): Ellipsoidal, 23–25 μm long, with a single longitudinal furrow which may close tightly throughout its entire length, not gapping at its ends.

Emended Description (Nichols et al. 1973, p. 248): Monocolpate pollen, amb elongate-ellipsoidal; colpus tapered, not expanded or gapping at ends. Exine tectate in structure, but exine stratification often obscure; sometimes exhibiting columellae in optical section. Exine psilate to finely foveolate or scrobiculate, diameter of foveolae about 0.5 μm. Size variable in different species, *ca.* 20–50 microns.

Remarks In accordance with the recommendations of Nichols et al. (1973), small monocolpate pollen with tapered sulcus, tectate exine and exine possessing psilate, foveolate, scrobiculate, foraminate, punctate or pitted ornamentation have been included under *Arecipites*. Distinctly reticulate forms with tapered sulcus have been included in *Liliacidites*.

Botanical Affinity *Phoenix dactylifera* (Arecaceae).

Species Recorded from India

Arecipites bellus Sah & Kar 1970 (Plate 1, Fig. 8)

Description (Sah and Kar 1970, p. 128, 129): Pollen grains oval with equally rounded lateral ends, 58–66 × 48–55 μm. Monocolpate, colpus extending end to end. Exine punctate, about 2 μm thick, sexine thicker than nexine. Puncta more or less 0.5 μm across, closely placed, uniformly distributed.

Occurrence Early Eocene, Laki Series, Kutch, Gujarat (Sah and Kar 1970); Early-Middle Eocene, Naredi and Harudi formations, Kutch, Gujarat (Kar 1978); Middle-Late Eocene subsurface sediments, Rataria, Kutch, Gujarat (Kar and Saxena 1981); Eocene, Naredi and Harudi formations, Kutch, Gujarat (Kar 1985); Palaeocene- Eocene subsurface sediments, Kapurdi, Barmer, Rajasthan (Tripathi 1995); Late Palaeocene-Middle Eocene subsurface sediments, Tiruchirapalli, Tamil Nadu (Saxena and Khare 2004).

Arecipites indicus Venkatachala & Rawat 1972 (Plate 1, Fig. 9)

Description (Venkatachala and Rawat 1972, p. 298): Pollen oval with slightly rounded lateral ends; monosulcate, 26.4 × 19.8 μm. Sulcus long, almost reaching the lateral margins, uniformly broad, slightly broadened at the ends. Exine about 1–1.3 μm thick, sexine thicker than nexine, finely punctate.

Occurrence Palaeocene-Eocene, Cauvery Basin, Tamil Nadu (Venkatachala and Rawat 1972); Early Eocene, Kadi Formation, Cambay Basin, Gujarat (Rawat et al. 1977).

Arecipites intrapunctatus Kar & Saxena 1981 (Plate 1, Fig. 10)

Description (Kar and Saxena 1981, p. 112): Pollen grains oval, 62–75 × 40–51 μm in size. Monocolpate, colpus end to end, distinct to indistinct. Exine 1–2 μm thick, intrapunctate.

Occurrence Middle-Late Eocene subsurface sediments, Rataria, Kutch, Gujarat (Kar and Saxena 1981; Kar 1985); Palaeocene, Tura Formation, Nongwal Bibra area, East Garo Hills, Meghalaya (Saxena et al. 1996); Early Eocene, Tarkeshwar Formation, Cambay Basin, Gujarat (Samant and Phadtare 1997); Late Palaeocene-Early Eocene, Palana Formation, Bikaner, Rajasthan (Kar and Sharma 2001).

Arecipites keralensis Rao & Ramanujam 1978 (Plate 1, Fig. 11)

Description (Rao and Ramanujam 1978, p. 412): Pollen grains heteropolar, ellipsoidal to rhombic, 38–45 × 19–24 μm in size, monosulcate, sulcus long, narrow all along, margins slightly thickened, ends blunt. Exine up to 2.5 μm thick, sexine thicker than nexine, surface finely punctate, puncta locally in regular longitudinal rows.

Occurrence Miocene, Quilon beds, Padappakkara, Kerala (Rao and Ramanujam 1978).

Arecipites magnus Ramanujam 1987 (Plate 1, Fig. 12)

Description (Ramanujam 1987, pp. 28, 29): Pollen grains heteropolar, amb broadly oval, 39.5–43 × 31–33 μm, monosulcate, sulcus long, uniformally broad with rounded end and incrassate margin. Exine up to 2 μm thick, punctitectate, columella fine, surface finely punctate.

Occurrence Miocene, Warkalli Beds, Kerala (Ramanujam 1987).

Arecipites wodehousei Samant & Phadtare 1997 (Plate 1, Fig. 13)

Description (Samant and Phadtare 1997, p. 16): Pollen grains heteropolar, almost circular, 40–41 μm in diameter, generally isolated, rarely in tetrads, monosulcate; sulcus long but not reaching to the margins, distinctly closed, slightly crassimarginate; exine about 2 μm thick, semitectate, columellate, sexine and nexine of same thickness, micropunctate.

Occurrence Early Eocene, Tarkeshwar Formation, Rajpardi, Cambay Basin, Gujarat (Samant and Phadtare 1997).

Genus *Arengapollenites* Kar 1985

Type Species *Arengapollenites achinatus* Kar 1985

Generic Description (Kar 1985, p. 70): Pollen grains monocolpate, oval-subcircular in shape, achinate, spines so arranged on margins of colpus that they interlock the aperture on invagination, interspinal exine laevigate.

Botanical Affinity *Arenga* (Arecaceae).

Species Recorded from India

Arengapollenites achinatus Kar 1985 (Plate 1, Fig. 14)

Description (Kar 1985, p. 70): Pollen grains monocolpate, colpus distinct, extending end to end, equally broad, oval-subcircular in shape, spinose, spines with broad base and pointed tip, sparsely placed except in apertural region where they are closely placed on two margins in alternate fashion so as to close the aperture on invagination. Exine 1–2 μm thick, interspinal exine laevigate.

Occurrence Early Eocene, Naredi Formation, Kutch, Gujarat (Kar 1985); Early Eocene, Rajpardi lignite mine, Cambay Basin, Gujarat (Kar and Bhattacharya 1992); Early Eocene, Tarkeshwar Formation, Rajpardi, Cambay Basin, Gujarat (Samant and Phadtare 1997); Late Palaeocene-Middle Eocene subsurface sediments, Tiruchirapalli, Tamil Nadu (Saxena and Khare 2004).

Arengapollenites clavatus Samant & Phadtare 1997 (Plate 1, Fig. 15)

Description (Samant and Phadtare 1997, p. 17): Pollen grains heteropolar, subprolate to prolate, 39–49 × 26–35 μm in size, monosulcate; sulcus generally widely open, broad on either ends, extending from end to end, beset with alternately arranged clavae; exine about 1 μm thick (excluding clavae), sexine and nexine undifferentiable, clavate, clavae about 4–5 μm long, spindle-shaped, blunt tipped with nexinal thickening at base, interclavate exine is smooth.

Occurrence Early Eocene, Tarkeshwar Formation, Rajpardi, Cambay Basin, Gujarat (Samant and Phadtare 1997).

Arengapollenites ovatus Kar & Bhattacharya 1992 (Plate 2, Fig. 1)

Description (Kar and Bhattacharya 1992, p. 254): Pollen generally oval, sometimes circular, 40–50 × 34–54 μm, colpus distinct, extended from end to end. Exine up to 2 μm thick, spinose, spines sparsely placed, in apertural region spines arranged in alternate fashion in a manner of sharply toothed jaws, inter-spinal space laevigate.

Occurrence Early Eocene, Rajpardi lignite mine, Cambay Basin, Gujarat (Kar and Bhattacharya 1992).

Genus *Asperitricolporites* Venkatachala & Rawat 1972

Type Species *Asperitricolporites pilosus* Venkatachala & Rawat 1972

Generic Description (Venkatachala and Rawat 1972, p. 314): Pollen grains with more or less oval amb in equatorial view, tricolporate. Colpi long; ora circular to lalongate. Exine tectate, spinulose with fine sharp tips.

Species Recorded from India

Asperitricolporites pilosus Venkatachala & Rawat 1972 (Plate 13, Fig. 10)

Description (Venkatachala and Rawat 1972, p. 314): Pollen oval in equatorial view; 27–23.1 μm; tricolporate. Colpi long, narrow reaching almost up to the poles; ora circular to slightly lalongate. Exine about 2.7 μm thick; tectate; sexine thicker than nexine, suprategillar ornamentation echinate, spinules small, about 1 μm long, with pointed to blunt tips; surface pseudoreticulate.

Occurrence Palaeocene-Eocene, Cauvery Basin, Tamil Nadu (Venkatachala and Rawat 1972).

Genus *Assamiapollenites* Singh 1975

Type Species *Assamiapollenites brownii* (Biswas 1962) Singh 1975

Generic Description (Singh 1975, p. 102): Pollen grains subcircular to circular in outline, non-aperturate, medium-sized. Exine thick, ornamented with pila, bacula, grana or some other thick projections forming a reticulum. Lumina of the reticulum quite big, rounded to pentagonal. Extrema lineamenta more or less rounded and sometimes possesses a shallow depression near the equator.

Species Recorded from India

Assamiapollenites brownii (Biswas 1962) Singh 1975 (Plate 1, Fig. 1)

Description (Biswas 1962, p. 42): Exine thick, coarsely reticulate, with rod-like projections that have rounded tips; diameter 42 μm.

Occurrence Early Eocene, Tura Formation, Garo Hills, Meghalaya (Biswas 1962).

Remarks *Nonaperturipites brownii* Biswas (1962, p. 42, Plate 7, Fig. 40) was transferred to *Assamiapollenites brownii* by Singh (1975).

B

Genus *Bacubrevitricolpites* Rao & Ramanujam 1982

Type Species *Bacubrevitricolpites rotundus* Rao & Ramanujam 1982

Generic Description (Rao and Ramanujam 1982, p. 72): Pollen grains subprolate equatorially, amb rounded, tricolpate, brevicolpate, colpi narrow, exine intectate, beset with numerous prominent baculae, heads of bacules generally rounded in surface view.

Species Recorded from India

Bacubrevitricolpites rotundus Rao & Ramanujam 1982 (Plate 9, Fig. 4)

Description (Rao and Ramanujam 1982, p. 72): Pollen grains isopolar, amb rounded, subprolate equatorially, polar diameter 21–26.5 μm; tricolpate, brevicolpate, colpi narrow, margins thin, ends blunt. Exine 2–3 μm thick, intectate, baculate, bacules fine, densely distributed all over, up to 2.2 μm high, heads of bacules usually rounded in surface view.

Occurrence Miocene, Quilon Formation, Kerala (Rao and Ramanujam 1982).

Genus *Barringtoniapollenites* Kar & Sharma 2001

Type Species *Barringtoniapollenites retipilatus* Kar & Sharma 2001

Generic Description (Kar and Sharma 2001, p. 133): Pollen grains oval in equatorial and subcircular in polar views, trisyncolpate, exine pilate-baculate, sometimes providing negative reticulum in surface view.

Botanical Affinity and Ecology *Barringtonia* (Barringtoniaceae); trees of coastal swamps.

Species Recorded from India

Barringtoniapollenites retibaculatus Kar & Sharma 2001 (Plate 11, Fig. 11)

Description (Kar and Sharma 2001, p. 134): Pollen grains oval-subcircular in equatorial view, size range 60–65 × 30–52 μm, exine 1–2 μm thick, nexine as thick as sexine, retibaculate, bacula closely placed, trisyncolpate.

Occurrence Late Palaeocene-Early Eocene, Palana Formation, Bikaner, Rajasthan (Kar and Sharma 2001, p. 133).

Barringtoniapollenites retipilatus Kar & Sharma 2001 (Plate 11, Fig. 12)

Description (Kar and Sharma 2001, p. 134): Pollen oval-subcircular, size range 50–55 × 35–46 μm, exine thick, sexine and nexine well-developed, retipilate, pila 2–3 μm long, with swollen tip and narrow base, pila closely placed to form pseudoreticulation, trisyncolpate, colpi extending almost from one end to the other in equatorial view.

Occurrence Late Palaeocene-Early Eocene, Palana Formation, Bikaner, Rajasthan (Kar and Sharma 2001).

Genus *Bithnokiapollis* Kar & Sharma 2001

Type Species *Bithnokiapollis striatus* Kar & Sharma 2001

Generic Description (Kar and Sharma 2001, p. 133): Pollen grains oval in equatorial and subtriangular in polar views, size range 35–72 × 31–64.5 μm, exine thin, tectate, columellate, finely reticulate, striato-reticulate, retipilate, tricolpate-tricolporoidate, colpi funnel-shaped in polar view, slit-like, almost extending end to end in equatorial view.

Species Recorded from India

Bithnokiapollis retipilatus Kar & Sharma 2001 (Plate 9, Fig. 5)

Description (Kar and Sharma 2001, p. 133): Pollen grains generally found in equatorial view, size range 35–38 × 31–57 μm, exine 1–2.5 μm thick, retipilate, pila 2–3 μm long, closely placed, with swollen tips and narrow base, pila aligned in rows to form pseudoreticulation, tricolpate-colporoidate, colpi long, well-defined, extending from pole to pole in equatorial view.

Occurrence Late Palaeocene-Early Eocene, Palana Formation, Bikaner, Rajasthan (Kar and Sharma 2001).

Bithnokiapollis striatus Kar & Sharma 2001 (Plate 9, Fig. 6)

Description (Kar and Sharma 2001, p. 133): Pollen grains mostly found in polar view, subcircular to subtriangular, size range 36–72 × 33–64.5 μm, exine thin, tectate, columellate, finely reticulate, columellae aligned in rows to form striations, striato-reticulate, lumina small, mesh size 0.5 μm, tricolpate, colpi distinct.

Occurrence Late Palaeocene-Early Eocene, Palana Formation, Bikaner, Rajasthan (Kar and Sharma 2001).

Genus *Bombacacidites* Couper 1960

Type Species *Bombacacidites bombaxoides* Couper 1960

Generic Description (Couper 1960, p. 53): Isopolar, tricolporate, colpi short, planaperturate (apertures midway in sides of grains in polar view); peroblate, triangular amb clearly sculptured.

Emended Description (Krutzsch 1970, p. 278): Shape flat, lenticular to spheroidal, equatorial outline typically over-round, in part subcircular or convexly triangular. Three equatorial germinals in the over-rounded forms in the centre of sides, in other aspects, in part, also apparently displaced towards the corners. Exocolpi short, perpendicular and symmetrical with respect to the equator. Endoaperture distinctly thickened or only weakly so. Surface always clearly reticulate, but in the corner regions often with finer meshes. Sometimes (secondarily) with fine sculpture such as spinules, granulate, pilate etc., small to medium-large in size, rarely large forms.

Botanical Affinity and Ecology Bombacaceae; found in varied ecological conditions.

Species Recorded from India

Bombacacidites assamicus Sah & Dutta 1968 (Plate 13, Fig. 11)

Description (Sah and Dutta 1968, p. 189): Size range 30–36 µm; amb roundly triangular; planaperturate, 3-colporate, longicolpate; exine finely reticulate.

Occurrence Miocene, Tipam Sandstone, Assam (Sah and Dutta 1968).

Bombacacidites minutus Rao & Ramanujam 1982 (Plate 13, Fig. 13)

Description (Rao and Ramanujam 1982, p. 77): Pollen grains isopolar, amb subtriangular, spheroidal equatorially, polar diameter 18.5–23 µm, planaperturate, tricolporate, brevicolpate, colpae narrow, margin thin, ends blunt, ora slightly lalongate. Exine 1.5 µm thick, sexine as thick as nexine, columella fine, surface finely reticulate.

Occurrence Miocene, Quilon Beds, Kerala (Rao and Ramanujam 1982); Miocene, Warkalli Beds, Kerala (Ramanujam 1987).

Bombacacidites triangulatus Kar 1985 (Plate 13, Fig. 14)

Description (Kar 1985, p. 137): Pollen grains triangular in polar view, 56–80 × 55–79 µm. Tricolpate, colpi situated midway between sides, margin thickened, brevicolpate, pores not traceable. Exine 1.5–2.5 µm thick, sexine thicker than nexine, reticulate, reticulum broader in middle and shorter at apices.

Occurrence Oligocene, Maniyara Fort Formation, Kutch, Gujarat (Kar 1985); Early Miocene, Alleppey, Kerala (Rao et al. 1993); Late Palaeocene-Early Eocene, Palana Formation, Bikaner, Rajasthan (Kar and Sharma 2001).

C

Genus *Calophyllumpollenites* Sah & Kar 1974

Type Species *Calophyllumpollenites rotundus* Sah & Kar 1974

Generic Description (Sah and Kar 1974, p. 172): Pollen grains subcircular-circular with three constrictions due to apertures. Tricolporate, colpi long, pore margin more or less thickened. Exine thick, more or less laevigate to finely reticulate.

Species Recorded from India

Calophyllumpollenites rotundus Sah & Kar 1974 (Plate 13, Fig. 15)

Description (Sah and Kar 1974, pp. 172, 173): Pollen grains subcircular-circular in polar view; 36–48 μm, 3 colporate, colpi long, pore margin thickened. Exine thick, laevigate to finely reticulate.

Occurrence Early Eocene, Palana lignite, Bikaner, Rajasthan (Sah and Kar 1974).

Genus *Caprifoliipites* Wodehouse 1933

Type Species *Caprifoliipites viridi-fluminis* Wodehouse 1933

Generic Description (Wodehouse 1933. p. 481): Pollen grains very small, ellipsoidal, tricolporate with furrows long and pointed, with conspicuous internally projecting furrow rims and pore rims; exine coarsely reticulate.

Species Recorded from India

Caprifoliipites descretus Venkatachala & Rawat 1972 (Plate 14, Fig. 1)

Description (Venkatachala and Rawat 1972, p. 309, 310): Pollen grain with roundly triangular amb in polar view; 19.2–20.5 × 19.2 20.5 μm; tricolporate. Colpi short, brevicolpate, up to 5.3 μm deep in polar view, wide in equatorial region, tapering towards poles; ora lalongate. Exine about 1.3 μm thick, tectate, sexine thicker than nexine, nexinous thickening below the ora prominent, sexine formed of thin pila, with slightly rounded heads, pila uniformly placed; surface finely reticulate, lumina more or less circular, about 1 μm in diameter, uniformly wide, muri as thick as lumina.

Occurrence Palaeocene-Eocene, Cauvery Basin, Tamil Nadu (Venkatachala and Rawat 1972).

Caprifoliipites wodehousei Ramanujam 1966 (Plate 14, Fig. 4)

Description (Ramanujam 1966, p. 168): Pollen grains isopolar, prolate spheroidal, 3-zonicolpate, equatorial diameter 23.8 μm. Amb lobed, lobes bulging, fossaperturate, colpi long but not reaching poles, ends sharply pointed, margin often faint, ora rounded, conspicuous, projecting from general surface. Exine 1.8 μm thick, sexine as thick as nexine, columellae distinct, surface reticulate, meshes very small, microluminate, muri look beaded under high magnification.

Occurrence Miocene-Pliocene, Cuddalore Series, Neyveli, Tamil Nadu (Ramanujam 1966).

Genus *Chiranthodendronpollenites* Kar & Sharma 2001

Type Species *Chiranthodendronpollenites bikanerensis* Kar & Sharma 2001

Generic Description (Kar and Sharma 2001, p. 138): Pollen grains subcircular in polar and oval in equatorial views, size range 52–68 × 57–62 μm, exine 1–2 μm thick, columellate, reticulate, scrobiculate, tricolporate, colpi long, funnel-shaped in polar view, slit-like in equatorial view, pore distinct.

Species Recorded from India

Chiranthodendronpollenites bikanerensis Kar & Sharma 2001 (Plate 14, Fig. 5)

Description (Kar and Sharma 2001, p. 138): Pollen grains mostly found in polar view, a few specimens also found in equatorial view, exine 1–2 μm thick, tectate, columellate, uniformly reticulate, scrobiculate, mesh 0.5 μm wide, tricolporate, colpi long, margin thickened, funnel-shaped in polar and slit-like in equatorial views, pores well-developed, lalongate in polar view and poorly developed in equatorial view, margin thin.

Occurrence Late Palaeocene-Early Eocene, Palana Formation, Bikaner, Rajasthan (Kar and Sharma 2001).

Genus *Clavadiporopollenites* Ambwani & Singh 1996

Type Species *Clavadiporopollenites raneriensis* Ambwani & Singh 1996

Generic Description (Ambwani and Singh 1996, p. 140): Pollen grains circular in equatorial plane, concave at poles; 20–25 × 45–55 μm in size, body tire-shaped excluding apertural projections, diporate, pores opposite to each other (juxtaposed twin), circular, annulated, surrounded by 3–7 clavae. Exine microreticulate to foveolate having verrucate, gemmate and clavate sculptural elements.

Botanical Affinity Arecaceae.

Species Recorded from India

Clavadiporopollenites raneriensis Ambwani & Singh 1996 (Plate 23, Fig. 3)

Description (Ambwani and Singh 1996, p. 140): Pollen grains subcircular, body tire-shaped excluding apertural projections, 20–25 × 45–55 μm in size; diporate, pores opposite to each other (juxtaposed twin), circular, 22–27 μm wide, sunken, annulated, surrounded by finger-like projections (clavae), 3–7 in number around each pore, arranged in a whorl. Exine microreticulate to foveolate, lumina less than 1 μm wide, muri 1–2 μm thick, muri fused and raised at places forming verrucate, gemmate, clavate sculptural elements. Verrucae and gemmae heads 5–7 μm wide, clavae 12–22 in number, 8–15 μm long, 5–7 μm wide, tips sometimes bifurcated.

Occurrence Palaeocene subsurface sediments, Bikaner, Rajasthan (Ambwani and Singh 1996); Late Palaeocene, Akli Formation, Barmer, Rajasthan (Tripathi et al. 2003; Tripathi and Srivastava 2010).

Genus *Clavapalmaedites* Rao & Ramanujam 1978

Type Species *Clavapalmaedites hammenii* Rao & Ramanujam 1978

Generic Description (Rao and Ramanujam 1978, p. 414): Grains heteropolar, ellipsoidal-oval, or oblong, monosulcate, sulcus long, margins thin or slightly incrassate, membrane and sulcus margins densely beset with pila or clava. Exine densely and coarsely clavate-pilate; clava/pila processes free, heads rounded to triangular, locally grouped to give a reticuloid look in surface view.

Botanical Affinity Arecaceae

Species Recorded from India

Clavapalmaedites hammenii Rao & Ramanujam 1978 (Plate 2, Fig. 2)

Description (Rao and Ramanujam 1978, p. 414): Pollen grain ellipsoidal to oval, 25–31 × 16–21 μm, monosulcate, sulcus long, extending all along long axis of grain, uniformly wide, margins incrassate, ends rounded or blunt, sulcus membrane and margins beset with clava or pila. Exine 3.8–4.5 μm thick, sexine much thicker than nexine, clavate-pilate, clava up to 3.5 μm high, heads free, rounded to angular with a tendency towards local reticuloid grouping.

Occurrence Miocene, Quilon Beds, Padappakkara, Kerala (Rao and Ramanujam 1978); Miocene, Warkalli Beds, Kerala (Ramanujam 1987); Early Eocene, Tarkeshwar Formation, Rajpardi, Cambay Basin, Gujarat (Samant and Phadtare 1997); Early Eocene, Surat lignite, Cambay Basin, Gujarat (Samant and Tapaswi 2001).

Clavapalmaedites heteroclavatus Kulkarni et al. 1985 (Plate 2, Fig. 3)

Description (Kulkarni et al. 1985, p. 298): Pollen grains oblate, heteropolar, ellipsoidal in distally polar view, 18–21 × 29–33 μm, monosulcate, sulcus tenuimarginate, extended up to either end of the longest axis, beset with prominent clavae; sexine thicker than nexine, exine about 3.5 μm thick, studded with evenly distributed heterogeneous clavae, clavae along the sulcus margin smaller than those on distal side, 1.5–2.5 μm in size, heads rounded, interpilar exine psilate.

Occurrence Miocene, Ratnagiri Beds, subsurface sediments, Ratnagiri, Maharashtra (Kulkarni et al. 1985).

Genus *Clavaperiporites* Ramanujam 1966

Type Species *Clavaperiporites jacobii* Ramanujam 1966

Generic Description (Ramanujam 1966, p. 190): Pollen grains oblate spheroidal or rounded, pantoporate (periporate), oligoporate, pores 5–10 in number. Sexine beset with clavae placed on baculate platform. Clavae polygonal or triangular in surface view, aligned in such a way as to form a loose reticulum.

Botanical Affinity Linaceae.

Species Recorded from India

Clavaperiporites clavatus Navale & Misra 1979 (Plate 26, Fig. 14)

Description (Navale and Misra 1979, p. 236): Pollen grain prolate or circular, 40–48 × 40–46 μm. Pores 5–10, panporate, faintly visible due to ornamentation, each pore 2–3 μm in diameter, tenuimarginate, margin often irregular. Exine 5–6.5 μm thick (including sculptural elements), sexine thicker than nexine, surface ornamented with densely packed clavae, clavae heads generally triangular, occasionally polygonal, arranged in somewhat crotonoid pattern, forming a loose reticulum, meshes polygonal, lumina smooth.

Occurrence Miocene, Neyveli lignite, Tamil Nadu (Navale and Misra 1979).

Clavaperiporites heteroclavatus Samant & Phadtare 1997 (Plate 26, Fig. 15)

Description (Samant and Phadtare 1997, p. 20): Pollen grains spheroidal, 45–60 μm in diameter, pantoporate; pores generally distinct, about 1–2 μm in diameter, margins smooth, sometimes irregular due to surrounding clavae, tenuimarginate; exine about 5–6 μm thick, rarely up to 10 μm thick, intectate, sexine thicker than nexine, nexine smooth, continuous, sometimes appears undulating; heteroclavate, some specimens on surface focusing show clavae heads forming an indistinct croton pattern, clavae heads circular to subcircular.

Occurrence Early Eocene, Tarkeshwar Formation, Rajpardi, Cambay Basin, Gujarat (Samant and Phadtare 1997).

Clavaperiporites homoclavatus Samant & Phadtare 1997 (Plate 10, Fig. 5)

Description (Samant and Phadtare 1997, p. 19): Pollen grains spheroidal, 36–55 μm in diameter, periporate; pores distinct, deeply placed, circular, 2–3 μm in diameter, margins irregular, surrounded by clavae; exine 4–6 μm thick, intectate, sexine thicker than nexine, nexine continuous, generally smooth, clavate, homoclavate, clavae bases fused locally forming reticulate pattern (non-crotonoid).

Occurrence Early Eocene, Tarkeshwar Formation, Rajpardi, Cambay Basin, Gujarat (Samant and Phadtare 1997).

Clavaperiporites jacobii Ramanujam 1966 (Plate 10, Fig. 6)

Description (Ramanujam 1966, p. 190): Pollen grains oblate spheroidal or rounded, 28–35 μm. Pores 5–10 in number, pantoporate, each pore 2 μm in diameter, tenuimarginate, pore margin often irregular. Exine 4 μm thick including sculpture, surface densely clavate, clavae heads polygonal or triangular forming a loose reticulum, meshes polygonal, lumina smooth.

Occurrence Miocene-Pliocene, Cuddalore Series, Neyveli, Tamil Nadu (Ramanujam 1966); Miocene, Quilon Beds, Kerala (Rao and Ramanujam 1975); Early-Middle Miocene, Warkalli Beds, Kerala (Ramanujam and Rao 1977, Ramanujam 1987); Neogene, Ratnagiri Beds, Sindhudurg, Maharashtra (Saxena and Misra 1990); Early Miocene subsurface sediments, Alappuzha (Alleppey), Kerala (Rao et al. 1993).

Clavaperiporites ramanujamii Samant & Phadtare 1997 (Plate 10, Fig. 8)

Description (Samant and Phadtare 1997, p. 20): Pollen grains spheroidal, 37–42 μm in diameter, pantoporate; pores endexinous, about 2 μm in diameter, distinctly tenuimarginate, margins irregular, surrounded by 8 relatively small clavae; exine about 3 μm thick, semitectate, sexine (about 2.5 μm) thicker than nexine (about 0.5 μm), homoclavate, clavae heads free, about 6 triangular clavae heads form distinct crotonoid arrangement, along subsurface level clavae base unite laterally forming a distinct reticulate pattern.

Occurrence Early Eocene, Tarkeshwar Formation, Rajpardi, Cambay Basin, Gujarat (Samant and Phadtare 1997)

Genus *Clavasyncolpites* Rao & Ramanujam 1982

Type Species *Clavasyncolpites gracilis* Rao & Ramanujam 1982

Generic Description (Rao and Ramanujam 1982, p. 73): Pollen grains with triangular-rounded triangular amb, tricolpate, syncolpate. Exine intectate, clavate, clavae locally seen in a reticuloid alignment.

Species Recorded from India

Clavasyncolpites gracilis Rao & Ramanujam 1982 (Plate 11, Fig. 13)

Description (Rao and Ramanujam 1982, p. 73): Pollen grains isopolar, amb triangular to rounded triangular, sides convex, polar diameter 22–31 μm; tricolpate, syncolpate, colpi wide and gaping at equatorial region, margins slightly thickened and beset with clavate-baculate processes. Exine 2.5–3.5 μm thick, sexine thicker than nexine, intectate, clavate, clavae 1.5–2 μm high, locally mixed with bacules, densely and uniformly distributed all over, here and there seen in reticuloid alignment.

Occurrence Miocene, Quilon Formation, Kerala (Rao and Ramanujam 1982).

Genus *Compositoipollenites* Potonié 1951 ex Potonié 1960

Type Species *Compositoipollenites rhizophorus* (Potonié 1934) Potonié 1951

Generic Description (Potonié 1960, p. 105): Pollen with spherical shape with 3 pores (and colpi), exine all over ornamented with spines which are often more or less irregular and commonly curved; spines 3.4 μm long in type specimen; between their bases a negative reticulum can be observed.

Botanical Affinity Asteraceae.

Species Recorded from India

Compositoipollenites alleppeyensis Rao 1995 (Plate 14, Fig. 6)

Description (Rao 1995, pp. 328, 331): Pollen grains subcircular in shape. Size range 72–90 × 65–72 μm (including spines). Tricolporate, zonaperturate, colpi not distinct due to thickening of annulus around pore, pore distinct. Exine 2.5–3 μm thick, sexine and nexine not differentiated, intectate, spines sparsely placed, 4–6 μm long, 2–3 μm wide, 16 spines present, bacula present in between spines, bacula closely placed. Surface showing fine negative reticulum.

Occurrence Eocene-Early Miocene subsurface sediments, Alappuzha (Alleppey), Kerala (Rao 1995).

Genus *Coneopollis* Venkatachala & Rawat 1973

Type Species *Coneopollis decorus* Venkatachala & Rawat 1973

Generic Description (Venkatachala and Rawat 1973, p. 251): Pollen grains spherical in polar view; tricolporate; 25.5 × 25.5 μm in size, brevicolpate. Pores broadly oval with thickened margin. Colpi small, wedged-shaped, exine thick, sexine thicker than nexine, tegillate, coarsely reticulate with broader lumina, muri narrow, polygonal, nexine thickened along the pores forming a broad collar.

Emended Description (Venkatachala and Rawat in Thanikaimoni et al. 1984, p. 16): Pollen grains more or less circular in polar view, tricolporate with costae-pori; ectoaperture with tapering ends; sexine reticulate, tectate, thicker than nexine.

Species Recorded from India

Coneopollis decorus Venkatachala & Rawat 1973 (Plate 14, Fig. 8)

Description (Venkatachala and Rawat 1973, p. 251): Pollen grains spheroidal in polar view, tricolporate; 25.5 × 25.5 μm, brevicolpate. Pores broadly oval with thickened margin, wedged-shaped. Exine up to 2.5 μm thick; sexine thicker than nexine, tegillate, coarsely reticulate with broader polygonal lumina and narrow muri. Nexine thicker along the pores forming a broad collar.

Occurrence Oligocene-Miocene, Cauvery Basin, Tamil Nadu (Venkatachala and Rawat 1973).

Coneopollis reticulatus (Venkatachala and Rawat 1973) Venkatachala & Rawat in Thanikaimoni et al. 1984 (Plate 14, Fig. 9)

Description (Venkatachala and Rawat 1973, p. 253): Pollen grain spheroidal, amb more or less circular, triporate, 33.66 × 34.4 μm. Pores equatorially placed, circular to oval in shape, about 3.3 μm wide, bordered by 1.3 μm wide sexinous thickenings. Exine about 2 μm thick, sexine thicker than nexine, tectate, columella distinct with rounded distal ends, surface reticulate.

Emended Description (Venkatachala and Rawat in Thanikaimoni et al. 1984, pp. 16, 17): Pollen grains more or less circular in polar view, tricolporate, E = 30–35 μm; ectoaperture with tapering ends, t = 12–15 μm; endoaperture more or less circular, d = 3–3.5 μm; exine 2 μm in thickness, sexine thicker than nexine, tectate, tectum perforate, lumina 1–2 μm, muri 0.5 μm, columella distinct, costae 1.5 μm thick.

Occurrence Oligocene-Miocene, Cauvery Basin, Tamil Nadu (Venkatachala and Rawat 1973).

Remarks *Maculoporites reticulatus* Venkatachala & Rawat (1973, p. 253, Plate 5, Fig. 28) was originally described as triporate but in fact it is tricolporate. This taxon was therefore merged with *Coneopollis*.

Genus *Coramandalipollis* Venkatachala & Rawat 1973

Type Species *Coramandalipollis pauciornatus* (Venkatachala and Rawat in Thanikaimoni et al. 1984)

Generic Description (Venkatachala and Rawat 1973, p. 247): Pollen grains spheroidal, circular to roundly triangular in polar view, tricolporate, brevicolpate; ora more distinct than colpi, with thick margo. Sexine equal or thicker than nexine, variously ornamented.

Restated Description (Venkatachala and Rawat in Thanikaimoni et al. 1984, p. 18): Pollen grains more or less circular in polar view, tricolporate, ectoaperture short, costaepori distinct, sexine reticulate, nexine more or less endosculptured.

Remarks Please refer Thanikaimoni et al. (1984, p. 18).

Species Recorded from India

Coramandalipollis pauciornatus Venkatachala & Rawat 1973
(Venkatachala and Rawat in Thanikaimoni et al. 1984) (Plate 14, Fig. 10)

Description (Venkatachala and Rawat 1973, p. 247): Pollen grains with rounded amb, tricolporate, 17.2–17.8 × 18.5 µm in size; colpi very short, about 4 µm deep in polar view; margo present, pores lalongate, encircled by an annulus, about 2 µm wide, Exine about 1.3 µm thick, sexine as thick as nexine or thicker, scabrate with vermiculate-pseudoreticulate ornamentation.

Occurrence Eocene subsurface sediments, Cauvery Basin, Tamil Nadu (Venkatachala and Rawat 1973).

Genus *Couperipollis* Venkatachala & Kar 1969

Remarks Please refer to the remarks under *Neocouperipollis* (Venkatachala and Kar 1969) Kar & Kumar 1987

Genus *Crotonisulcites* Rao & Ramanujam 1978

Type Species *Crotonisulcites grandis* Rao & Ramanujam 1978

Generic Description (Rao and Ramanujam 1978, p. 414): Pollen grains heteropolar, amb ovoid, monosulcate, sulcus long, locally broader, margins thickened, exine tectate, heavily ornamented with free pilate-clavate processes with triangular heads, aligned in a reticuloid pattern of sculpture.

Species Recorded from India

Crotonisulcites grandis Rao & Ramanujam 1978 (Plate 2, Fig. 4)

Description (Rao and Ramanujam 1978, p. 414): Pollen grains ovoid, 45–56 × 37.5–43 µm, monosulcate, sulcus long, 8.5–12 µm broad, more or less dumble-shaped, sulcus margin incrassate and beset with clava. Exine 5–6.5 µm thick, sexine much thicker than nexine, tectate, pilate-clavate, pila and clava processes 3.75–4–5 µm high, densely and uniformly placed all over, heads triangular, free heads of these processes aligned in a reticuloid pattern resembling crotonoid pattern of sculpture.

Occurrence Miocene, Quilon Beds, Padappakkara, Kerala (Rao and Ramanujam 1978).

Genus *Cruciferoipollenites* Navale & Misra 1979

Type Species *Cruciferoipollenites elongatus* Navale & Misra 1979

Generic Description (Navale and Misra 1979, p. 226): Pollen grain isopolar, prolate, size range 47–120 × 33–110 μm. Tricolpate to tricolporoidate, colpi long, generally uneven, distinct and extending almost up to the poles. Exine 3–4 μm thick, sexine thicker than nexine. Surface tegillate, granulate to microreticulate, grana comparatively denser at the poles.

Species Recorded from India

Cruciferoipollenites elongatus Navale & Misra 1979 (Plate 9, Fig. 7)

Description (Navale and Misra 1979, p. 227): Isopolar pollen grain, prolate, mostly seen in equatorial compression while in polar compression they are seldom intact. Size range 47–56 × 33–54 μm. Tricolpate to tricolporoidate, colpi quite long, margins generally uneven, distinct, extending up to the pole, mesocolpi convex towards periphery. Exine 3.5–4 μm thick, thinning gradually towards colpi, sexine nearly double the thickness of nexine, tegillate, surface structure finely microreticulate, lumina of varying shapes.

Occurrence Miocene, Neyveli lignite, Tamil Nadu (Navale and Misra 1979); Miocene Cuddalore Formation, Neyveli Lignitefield, Tamil Nadu (Singh et al. 1992).

Genus *Cryptopolyporites* Venkatachala & Kar 1969

Type Species *Cryptopolyporites cryptus* Venkatachala & Kar 1969

Generic Description (Venkatachala and Kar 1969, p. 174): Pollen grains circular in polar view. Panporate, pores hidden, not easily recognizable, covered with baculate processes as in non-porate region.

Botanical Affinity and Ecology Gonystylaceae; fresh water swamp forest.

Species Recorded from India

Cryptopolyporites cryptus Venkatachala & Kar 1969 (Plate 10, Fig. 9)

Description (Venkatachala and Kar 1969, p. 175): Pollen grains circular in overall shape, 30–50 μm. Panporate, pores are concealed by the ornamentation pattern of the exine; aperturate as well as non-aperturate regions ornamented by closely set bacula.

Occurrence Early Eocene, Laki Series, Matanomadh, Kutch, Gujarat (Venkatachala and Kar 1969); Early Eocene, Naredi Formation, Kutch, Gujarat (Kar 1985); Early Eocene, Tarkeshwar Formation, Rajpardi, Cambay Basin, Gujarat (Samant and Phadtare 1997).

Cryptopolyporites indicus Saxena & Khare 2004 (Plate 9, Fig. 10)

Description (Saxena and Khare 2004, p. 87): Pollen grains subcircular in shape, size 41–44 × 37–39 μm, polyporate, periporate, pores circular, up to 2 μm in diameter, uniformly distributed. Exine 2 μm thick, sexine as thick as nexine, punctate to granulate.

Occurrence Late Palaeocene-Middle Eocene subsurface sediments, Tiruchirapalli, Tamil Nadu (Saxena and Khare 2004).

Genus *Ctenolophonidites* von Hoeken-Klinkenberg 1966

Type Species *Ctenolophonidites costatus* (von Hoeken-Klinkenberg 1964) von Hoeken-Klinkenberg 1966

Generic Description (von Hoeken-Klinkenberg 1966, p. 42): Stephanocolpate pollen grains with endexinous thickenings between the furrows. These thickenings fusing at poles forming an irregular ring-like pattern, as in recent Ctenolophonaceae.

Remarks *Ghoshiacolpites globatus* Sah & Kar (1970) has following diagnosis "Pollen grains circular to subcircular, polycolpate, colpi 5–9, mesocolpia broad with a ridge in the middle. Mesocolpial ridge well-developed, joining each other in middle region to form a starfish-like appearance." Diagnosis of this form suggests its inclusion in the genus *Ctenolophonidites.*

The genus *Ctenolophonidites* is commonly found in Indian Early Palaeogene sediments. Pollen grains of this genus are hexa to octacolporate and provided with exine thickenings which are united in apocolpial region. Some species of this genus resemble with pollen of extant *Ctenolophon* which are inhabitants of Central Africa and southeast Asia. First record of *Ctenolophon* pollen is from Late Cretaceous of Africa and occurrence of this pollen in Indian Palaeogene is indicative of its dispersal from Africa. These pollen grains, though described under different names, became widely distributed during Early and Middle Eocene and continued up to Miocene time (Ambwani et al. 1981; Thanikaimoni et al. 1984).

Botanical Affinity and Ecology *Ctenolophon* (Ctenolophonaceae); fresh water swamp forest.

Species Recorded from India

Ctenolophonidites erdtmanii Ramanujam & Rao 1973 (Plate 12, Fig. 7)

Description (Ramanujam and Rao 1973, pp. 211, 212): Pollen grains rounded in polar view, 40–50 μm in diameter, 7-colpate, radially symmetrical, isopolar. Colpi gaping, 12–16 μm long with incrassate margins. Colpi margins not much raised in polar view. One conspicuous irregular ring-like endexinous thickening around apocolpium of each hemisphere, interconnected only at one or two places by meridional mesocolpial ridges or radial costae, 2–3.5 μm thick. Majority of mesocolpial radial costae subequatorial as seen in polar view. More than one radial costae often in some mesocolpia. Additional thickenings present inside apocolpial ring, joining the latter locally. Unthickened wall finely granular.

Occurrence Miocene, Warkalli Beds, Varkala, Kerala (Ramanujam and Rao 1973).

Ctenolophonidites keralensis Ramanujam & Rao 1973 (Plate 12, Fig. 8)

Description (Ramanujam and Rao 1973, p. 212): Pollen grains stellate-spherical in polar view, 47–66 μm in diameter, suboblate equatorially, 7 or 8 colpate, radially symmetrical, isopolar. Colpi broad, gaping, 13–19 μm long, more or less spindle-shaped, with incrassate margins. One conspicuous but highly irregular, sinuous, ring-like endexinous thickening in apocolpium of each hemisphere, interconnected by meridional mesocolpial radial costae, 2.5–4.5 μm thick. Radial costae 2–4 in each mesocolpium. Additional incomplete ring-like sinuous thickening within main apocolpial ring; unthickened wall smooth.

Occurrence Miocene, Warkalli Beds, Varkala, Kerala (Ramanujam and Rao 1973).

Ctenolophonidites magnus Saxena & Khare 2004 (Plate 12, Fig. 9)

Description (Saxena and Khare 2004, P. 84): Pollen grains circular in polar view, size 61 μm, decacolpate, colpi distinct, 11 μm long, 5–6 μm broad. Exine 4.5 μm thick, granulate, irregular strips of endexinous thickenings developed in apocolpial region.

Occurrence Late Palaeocene-Middle Eocene subsurface sediments, Tiruchirapalli, Tamil Nadu (Saxena and Khare 2004).

Ctenolophonidites neyveliensis Ambwani et al. 1981 (Plate 12, Fig. 10)

Description (Ambwani et al. 1981, p. 103): Pollen grains stellate, spherical in polar view, 33–35 μm in diameter, radially symmetrical, 8–9 colpate, isopolar; colpi 5 × 1.5 μm, tenuimarginate; one conspicuous ring of endexinous thickening present around apocolpium of each hemisphere; radial costa arrow-head shaped present in each mesocolpium, about 5 μm thick; area between radial costa and outer margin of mesocolpium lamellate; unthickened wall of pollen grain profusely granulose, grana more or less 1–1.5 μm in diameter.

Occurrence Miocene, Neyveli lignite, Tamil Nadu (Ambwani et al. 1981).

Ctenolophonidites palaeoparvifolius Kar & Jain 1981 (Plate 12, Fig. 11)

Description (Kar and Jain 1981, p. 121): Pollen grains subcircular, 39–50 × 30–41 μm, mostly 7-colpate, colpi short. Exine 2–4 μm thick, sexine thicker than nexine, laevigate and intrabaculate. Exinal thickening uniformly spread like a blanket in apocolpate and mesocolpate regions.

Occurrence Neogene, Quilon and Varkala, Kerala (Kar and Jain 1981).

Ctenolophonidites saadii Ramanujam & Rao 1973 (Plate 12, Fig. 12)

Description (Ramanujam and Rao 1973, pp. 212, 213): Pollen grains rounded to stellate-rounded in polar view, 31–50 μm in diameter, suboblate equatorially, 5–7 colpate, radially symmetrical, isopolar. Colpae short, 10–13 μm long, 6.5 μm broad, with incrassate margins, ends blunt. Locally anastomosing irregular strips of endexinous thickening in apocolpium of each hemisphere. Pairs of radial costae in mesocolpia arching over colpus ends on either side and fusing; neighboring arcs of radial costae seldom join towards apocolpial region. Unthickened wall minutely granular.

Occurrence Miocene, Warkalli Beds, Varkala, Kerala (Ramanujam and Rao 1973); Late Palaeocene-Middle Eocene subsurface sediments, Tiruchirapalli, Tamil Nadu (Saxena and Khare 2004).

Ctenolophonidites stellatus Navale & Misra 1979 (Plate 12, Fig. 13)

Description (Navale and Misra 1979, p. 228): Pollen grains isopolar, spherical (in polar view), radially symmetrical; polar diameter 42–57 μm, 9-colpate, colpi 10–15 μm deep, funnel-shaped with distinct incrassate margins and ring-like endexinous thickening around apocolpium of each pole, interconnected with 2.5–4.5 μm thick radial costae going into each mesocolpium. Additional irregular thickenings are also present inside each apocolpial ring. Polar area inside apocolpial ring smooth or minutely punctate.

Occurrence Miocene, Neyveli Lignitefield, Tamil Nadu (Navale and Misra 1979); Palaeocene, Neyveli Formation, Neyveli, Tamil Nadu (Siddhanta 1986).

Genus *Cupanieidites* Cookson & Pike 1954

Type Species *Cupanieidites major* Cookson & Pike 1954

Generic Description (Cookson and Pike 1954, p. 210): Grains medium sized, tricolporate; amb triangular to almost circular; angulaperturate; arci present, either uniting to form more or less distinct polar islands, or not, reaching the poles, sexine as thick as nexine, distinctly to faintly reticulate.

Botanical Affinity Sapindaceae.

Species Recorded from India

Cupanieidites cooksoniae Ramanujam 1987 (Plate 14, Fig. 11)

Description (Ramanujam 1987, p. 34): Pollen grains isopolar, amb triangular, sides convex, 20–25 μm, angulaperturate, tricolporate, syncolpate, colpal margins incrassate, ora lalongate; exine up to 2 μm thick, tectate, surface coarsely granular to locally microreticulate

Occurrence Miocene, Warkalli Beds, Alappuzha (Alleppey), Kerala (Ramanujam 1987).

Cupanieidites flaccidiformis Venkatachala & Rawat 1972 (Plate 14, Fig. 12)

Description (Venkatachala and Rawat 1972, p. 321): Pollen grain with roundly triangular amb, sides convex; trisyncolporate; 23.7–25 × 26.44 μm. Arci prominent, not forming a polar island. Pore aspidate. Exine about 1 μm thick, sexine thicker than nexine, reticulate, flangy, equatorial flange developed, only restricted to the mesocolpium area.

Occurrence Palaeocene-Eocene, Cauvery Basin, Tamil Nadu (Venkatachala and Rawat 1972).

Cupanieidites granulatus Jain et al. 1973 (Plate 14, Fig. 13)

Description (Jain et al. 1973, p. 158): Pollen grains tricolporate, angulaperturate, syncolpate, triangular, 18–25 μm, sides slightly convex. Exine infragranulose.

Occurrence Palaeocene, Barmer Sandstone, Barmer, Rajasthan (Jain et al. 1973).

Cupanieidites punctatus Rao & Ramanujam 1982 (Plate 14, Fig. 14)

Description (Rao and Ramanujam 1982, p. 80): Pollen grains isopolar, amb triangular, sides convex, polar diameter 19.5–27 μm, zonaperturate, tricolporate, syncolpate, apocolpial area triangular, colpal margins thin, ora slightly incrassate, lalongate. Exine up to 2 μm thick, sexine as thick as nexine, punctitectate, columella indistinct, surface finely punctate.

Occurrence Miocene, Quilon Beds, Kerala (Rao and Ramanujam 1982).

D

Genus *Dakshinipollenites* Navale & Misra 1979

Type Species *Dakshinipollenites tripakshii* Navale & Misra 1979

Generic Description (Navale and Misra 1979, p. 229): Pollen grain isopolar, triangular to subtriangular (in polar view), subprolate to suboblate (equatorial view); 48–70 × 48–66 μm in size. Tricolporate, angulaperturate, colpi long, narrow and deep, apocolpi small. Ora faint to indistinct, lalongate, slit-like. Exine thin (1.5 μm) in the middle part of the mesocolpi where it is projected out like a hump, gradually thickens laterally, being thickest (3–4.5 μm) on the colpi margins, subsurface microgranulate to granulate, granulation more prominent towards colpi margins.

Emended Description (Singh and Misra 1991b, p. 205): Pollen isopolar, tricolpate, triangular to subtriangular in polar view, prolate-oblate spheroidal or subspheroidal in equatorial view; exine thin in middle of mesocolpium where ambitus shows a bulge, gradually thickening towards colpi and thickest at colpus margins, exine comprised of a perforate tectum, columellae layer and unstructured nexine, columellae shorter at mesocolpium, gradually becoming longer towards colpi margin; ornamentation finely granulate to finely reticulate as gaps between the granules appear like pits or lumina.

Species Recorded from India

Dakshinipollenites tripakshii Navale & Misra 1979 (Plate 14, Fig. 15)

Description (Navale and Misra 1979, p. 229): Isopolar pollen grain, subprolate to suboblate in equatorial view, triangular to subtriangular in polar view. Size range 48–70 × 48–66 μm. Tricolporate, angulaperturate, longicolpate, colpi deep and narrow, extending quite up to the poles, apocolpium small. Pore obscure, lalongate, like a slit. Exine thin (1.5–2 μm) in the middle part of the mesocolpi (clearly visible in the polar view) projecting outwards like a hump, gradually thickening laterally, being thickest (3–4.5 μm) at the colpi margins. Microgranulate to granulate, granulation becoming more prominent towards colpi margins.

Emended Description (Singh and Misra 1991b, p. 206): Pollen 40–90 μm, isopolar, triangular to subtriangular in polar view, prolate-oblate spheroidal to subspheroidal in equatorial view, tricolpate, colpi medium to long; exine comprised of a perforate tectum, columellae layer and unstructured smooth nexine; exine 1–4 μm in the middle of mesocolpium, gradually thickening towards colpi and 3–8 μm at colpus margin where exine 2–4 times thicker than at the equatorial bulge; nexine smooth, gradually thickens towards margin, sexine and nexine thickness is more or less equal; surface ornamentation finely granulate to finely reticulate.

Occurrence Miocene, Neyveli lignite, Tamil Nadu (Navale and Misra 1979; Singh and Misra 1991b).

Genus *Densiverrupollenites* Tripathi & Singh 1984

Type Species *Densiverrupollenites eocenicus* Tripathi & Singh 1984

Generic Description (Tripathi and Singh 1984, p. 154): Pollen grains spheroidal to subspheroidal in polar view; tricolporate, apertures subequatorially placed, brevicolpate, ora circular to slightly lalongate; exine verrucate, verrucae beset with gemmae.

Botanical Affinity Alangiaceae.

Species Recorded from India

Densiverrupollenites eocenicus Tripathi & Singh 1984 (Plate 15, Fig. 1)

Description (Tripathi and Singh 1984, p. 155): Pollen grains spheroidal to subspheroidal; 75–95 μm in size, tricolporate, brevicolpate; apertures subequatorially placed; exine 1.5–2.5 μm thick, sexine as thick as nexine, verrucate to gemmate, verrucae/gemmae very closely placed.

Occurrence Late Eocene, Kopili Formation, Jaintia Hills, Meghalaya (Tripathi and Singh 1984); Late Eocene (Barail Group), Mokokchung District, Nagaland (Mandal 1997); Late Eocene, Kopili Formation, North Cachar Hills District, Assam (Trivedi and Saxena 2000).

Genus *Dermatobrevicolporites* Kar 1985

Type Species *Dermatobrevicolporites dermatus* (Sah and Kar 1970) Kar 1985

Generic Description (Kar 1985, p. 89): Pollen grains triangular-subtriangular, tricolporate, brevicolpate, pores large, margin thickened, exine thick, laevigate, finely intrastructured.

Species Recorded from India

Dermatobrevicolporites dermatus (Sah and Kar 1970) Kar 1985 (Plate 15, Fig. 2)

Description (Sah and Kar 1970, p. 140): Pollen grains subtriangular-subcircular, 30–38 × 28–35 µm, 3-orate, ora with thickened margin. Exine thick, more or less laevigate to finely intrastructured.

Restated Description (Kar 1985, pp. 89, 90): Pollen grains triangular-subtriangular, 30–38 × 28–35 µm, always found in polar view, tricolporate, colpi short, 10–15 µm in length, pore well-developed, 5–10 µm in diameter. Exine 2–4 µm thick, sexine thicker than nexine, more or less psilate or sometimes weakly intrastructured.

Occurrence Early Eocene, Naredi Formation, Kutch, Gujarat (Sah and Kar 1970; Kar 1985); Early Eocene Cambay Shale, Tarkeshwar, Surat, Gujarat (Samant and Tapaswi 2001).

Remarks Forms described as *Triorites dermatus* Sah and Kar (1970) was reinvestigated by Kar (1985) and was found to be tricolporate. As *Triorites* (Erdtman) Couper (1953) is triporate, this species was placed under a new genus by Kar (1985).

Dermatobrevicolporites exaltus Kar 1985 (Plate 15, Fig. 3)

Description (Kar 1985, p. 90): Pollen grains subtriangular in polar view, 40–48 µm. Tricolporate, brevicolpate, colpi 10–20 µm, pore mostly distinct, margin thickened, 4–8 µm thick and at apertural region 8–12 µm thick. Sexine as thick as nexine, generally laevigate, sometimes weakly intrastructured.

Occurrence Early Eocene, Naredi Formation subsurface sediments, Kutch, Gujarat (Kar 1985).

Dermatobrevicolporites globatus (Kar and Saxena 1981) Kar 1985 (Plate 15, Fig. 4)

Description (Kar 1985, p. 123): Pollen grains subcircular, 60–89 × 56–78 µm. Tricolporate, brevicolpate, colpi 8–12 µm long, pore distinct, margin thickened. Exine 1–2 µm thick, sexine thicker than nexine, laevigate to granulose.

Occurrence Middle-Late Eocene subsurface sediments, Rataria, Kutch, Gujarat (Kar and Saxena 1981; Kar 1985).

Remarks *Myricipites globatus* Kar and Saxena (1981) was transferred to the genus *Dermatobrevicolporites.*

Dermatobrevicolporites triangulus (Saxena 1979) Kar 1985 (Plate 15, Fig. 5)

Description (Saxena 1979, p. 137): Pollen grains triangular-subtriangular in polar view, 40–60 μm. Tricolporate, brevicolpate, pores distinct, pore margin thickened. Exine up to 4 μm thick, laevigate to finely intrastructured.

Occurrence Palaeocene, Matanomadh Formation, Kutch, Gujarat (Saxena 1979; Kar 1985).

Remarks Saxena (1979) described *Meliapollis triangulus* which is triangular to subtriangular in shape. As, the forms assigned to the genus *Meliapollis* are sub-circular in shape, this species was transferred to *Dermatobrevicolporites* by Kar (1985).

Genus *Dicolpopollis* Pflanzl 1956 emend. Potonié 1966

Type Species *Dicolpopollis kockelii* Pflanzl 1956

Emended generic description (Potonié 1966, p. 160): Microspores with two opposite sulculi that run close to and approximately parallel to the equator; the two sulculi are only slightly displaced towards the distal side in the type species. Equator of the spore partly of more or less axe-shaped in outline. Meridian somewhat oval; exine more or less microreticulate to microverrucate.

Botanical Affinity and Ecology *Calamus* (Rattan palms); Tropical.

Species Recorded from India

Dicolpopollis edavensis Rao & Ramanujam 1978 (Plate 8, Fig. 1)

Description (Rao and Ramanujam 1978, p. 418): Pollen grains isopolar, more or less barrel-shaped in polar view, 28–35 × 18–22 μm, dicolpate. Exine up to 1.8 μm thick, sexine as thick as nexine, punctitectate, columella distinct, surface punctate, punctae fine, aligned in transverse rows (at right angles to colpae).

Occurrence Miocene, Quilon Beds, Edavai, Kerala (Rao and Ramanujam 1978).

Dicolpopollis fragilis Salujha et al. 1972 (Plate 8, Fig. 2)

Description (Salujha et al. 1972, p. 279): Golden yellow, oval, measuring 36.8–51.2 × 27.2–44.8 μm; dicolpate, colpi 8.5 μm deep; exine over 1 μm thick, finely granulate, grana less than 1 μm wide.

Occurrence Palaeocene, Garo Hills, Meghalaya (Salujha et al. 1972); Kannur (Cannanore), Kerala (Ramanujam 1982).

Dicolpopollis microreticulatus Rao & Ramanujam 1978 (Plate 8, Fig. 3)

Description (Rao and Ramanujam 1978, p. 418): Pollen grains isopolar, polar view barrel-shaped with somewhat rounded ends, 29–38.5 × 22–30 μm, dicolpate, colpae fairly long, gradually tapering, margins thin. Exine 2 μm thick, columella distinct, surface very finely reticulate, meshes very small, angular.

Occurrence Miocene Quilon Beds, Padappakkara, Kerala (Rao and Ramanujam 1978); Miocene, Kannur (Cannanore), Kerala (Ramanujam 1982).

Dicolpopollis proprius Salujha et al. 1972 (Plate 8, Fig. 4)

Description (Salujha et al. 1972, p. 279): Golden yellow, oval, size 30–31.6 × 40–43 μm; sculptured at two ends; dicolpate, colpi 3.5–4.2 μm deep; exine 1–1.5 μm thick, faintly reticulate, muri less than 1 μm thick with equally wide lumina.

Occurrence Palaeocene, Garo Hills, Meghalaya (Salujha et al. 1972).

Dicolpopollis triangularis Kulkarni et al. 1985 (Plate 8, Fig. 5)

Description (Kulkarni et al. 1985, p. 304): Pollen grains heteropolar, triangular laterally, 32–33.5 μm in diameter, dicolpate, colpi slightly crassimarginate, open, longicolpate, sexine thicker than nexine, exine 1.7 μm thick, scrobiculate to microreticulate.

Occurrence Miocene, Ratnagiri Beds, Maharashtra (Kulkarni et al. 1985).

Genus *Dipterocarpuspollenites* Kar 1992

Type Species *Dipterocarpuspollenites retipilatus* (Kar and Jain 1981) Kar 1992

Generic diagnosis (Kar 1992, pp. 81, 82): Pollen grains subcircular and oval in polar and equatorial views respectively. Size range 51–69 μm. Tricolpate, very rarely tricolporoidate, colpi long. Exine 1.5–3.5 μm thick, pilate, pilae up to 3 μm long, 1–1.5 μm broad, closely placed, sometimes fused together, pila provide pseudoreticulate appearance in surface view.

Emended Description (Mandal and Rao 2001, p. 352): Pollen spheroidal in polar and subprolate in equatorial view. Tricolpate, longicolpate, sexine thicker than nexine. Exine columellate, with (strongly) perforate tectum, which tends to pushed up between each two columellae. Surface pseudoreticulate at lower focus and verrucate at top focus.

Botanical Affinity *Dipterocarpus* (Dipterocarpaceae).

Species Recorded from India

Dipterocarpuspollenites retipilatus (Kar and Jain 1981) Kar 1992 (Plate 9, Fig. 8)

Synonyms: (after Kar 1992)

1981 *Tricolpites baculatus* Kar & Jain, *The Palaeobotanist* 27: 120, Plate 3, Figs. 81–82, *non:Tricolpites baculatus* Jain et al. 1973 (now *Albertipollenites baculatus*)

1982 *Retitricolpites dipterocarpoides* Rao & Ramanujam, *The Palaeobotanist* 30: 69–70, Plate 1, Fig. 3

Description (Kar and Jain 1981, p. 119): Pollen grains subcircular in equatorial view, 55–66 μm. Tricolpate, longicolpate, colpi long, prominent, wide in polar view. Exine up to 2 μm thick, sexine thicker than nexine, reticulate, forming negative reticulum.

Restated Description (Mandal and Rao 2001, p. 352): Pollen grains rounded triangular in polar view, 45–50 × 49–60 μm. Tricolpate, longicolpate, colpi 19.5 μm deep, often gaping at equator. Exine 2 μm thick, uniformly thick, sexine thicker than nexine, tectate, columellae 1.5 μm long, with strongly perforate tectum, which tends to pushed up between each two columellae. Surface appears pseudoreticulate at low focus and verrucate at top focus.

Occurrence Miocene, Kerala Basin (Kar and Jain 1981; Rao and Ramanujam 1982).

Genus *Dorreenipites* Biswas 1962

Type Species *Dorreenipites platydesma* Biswas 1962

Generic Description (Biswas 1962, p. 37): Polar compression; exine thick, faintly reticulate in polar region, reticulate in subequatorial region, finely papillate near the equatorial (?) girdle; characteristically anisopolar, prominent thickening in the endexine at the base of the pores; 39 μm in size.

Emended Description (Misra et al. 1996, p. 335): Pollen heteropolar, rounded triangular to subcircular in polar view; trilatiporate, concavo-convex in equatorial view; trilatiporate (rarely 2-, 4-, 5-porate), pores borne on well-developed projections, circular, subequatorial, surrounded by endexinal (nexinal) thickening; exine comprising a perforate tectum, columellate layer (base of pila) and nexine; exine structure reticulate, occasionally psilate with imperforate tectum, reticulum usually finer and fainter towards poles varying from distinct to obscure or absent with a circular psilate area at poles.

Remarks Biswas (1962) instituted the genus *Dorreenipites* with insufficient morphological description. Ramanujam (1966) established a genus *Trilatiporites* that resembles *Dorreenipites* but did not designate its type species. Navale and Misra (1979), considering the priority, maintained *Dorreenipites* Biswas (1962) and considered *Trilatiporites* Ramanujam (1966) its junior synonym.

Botanical Affinity *Sclerosperma mannii* (Arecaceae).

Species Recorded from India

Dorreenipites bengalensis Mathur & Chopra 1987 (Plate 24, Fig. 6)

Description (Mathur and Chopra 1987, p. 124): Triporate pollen grains, free anisopolar, amb more or less rounded, diameter 38–45 μm; pores strongly subequatorial, grandiporate, 5–7 μm in diameter, situated on raised neck-like projections, 2–4 μm thick; surface near equatorial region microreticulate, in polar region completely smooth.

Occurrence Early Eocene subsurface sediments, Bengal Basin, West Bengal (Mathur and Chopra 1987).

Dorreenipites cooksoniae (Ramanujam 1966) Navale & Misra 1979 (Plate 24, Fig. 7)

Description (Ramanujam 1966, p. 184): Pollen grains markedly heteropolar, suboblate to oblate spheroidal, equatorial diameter 40–48 μm. Amb subtriangular, pores three, prominently nonequatorial, 3.3–6.6 μm in diameter, surrounded by a slightly thickened rim, all in one hemisphere, aligned in a triangular contour. Exine 4–6 μm thick, sexine slightly thicker than nexine, columellae fine, often faint, surface smooth both at meso- and apoporia.

Occurrence Miocene, Neyveli Lignite, Neyveli, Tamil Nadu (Ramanujam 1966).

Remarks *Trilatiporites cooksoniae* Ramanujam (1966) was transferred to the genus *Dorreenipites* by Navale and Misra (1979).

Dorreenipites distinctus Navale & Misra 1979 (Plate 24, Fig. 8)

Description (Navale and Misra 1979, p. 235): Pollen heteropolar, amb rounded triangular to subcircular, 44–58 × 48–54 μm in size. 2–4 grandiporate, latiporate, pore more or less circular, 5–7 μm in diameter, surrounded by 2–2.5 μm thick annulus, non-equatorial, all in one hemisphere. Exine 4–4.5 μm thick, sexine equal to or slightly thicker than nexine, tegillate, pila distinct, surface evenly granulate to microreticulate.

Occurrence Miocene, Neyveli Lignite, Neyveli, Tamil Nadu (Navale and Misra 1979).

Dorreenipites erdtmanii (Ramanujam 1966) Navale & Misra 1979 (Plate 24, Fig. 9)

Description (Misra et al. 1996, p. 337): Pollen size 30–50 μm in maximum diameter, trilatiporate, pore 3–7 μm in diameter; exine 3–5 μm thick, comprising a perforate tectum, columellate layer and smooth nexine; exine structure reticulate at both hemispheres.

Occurrence Miocene, Neyveli Lignite, Neyveli, Tamil Nadu (Ramanujam 1966).

Remarks *Trilatiporites erdtmanii* Ramanujam 1966 and *Trilatiporites noremii* Ramanujam 1966 were transferred to *Dorreenipites erdtmanii* by Navale and Misra (1979).

Dorreenipites kutchensis (Venkatachala and Kar 1969) Navale & Misra 1979 (Plate 24, Fig. 10)

Description (Venkatachala and Kar 1969, p. 172): Pollen grains mostly roundly triangular, 30–35 × 27–33 μm. Triporate, pores more or less subequatorial in position, margin of pores not appreciably thickened. Exine thick, sexine thicker than nexine, finely microreticulate.

Occurrence Early Eocene, Laki Series, Matanomadh, Kutch, Gujarat (Venkatachala and Kar 1969).

Remarks *Trilatiporites kutchensis* Venkatachala & Kar (1969) was transferred to the genus *Dorreenipites* by Navale and Misra (1979).

Dorreenipites medius Navale & Misra 1979 (Plate 24, Fig. 11)

Description (Navale and Misra 1979, p. 236): Heteropolar pollen grains, amb rounded triangular to subcircular, 55–64 × 55–58 µm in size. Trilatiporate, pores large, 6–7 µm in diameter, circular to elliptical, surrounded by 2 µm thick rim, non-equatorial to prominently non-equatorial, all in one hemisphere. Exine 3.5–4.5 µm thick, sexine thicker or equal to nexine, surface tegillate, pila distinct, granulate to microgranulate.

Occurrence Miocene, Neyveli Lignite, Neyveli, Tamil Nadu (Navale and Misra 1979).

Dorreenipites neyveliensis Misra et al. 1996 (Plate 24, Fig. 12)

Description (Misra et al. 1996, p. 339): Pollen rounded triangular to subcircular in polar view, 34–83 µm in maximum diameter; pores 4–10 µm in diameter; exine 2–7 µm thick, comprising a perforate tectum, columellate layer and nexine; exine structure psilate at convex pole and reticulate at apoporium.

Occurrence Miocene, Cuddalore Formation, Neyveli Lignitefield (Misra et al. 1996).

Dorreenipites platydesma Biswas 1962

Description (Biswas 1962, p. 37): As Generic description.

Occurrence Lower-Middle Eocene, Tura Formation, Tura, Garo Hills, Meghalaya (Biswas 1962).

Dorreenipites psilatus Misra et al. 1996 (Plate 24, Fig. 13)

Description (Misra et al. 1996, p. 343): Pollen rounded triangular to subcircular in polar view, 29–60 µm in maximum diameter; pores 3–6 µm in diameter; exine 2–5 µm thick, comprising an imperforate tectum and smooth nexine separated by a thin gap (random columellae developed by degradation); surface of the pollen smooth.

Occurrence Miocene, Cuddalore Formation, Neyveli Lignitefield (Misra et al. 1996).

Dorreenipites sellingii (Ramanujam 1966) Navale and Misra 1979 (Plate 24, Fig. 14)

Description (Ramanujam 1966, p. 184): Pollen grains markedly heteropolar, prominently flattened along poles, equatorial diameter 45 μm. Amb roundly triangular. Pores three, prominently non-equatorial, 4–7 μm in diameter, rounded or elliptical, all in one hemisphere, markedly aspidate. Exine 4.6 μm thick, crassisexinous, tegillate, columellate, columellae closely placed giving a radially striated look at periphery, surface coarsely granular to microreticulate at mesocolpia.

Occurrence Miocene, Neyveli lignite, Tamil Nadu (Ramanujam 1966); Miocene, Cuddalore Formation, Neyveli Lignitefield (Misra et al. 1996).

Remarks *Trilatiporites sellingii* Ramanujam (1966) was trans ferred to the genus *Dorreenipites* by Navale and Misra (1979).

Dorreenipites undulatus Misra et al. 1996 (Plate 24, Fig. 15)

Description (Misra et al. 1996, p. 339): Pollen rounded triangular to subtriangular in polar view, 36–82 μm in maximum diameter; pores 5–10 μm in diameter; exine 4–9 μm thick, comprising a perforate tectum, columellate layer and undulating nexine; psilate circular area present at poles of both hemispheres.

Occurrence Miocene, Cuddalore Formation, Niyveli Lignitefield (Misra et al. 1996).

Genus *Dracaenoipollis* Sah & Kar 1970

Type Species *Dracaenoipollis circularis* Sah & Kar 1970

Generic Description (Sah and Kar 1970, p. 129): Pollen grains circular subcircular, size range 22–36 × 19–34 μm. Aperture single, circular-subcircular. Exine more or less laevigate, weakly granulose or intrastructured.

Species Recorded from India

Dracaenoipollis circularis Sah & Kar 1970 (Plate 22, Fig. 1)

Description (Sah and Kar 1970, p. 130): Pollen grains circular to subcircular, 22–36 × 19–34 μm. Aperture single, circular to subcircular, distinct. Exine more or less laevigate, weakly granulose or intrastructured.

Occurrence Early Eocene, Laki Series, Kutch, Gujarat (Sah and Kar 1970).

E

Genus *Echidiporites* Mandal 1997

Type Species *Echidiporites indicus* Mandal 1997

Generic Description (Mandal 1997, p. 102): Pollen grains drum-shaped, biporate, pores big, simple, placed at either side of pole; exine thin, tectate, dominantly spinose, mixed with few bacula and coni, interspinal areas finely granular, sexine thicker than nexine, uniformly thick.

Species Recorded from India

Echidiporites indicus Mandal 1997 (Plate 23, Fig. 4)

Description (Mandal 1997, p. 102): Pollen drum-shaped, 50–60 × 65–78 μm, diporate, pores on either side of pole, big, 15–17 μm in diameter, without thickening, pore margin granular, free of spines; exine thin, 1.5 μm, tectate, nexine thinner (0.5 μm) than sexine, surface sculpture dominantly spinose, mixed with bacula and coni, closely and irregularly placed. Spine tips blunt, rarely branched and varying in size, 1.5–5 μm long. Interspinal space finely granular.

Occurrence Late Eocene, Barail Group, Moriani-Mokokchung Road, Mokokchung District, Nagaland (Mandal 1997).

Genus *Echimonoporopollis* Saxena et al. 1991

Type Species *Echimonoporopollis grandiporus* Saxena et al. 1991

Generic Description (Saxena et al. 1991, p. 46): Pollen grains spherical, subspherical or ovoid, small to medium sized. Monoporate. Exine generally thin, spinose, spines vary in shape and size.

Species Recorded from India

Echimonoporopollis Circularis Samant & Tapaswi 2001 (Plate 22, Fig. 2)

Description (Samant and Tapaswi 2001, p. 124): Pollen grains spherical; 27–29 μm in diameter; monoporate, pore 10 μm in diameter, tenuimarginate with slightly irregular margins, spines extending up to spore margins; exine less than 1 μm thick (excluding spines) sexine and nexine undifferentiable; sculpture spinose, spines about 1–1.5 μm long, with curved tips and bulbous base, scarcely placed and evenly distributed, interspinal area smooth.

Occurrence Early Eocene, Surat lignite, Cambay Shale, Surat, Gujarat (Samant and Tapaswi 2001).

Echimonoporopollis grandiporus Saxena et al. 1991 (Plate 22, Fig. 3)

Description (Saxena et al. 1991, p. 48): Pollen grains spherical to subspherical in shape. Size range 30–42 × 29–40 μm (excluding spines). Monoporate, pores distinct, circular to oval, 7–13 μm in diameter, pore margin unthickened. Exine up to 1.5 μm thick, sexine as thick as nexine, spinose. Spines 3–6 μm long, evenly distributed, bases of spines bulbous, tips of spines mostly pointed but may also be blunt. Interspinal area psilate, punctate or microreticulate.

Occurrence Palaeocene Neyveli Formation subsurface sediments, Tiruchirapalli, Tamil Nadu (Saxena et al. 1991); Late Palaeocene-Middle Eocene subsurface sediments, Tiruchirapalli, Tamil Nadu (Saxena and Khare 2004).

Echimonoporopollis neyveliensis Saxena et al. 1991 (Plate 22, Fig. 4)

Description (Saxena et al. 1991, p. 48): Pollen grains spherical to ovoid in shape. Size range 41–46 × 38–45 μm (excluding spines). Monoporate, pore distinct, circular, 10–14 μm in diameter, pore margin unthickened, Exine 0.5–1 μm thick, sexine as thick as nexine, spinose. Spines uniformly distributed, robustly built, broad at base and gradually taper towards the tip, spine bases not bulbous, length of spines 6.2–9.6 μm. Interspinal area punctate-microreticulate.

Occurrence Palaeocene, Neyveli Formation subsurface sediments, Tiruchirapalli, Tamil Nadu (Saxena et al. 1991); Late Palaeocene-Middle Eocene subsurface sediments, Tiruchirapalli, Tamil Nadu (Saxena and Khare 2004).

F

Genus *Favitricolporites* Sah 1967

Type Species *Favitricolporites eminens* Sah 1967

Generic Description (Sah 1967, p. 85): Pollen grains 3-zonaperturate, colporate, triangular to rounded triangular; exine stratification well-defined, ornamentation reticulate.

Species Recorded from India

Favitricolporites complex Sah & Dutta 1968 (Plate 15, Fig. 6)

Description (Sah and Dutta 1968, p. 190): Size range 53–59 μm; amb triangular; 3-zonaperturate, colporate, brevicolpate, grandorate, crassimarginate; exine thick, undifferentiated, ornamentation finely reticulate.

Occurrence Oligocene, Barail Group, Assam (Sah and Dutta 1968).

Genus *Foveotricolpites* Pierce 1961

Type Species *Foveotricolpites sphaeroides* Pierce 1961

Generic Description (Pierce 1961, pp. 24, 49): Tricolpate, colpi elongate with conspicuously thickened aperture margins; subprolate, 18 × 21.5 μm; three polar and one equatorial plane of symmetry; exine *ca.* 1.5 μm thick, tectate, foveolate, foveolae small, closely placed.

Species Recorded from India

Foveotricolpites alveolatus Mandal & Rao 2001 (Plate 9, Fig. 9)

Description (Mandal and Rao 2001, p. 353): Pollen grains subprolate, 60–71 μm. Tricolpate, colpus 50.5 μm long, wide. Exine 3.5–4.5 μm thick, nexine about 1 μm, sexine psilate, pila more or less 3 μm long, capita 2.5–3 μm broad, pila heads generally fused forming 2.5 μm thick perforated tectum. Surface foveolate, lumina circular, more or less 1 μm, 2–3 μm apart, inter-foveolate space smooth.

Occurrence Palaeocene, Therria Formation, Meghalaya (Tripathi and Singh 1985).

Remarks *Foveotricolpites alveolatus* Couper (1953) in Tripathi & Singh (1985) *Non: Foveotricolpites alveolatus* Couper (1953) was transferred to *Foveotricolpites alveolatus* by Mandal and Rao (2001).

Foveotricolpites piercei Rao & Ramanujam 1982 (Plate 9, Fig. 11)

Description (Rao and Ramanujam 1982, p. 71): Pollen grains isopolar, amb rounded, lobes distinct, polar diameter 16–33 μm; zonaperturate, tricolpate, brevicolpate (colpoidate), colpal margins thin, ends rounded or blunt. Exine 1.8 μm thick, sexine thicker than nexine, tectate, tectum perforated, surface foveolate, foveolae of uniform size all over, locally coalescing, up to 3 μm in diameter.

Occurrence Miocene, Quilon Formation, Kerala (Rao and Ramanujam 1982).

Foveotricolpites prolatus Rao & Ramanujam 1982 (Plate 9, Fig. 12)

Description (Rao and Ramanujam 1982, p. 71): Pollen grains isopolar, amb triangular to rounded triangular, lobes distinct, prolate equatorially, 26–38 × 17.5–32 μm; tricolpate, longicolpate, colpi gaping wide at equator, margins uneven, thin, ends pointed. Exine 1.8 μm thick, sexine slightly thicker than nexine, tectate, columella distinct, surface foveo-reticulate, foveolae small, closely placed.

Occurrence Miocene, Quilon Formation, Kerala (Rao and Ramanujam 1982).

Foveotricolpites sijuensis (Baksi 1962) Mandal & Rao 2001 (Plate 9, Fig. 13)

Description (Baksi 1962, p. 16): Equatorial compression; larger diameter 32 μm, surface ornamentation granular to punctate; tricolpate, colpi wide.

Occurrence Lower- Middle Eocene, Meghalaya (Baksi 1962).

Remarks *Tricolpopites sijuensis* Baksi (1962) was transferred to *Foveotricolpites sijuensis* Baksi (1962) by Mandal and Rao (2001).

Foveotricolpites simplex (Rao and Ramanujam 1982) Mandal and Rao 2001 (Plate 9, Fig. 14)

Description (Rao and Ramanujam 1982, p. 71): Pollen grains isopolar, amb triangular, sides convex, polar diameter 25–42 μm; tricolpate, brevicolpate, margins thin, ends pointed. Exine 1.8 μm thick, sexine as thick as nexine, Surface reticulate, brochi small, more or less 1 μm in diameter, finer at mesocolpia.

Occurrence Miocene, Quilon Formation, Kerala (Rao and Ramanujam 1982).

Remarks *Retibrevitricolpites simplex* Rao and Ramanujam (1982) was transferred to *Foveotricolpites simplex* (Rao and Ramanujam 1982) by Mandal and Rao (2001).

Foveotricolpites tamilensis Saxena & Khare 2004 (Plate 9, Fig. 15)

Description (Saxena and Khare 2004, p. 80): Pollen grain oval in equatorial view, size range 56–62 × 48–52.5. Tricolpate, colpi narrow, about 2/3rd of the polar axis. Exine 2–5 μm thick, tectate, columella distinct, sculpture foveo-reticulate, foveolae small and densely placed.

Occurrence Late Palaeocene-Middle Eocene subsurface sediments, Tiruchirapalli, Tamil Nadu (Saxena and Khare 2004).

G

Genus *Gangapollenites* Mathur & Chopra 1987

Remarks *Gangapollenites* Mathur and Chopra (1987) was transferred to *Pseudonothofagidites* by Samant and Phadtare (1997).

Genus *Gemmatriporopollis* Saxena & Khare 1996

Type Species *Gemmatriporopollis triangulus* Saxena & Khare 1996

Generic Description (Saxena and Khare 1996, p. 130): Pollen grains isopolar, subtriangular to subcircular in shape. Triporate pores subcircular, large and equatorially placed. Exine gemmate, inter-gemmate area granulate or finely baculate/conate to microreticulate.

Species Recorded from India

Gemmatriporopollis triangulus Saxena & Khare 1996 (Plate 25, Fig. 1)

Description (Saxena and Khare 1996, p. 130): Pollen grains isopolar, subtriangular to subcircular, size range 44–49 × 42–48 μm. Triporate, pores distinct, subcircular in shape, 8–12 μm in diameter. Exine gemmate, 2–3 μm thick (excluding gemma); gemma large-sized, 3–5 μm in height, uniformly distributed, inter-gemmate area granulate, finely baculate/connate, finely pitted in apocolpial region.

Occurrence Palaeocene-Eocene subsurface sediments, Tiruchirapalli, Tamil Nadu (Saxena and Khare 1996).

Genus *Graminidites* Cookson 1947 ex Potonié 1960

Type Species *Graminidites media* Cookson 1947 ex Potonié 1960

Generic Description (Potonié 1960, p. 111): More or less spherical spores with one pore which is surrounded by annulus, exine thin, finely granulated (infragranulate to granulate). Larger than *Sparganiaceaepollenites* and with thinner and less sculptured exine, hence with more secondary folds.

Emended Description (Krutzsch 1970, p. 12): Monoporate pollen with a distinctly delimited, often not wide pore and almost usually with (only rarely without) a distinct annulus, surface smooth or just finely structured (chagrinate-punctate or finely granulate). Important diagnostic characters for the separation of species are size ratio, pore dimension, annulus width, wall thickness and structure, surface and form.

Botanical Affinity Poaceae.

Species Recorded from India

Graminidites assamicus Sah & Dutta 1968 (Plate 22, Fig. 5)

Description (Sah and Dutta 1968, p. 191): Size range 43–47 µm; amb oval-elliptical; monoporate, pore ulcerate; exine well-differentiated, sexine as thick as nexine, ornamentation psilate.

Occurrence Oligocene-Miocene, Barail and Tipam Groups, Digboi, Assam (Sah and Dutta 1968).

Graminidites granulatus Kar 1985 (Plate 22, Fig. 6)

Description (Kar 1985, p. 137): Pollen grains originally subcircular, but due to irregular folding may be oval, 55–63 × 53–61 µm. Monoporate, pores subcircular, 5–7 µm in diameter, annulus well-developed. Exine 1–1.5 µm thick, granulose, grana about 5 µm high, closely placed, 2–4 µm apart, evenly distributed.

Occurrence Oligocene, Maniyara Fort Formation, Kutch, Gujarat (Kar 1985).

Graminidites maximus Tripathi & Singh 1985 (Plate 22, Fig. 7)

Description (Tripathi and Singh 1985, p. 181): Pollen grains semicircular to ovoid in shape, 82–105 µm in size (holotype 102 × 80 µm), monoporate. Pore more or less circular in outline, annulate and 12–15 µm across (including annulus). Exine about 1 µm thick, undifferentiated, laevigate, associated with a few irregular folds.

Occurrence Palaeocene, Therria Formation, Jaintia Hills, Meghalaya (Tripathi and Singh 1985).

Graminidites pliocenicus Singh & Saxena 1981 (Plate 22, Fig. 8)

Description (Singh and Saxena 1981, p. 176): Subcircular, 40–45 × 36–40 µm. Monoporate, ulcus circular, 4–5 µm in diameter, surrounded by a well-defined annulus. Exine 1 µm thick, laevigate, with prominent but irregular folds.

Occurrence Early Miocene, Kasauli Formation, Solan, Himachal Pradesh (Singh and Saxena 1981).

Genus *Grevilloideaepites* Biswas 1962

Type Species *Grevilloideaepites eocenicus* Biswas 1962

Description of genus (Biswas 1962, p. 39): Exine thick, reticulate; long diameter of the body 50 μm and of the flanges 18 μm; ora 9 μm across.

Emended generic description (Singh and Misra 1991b, p. 207): Pollen isopolar, triangular to subtriangular in polar view, triorate; ora circular, unthickened, present at the extremity of three equatorial arms; arms distinctly demarcated from the central body; central body rounded to subtriangular, open at corners, with or without nexinal undulations or bands; nexinal bands start from base of arms towards central body, bands of an arm do not fuse with the bands of other arm; exine thin, comprised of an incomplete tectum, columellae layer and nexine; surface ornamentation finely granulate to finely reticulate.

Remarks Without assigning any reason Kar (1985, p. 48) considered *Grevilloideaepites* as invalid but Jansonius and Hills (1987) considered this genus as valid, a view with which Singh and Misra (1991b, p. 207) agreed. These authors considered *Triangulorites* (Kar 1985) as junior synonym of *Grevilloideaepites* proposed by Biswas (1962).

Botanical Affinity Proteaceae.

Species Recorded from India

Grevilloideaepites eocenicus Biswas 1962 Emend. Singh & Misra 1991b
(Plate 25, Fig. 2)

Emended Description (Singh and Misra 1991b, p. 209): Pollen isopolar, 50–75 μm (length of an arm and central body), triorate; ora 5–10 μm in diameter; equatorial arm 10–30 μm long and 5–18 μm broad, slightly constricted in the middle; central body rounded, 38–55 μm in size, with or without nexinal bands or undulations; exine thin (1–3 μm), comprised of an incomplete tectum, columellate layer and nexine; nexine more or less equal to sexine, slightly thicker at the base of arms, getting fainter in arms; ornamentation finely reticulate, lumina very small (in other focus appears to be infragranulate).

Occurrence Early Eocene, Tura Formation, Tura-Dalu Road, Garo Hills, Meghalaya (Biswas 1962); Early Eocene, Laki Series, Kutch, Gujarat (Sah and Kar 1970); Palaeocene-Early Eocene, Matanomadh and Naredi formations, Matanomadh (Kar 1985); Palaeocene-Eocene, Neyveli Formation, Neyveli, Tamil Nadu (Siddhanta 1986); Palaeocene, Matanomadh Formation, Matanomadh, Kutch, Gujarat (Saxena 1979); Palaeocene, Lakadong Sandstone, Khasi Hills, Meghalaya (Kar and Kumar 1986); Palaeocene-Eocene subsurface sediments, Kapurdi, Barmer, Rajasthan (Tripathi 1995); Late Palaeocene-Early Eocene, Palana Formation, Bikaner, Rajasthan (Kar and Sharma 2001); Miocene, Cuddalore

Formation, Neyveli Lignite, Tamil Nadu (Singh and Misra 1991b); Late Palaeocene Akli Formation, Barmer, Rajasthan (Tripathi et al. 2003).

Remarks *Triorites bellus* Sah and Kar (1970) and *Triangulorites bellus* Kar (1985) were transferred to *Grevilloideaepites eocenicus* Biswas (1962) by Singh and Misra (1991b).

Grevilloideaepites pachyexinous (Kar and Kumar 1986) Singh & Misra 1991b (Plate 25, Fig. 4)

Description (Kar and Kumar 1986, p. 203): Pollen grains triporate, triangular, interapical margin straight to slightly convex, 45–55 μm. Apertures protruded, annuli straight, pore canal broadening towards inside with thickened margin. Exine 2–4 μm thick, baculate, bacula 2–3 μm long.

Occurrence Palaeocene, Lakadong Sandstone, Khasi Hills, Meghalaya (Kar and Kumar 1986).

Remarks *Triangulorites pachyexinus* Kar and Kumar (1986) was transferred to *Grevilloideaepites pachyexinus* by Singh and Misra (1991b).

Grevilloideaepites trilobatus Mathur & Chopra 1987 (Plate 25, Fig. 5)

Description (Mathur and Chopra 1987, p. 121): Triporate pollen, free, isopolar, amb rounded triangular with three sexinal inflations at the corners; body diameter 25–30 μm, inflations 8–10 × 5–6 μm; exine finely reticulate.

Occurrence Neogene subsurface sediments, Bengal Basin, west Bengal (Mathur and Chopra 1987).

H

Genus *Hippocrateaceaedites* Ramanujam 1966

Type Species *Hippocrateaceaedites vancampoae* Ramanujam 1966

Generic Description (Ramanujam 1966, p. 170): Pollen grains isopolar, oblate to suboblate, 3-zonicolporate. Amb subtriangular; in polar view, the sexine of each lobe is prolonged into characteristic knob-like process on either side of colpus. Colpi broad at equator, suddenly narrowing towards poles, surrounded by a thickened exine, longicolpate, ora prominent, rounded with a thick annulus interrupted at equator. Exine punctitegillate.

Emended Description (Rao and Ramanujam 1982, p. 77): Pollen grains oblate to suboblate, amb triangular to subtriangular, tricolporate, longicolpate, in polar view sexine seen prolonged into characteristic knob-like process on either side of the colpus. Colpal margin thickened, ora prominent, rounded to lalongate with a thickened rim interrupted at equator. Exine punctitectate to finely reticulate.

Species Recorded from India

Hippocrateaceaedites constrictus Sah & Kar 1974 (Plate 15, Fig. 7)

Description (Sah and Kar 1974, p. 170): Pollen grain triangular with marked constriction at apices, 3-colporate, pore margin thickened. Exine 3–5 μm thick, pilate to reticulate, pila 3–5 μm long with prominent bulbous head, closely placed, forming reticulate pattern.

Occurrence Early Eocene, Palana lignite, Palana, Rajasthan (Sah and Kar 1974).

Hippocrateaceaedites quilonensis Rao & Ramanujam 1982 (Plate 15, Fig. 8)

Description (Rao and Ramanujam 1982, p. 78): Pollen grains isopolar, amb triangular, sides convex, polar diameter 18–31 μm, zonaperturate, tricolporate, longicolpate, colpae provided with characteristic prolongation of sexine on either side at its equatorial zone in polar view, ora prominent, lalongate with a distinctly thickened rim. Exine up to 2.5 μm thick, sexine thicker than nexine, tectate, columella distinct, surface finely reticulate (microreticulate), meshes less than 1 μm in diameter.

Occurrence Miocene, Quilon Beds, Edavai, Padappakkara, Kerala (Rao and Ramanujam 1982); Miocene, Warkalli Beds, Kerala (Ramanujam 1987).

Hippocrateaceaedites vancampoae Ramanujam 1966 (Plate 15, Fig. 9)

Description (Ramanujam 1966, p. 171): Pollen grains oblate to suboblate, 30 × 40 μm. 3-zonicolporate, ora conspicuous, rounded, 4–7 μm in diameter, surrounded by a thick luminous, interrupted at equator. Exine 2.5–3.7 μm thick, sexine thicker than nexine, columellae distinct, surface punctitegillate.

Occurrence Miocene-Pliocene, Cuddalore Series, Neyveli, Tamil Nadu (Ramanujam 1966).

I

Genus *Icacinoipollenites* Navale & Misra 1979

Type Species *Icacinoipollenites spinulatus* Navale & Misra 1979

Generic Description (Navale and Misra 1979, p. 228): Pollen grain isopolar, subprolate to oblate, size range 41–68 × 30–56 μm. Tricolpate, brevicolpate, colpi slit-like, surrounded by lip-like thickening. Exine 2.5–5.5 μm thick, crassisexinous, tegillate, spinulate, spinules very small (1 μm). Surface microgranulate.

Species Recorded from India

Icacinoipollenites spinulatus Navale & Misra 1979 (Plate 10, Fig. 1)

Description (Navale and Misra 1979, p. 228): Isopolar pollen grain, subprolate to oblate (in equatorial view), 41–68 × 30–56 μm in size. Tricolpate, brevicolpate, slit-like colpi surrounded by lip-like thickening which is about 2.5–3 μm thick. Exine 3.5–4.5 μm thick, crassisexinous, sexine nearly 3–4 times thicker than nexine, tegillate, spinulate, spinules very small (1 μm long) but their bases are distinctly seen in surface view, surface microgranulate.

Occurrence Miocene, Neyveli lignite, Tamil Nadu (Navale and Misra 1979).

Genus *Incrotonipollis* Jansonius & Hills 1981

Type Species *Incrotonipollis burdwanensis* (Baksi et al. 1979) Jansonius & Hills 1981

Generic Description (Baksi et al. 1979, p. 232): Pollen grains inaperturate, exine sculpture exhibiting characteristic "croton pattern".

Botanical Affinity *Jatropha* (Euphorbiaceae).

Species Recorded from India

Incrotonipollis burdwanensis (Baksi et al. 1979) Jansonius & Hills 1981 (Plate 1, Fig. 2)

Description (Baksi et al. 1979, p. 232): Pollen grains inaperturate, often ruptured to appear as tricolpate, more or less circular in outline, 50–72 μm in size. Exine intectate, 5.5 μm thick, ectexine thicker than endexine, ectexine consists of muroid ridges separated by rounded foveoloid areas at the lower part. Upper part of ectexine consists of well-defined, regularly arranged excresences or warts forming a distinct "croton pattern". The warts are situated on top of the muroid ridges so that any foveoloid area is encircled by six warts arranged in hexagonal pattern with apices of the triangles directed towards the centre of the foveoloid area. The warts, triangular both in surface view and in optical section, are 4.5 µm high, 3.9 µm wide and 3 µm in diameter.

Occurrence Palaeocene, Bengal Basin, West Bengal and Palaeocene, Tura Formation, Meghalaya (Baksi et al. 1979).

Incrotonipollis neyvelii (Baksi et al. 1979) Jansonius & Hills 1981 (Plate 1, Fig. 3)

Description (Baksi et al. 1979, p. 233): Pollen grains inaperturate, 38–48 µm in size, subspheroidal, exine intectate, 3.0–3.5 µm thick, ectexine thicker than endexine, ectexine consists of muroid ridges separated by rounded foveoloid areas at the lower part. The upper part of the ectexine consists of well-defined, regularly arranged clavae forming a distinct croton pattern. The clavae are on top of the muroid ridges so that any muroid area is encircled by 5–7 clavae arranged in a polygonal pattern. The clavae are subtriangular in surface view and 1.5 µm in diameter and 1.5–2.0 µm high. Sometimes there are suggestions of columellae below the muroid ridges and clavae.

Occurrence Eocene, Neyveli Lignite, Tamil Nadu (Baksi et al. 1979); Palaeocene-Eocene, Neyveli Formation, Neyveli, Tamil Nadu (Siddhanta 1986).

Genus *Intectocolporites* Mandal & Guleria 2006

Type Species *Intectocolporites baculatus* Mandal & Guleria 2006

Generic Description (Mandal and Guleria 2006, p. 53): Pollen grains subprolate to prolate, tricolporate, exine intectate.

Species Recorded from India

Intectocolporites baculatus Mandal & Guleria 2006 (Plate 15, Fig. 10)

Description (Mandal and Guleria 2006, p. 53): Pollen grains radially symmetrical, Isopolar, mostly preserved in equatorial position, subprolate in equatorial view and triangular with convex sides in polar view, 27–30 × 28–34 µm (equatorial), 36–40 × 31–36 µm (polar) in size. Tricolporate, colpi 22–24 µm long, tenuimarginate; pore simple, elliptical, 5 × 3 µm. Exine 1.5–2 µm thick; nexine 0.5 µm at meso-colpia and 1 µm thick near apertures; sexine columellate, columellae baculate, free, closely placed, about 1 µm long, 0.5 µm broad. Surface granular at top focus.

Occurrence Early Eocene, Vastan lignite mine, Surat, Gujarat (Mandal and Guleria 2006).

Genus *Intrareticulites* Kar 1985

Type Species *Intrareticulites brevis* (Sah and Kar 1970) Kar 1985

Generic Description (Kar 1985, p. 40): Pollen grains subcircular-subtriangular in polar view. Tricolpate, colpi distinct, long. Exine intrareticulate, structural elements restricted to base of exine.

Restated Description (Mandal and Rao 2001, p. 354): Pollen grains subtriangular, 23 × 25.5–34 × 38 μm in polar view, amb lobate, one lobe angular, mesocolpia convex. Tricolpate, colpi long, 11 μm, pointed, rarely gaping at equator, inconspicuous thickening about 2 μm around colpi especially at polar end. Exine 2 μm at mesocolpia which gradually thickens to 2.5 μm at colpus margins; sexine thicker than nexine, mesocolpial sexine 1.5 μm, both sexine and nexine little thicker near colpi; columellate, columellae distinct, slender, closely placed, 0.5 μm long, longer near colpi; tectate, tectum smooth, 1 μm thick uniformly. Surface granular on top focus and faintly microreticulate in polar area with discontinuous striations in equatorial areas on low focus; lumina nearly circular, about 0.5 μm, muri low.

Botanical Affinity Dipterocarpaceae.

Species Recorded from India

Intrareticulites brevis (Sah and Kar 1970) Kar 1985 (Plate 10, Fig. 2)

Description As above by Mandal and Rao (2001, p.354)

Occurrence Early Eocene, Naredi Formation, Kutch, Gujarat (Sah and Kar 1970; Kar 1985).

Remarks Venkatachala et al. (1989) treated *Intrareticulites brevis* as *Tricolpites brevis* Sah and Kar (1970). Based on reinvestigation of the specimen, Mandal and Rao (2001) found that observation made by Kar (1985) was correct.

Genus *Iridacidites* Ramanujam 1987

Remarks In the present work *Iridacidites* Ramanujam (1987) is treated as junior synonym of *Liliacidites*.

Genus *Iugopollis* Venkatachala & Rawat 1972

Type Species *Iugopollis tetraporites* Venkatachala & Rawat 1972

Generic Description (Venkatachala and Rawat 1972, p. 322): Pollen grains prolate, broadly oval to rectangular with rounded corners or elliptical in equatorial view. Tetraporate, pores circular to elliptical, annulate, annulus not prominent. Equatorial band bordering the apertures and forming an equatorial zone distinctly darker, closely ornamented as opposed to the polar apocolpial area. Exine tectate, sexine thicker than nexine, baculate.

Description (Venkatachala and Rawat in Thanikaimoni et al. 1984, p. 33): Pollen grains elliptic in meridional view, tetracolporate, ectoaperture very narrow and short, endoaperture perpendicular to the ectoaperture, sexine reticulate except at the equatorial region where the underlying nexine is as a dark more or less continuous band.

Botanical Affinity *Sonneratia alba* (Sonneratiaceae).

Species Recorded from India

Iugopollis tetraporites Venkatachala & Rawat 1972 (Plate 20, Fig. 8)

Description (Venkatachala and Rawat in Thanikaimoni et al. 1984, p. 33): same as emended description.

Occurrence Palaeocene-Early Eocene, Cauvery Basin, Tamil Nadu (Venkatachala and Rawat 1972); Eocene, Neyveli lignite, Tamil Nadu (Venkatachala 1973).

J

Genus *Jacobipollenites* Ramanujam 1966

Type Species *Jacobipollenites magnificus* Ramanujam 1966

Generic Description (Ramanujam 1966, p. 181): Pollen grains spheroidal, monoporate. Pore large, 5–10 μm across, rounded or slightly oval, appearing as a crude and rugged cavity in the sculptured wall. Exine retipilate.

Emended generic description (Singh and Misra 1991b, p. 211, 213): Pollen occurring as tetragonal tetrads, dyads and monads; each pollen subspherical to spherical, heteropolar, monoporate; pore large appearing as a rugged cavity in the structural wall, more or less circular (variable if compressed), thickened to annulate; exine comprised of a perforate tectum, columellate layer (base of pila) and nexine, free spines interspersed with pila; surface ornamentation reticulate, homo- to heterobrochate, lumina with or without spines.

Species Recorded from India

Jacobipollenites arthugalensis Rao 1990 (Plate 22, Fig. 9)

Description (Rao 1990, p. 251): Pollen spheroidal, size 55–60 × 45–55 μm; monoporate, pore distinct, 10–12 μm wide, thick annulus present around pore; exine up to 4 μm thick, retipilate, surface showing retipilariate ornamentation.

Occurrence Eocene-Early Miocene, subsurface sediments, Alappuzha (Alleppey), Kerala (Rao 1990).

Jacobipollenites crassimurus Singh & Misra 1991b (Plate 22, Fig. 10)

Description (Singh and Misra 1991b, p. 214): Pollen occurs as tetrads (187–195 μm) and monads (80–90 μm); each pollen heteropolar, spherical to sub-spherical, monoporate; 20–28 μm, more or less circular, annulate, outer pore margin irregular, merged with surface ornamentation; exine (10–15 μm), comprised of a perforate tectum, columellate layer (base of pila) and nexine; nexine 2–3 μm, more or less four times thinner than sexine, spines interspersed with pila, swollen heads of pila (caput) robust, variable in diameter and fused at several places; ornamentation reticulate, heterobrochate, lumina 2–15 μm across, irregular, with free spines; muri thick at few places, more or less equal to the size of lumina, distinctly thickened at joints, headed, occasionally duplipilate.

Occurrence Miocene, Cuddalore Formation, Neyveli Lignitefield, Tamil Nadu (Singh and Misra 1991b).

Jacobipollenites distinctus Singh & Misra 1991b (Plate 22, Fig. 11)

Description (Singh and Misra 1991b, pp. 213, 214): Pollen occurs as tetrads (70–105 × 70–120 μm), dyads (60–85 μm) and monads (37–70 μm); each pollen heteropolar, rounded to subcircular, monoporate; pore 6–14 μm, more or less circular, annulate, outer pore margin irregular, merged with the surface pattern; exine (5–7 μm) comprised of a perforate tectum, columellate layer (base of pila) and nexine; nexine 1.5–3 μm, 2–3 times thinner than sexine; spines interspersed with pila; surface ornamentation reticulate, heterobrochate, lumina more or less polygonal, with or without free spines.

Occurrence Miocene, Cuddalore Formation, Neyveli Lignitefield, Tamil Nadu (Singh and Misra 1991b, Singh et al. 1992).

Jacobipollenites magnificus Ramanujam 1966 (Plate 22, Fig. 12)

Description (Ramanujam 1966, p. 181): Pollen grains spheroidal, monoporate, 38.5–48 μm. Pore rugged, 5–10 μm across, margin often wavy, indistinct. Exine 4 μm thick, crassisexinous, retipilate, homobrochate, brochi polygonal, lumina large, muri beaded, pila at angles of brochi, larger.

Occurrence Miocene-Pliocene, Cuddalore Series, Neyveli, Tamil Nadu (Ramanujam 1966)

Jacobipollenites magniporus Singh & Misra 1991b (Plate 22, Fig. 13)

Description (Singh and Misra 1991b, p. 213): Pollen occurs as tetrads (105–176 × 110–176 μm), dyads (80–132 μm) and monads (50–100 μm); each pollen heteropolar, subspherical to spherical, monoporate; pore 17–25 μm, more or less circular, annulate, outer pore margin irregular, merged with the surface ornamentation; exine (6–10 μm) comprised of a perforate tectum, columellate layer (base of pila) and nexine; nexine 2–4 μm, 2–3 times thinner than sexine, smooth; spines interspersed with pila; surface ornamentation reticulate, heterobrochate, lumina more or less polygonal, 0.5–7 μm across with or without one to several free spines; muri beaded, pila heads not larger at the angles of muri.

Occurrence Miocene, Cuddalore Formation, Neyveli Lignitefield, Tamil Nadu (Singh and Misra 1991b).

Jacobipollenites ramanujamii Samant 2000 (Plate 22, Fig. 14)

Description (Samant 2000 p. 104): Pollen grains spheroidal; 25–30 μm in diameter, monoporate; pore almost circular, small, 4–6 μm in diameter, slightly crassimarginate, pore margins smooth, regular; exine 2–3 μm thick, nexine slightly thinner than sexine, nexine smooth to slightly undulating, columellate, semitectate; reticulate, simple to duplibaculate, heterobrochate, muri 1–1.5 μm thick, lumina about 1 μm wide, circular to polygonal.

Occurrence Early Eocene, Kharsalia Clay Formation, Cambay Basin, Gujarat (Samant 2000).

K

Genus *Kapurdipollenites* Tripathi 1994

Type Species *Kapurdipollenites gemmatus* Tripathi 1994

Generic Description (Tripathi 1994, p. 61): Pollen grains spherical to subspherical; monosulcate, zonisulcate, sulcus long, dividing the grain into two parts. Exine moderately thick, foveolate, imperfectly tectate, foveolate surface covered with densely or sparsely placed verrucae, gemmae or baculae.

Botanical Affinity Arecaceae.

Species Recorded from India

Kapurdipollenites baculatus Tripathi 1994 (Plate 2, Fig. 5)

Description (Tripathi 1994, p. 62): Pollen grains spherical to subspherical, 70–75 × 65–70 μm. Monosulcate, sulcus long, dividing the pollen into two equal halves. Exine about 2 μm thick, sexine thicker than nexine, foveolate with incomplete tectum, foveolate surface of exine possessing sparsely placed baculae and verrucae, baculae or verrucae 3–6 μm long and 2 μm wide.

Occurrence Early Eocene, subsurface sediments, Kapurdi, Barmer, Rajasthan (Tripathi 1994).

Kapurdipollenites clavatus Tripathi & Srivastava 2010 (Plate 2, Fig. 6)

Description (Tripathi and Srivastava 2010, p. 16): Pollen grains subcircular in shape, size range 62–68 μm (excluding processes), monosulcate. Sulcus long, meridional, running all along the peripheral region of the pollen. Exine 4–5 μm thick, reticulate, muri 2–2.5 μm thick, lumina 1–1.5 μm across. Exine provided with 8–12 clavae which are 7–15 μm long and 6–8 μm wide at the distal side.

Occurrence Late Palaeocene, Akli Formation, Giral, Barmer, Rajasthan (Tripathi and Srivastava 2010).

Kapurdipollenites gemmatus Tripathi 1994 (Plate 2, Fig. 7)

Description (Tripathi 1994, p. 62): Pollen grains spherical to subspherical, size range 68–75 μm; monosulcate, sulcus long, peripheral. Exine moderately thick, foveolate, incompletely tectate, perforate surface of exine covered with closely placed verrucae and gemmae.

Occurrence Early Eocene, subsurface sediments, Kapurdi, Barmer, Rajasthan (Tripathi 1994).

Kapurdipollenites ovatus Tripathi & Srivastava 2010 (Plate 2, Fig. 8)

Description (Tripathi and Srivastava 2010, p. 16, 17): Pollen grains oval in shape, size range 60–72 μm (excluding ornamentation), monosulcate. Sulcus long, meridional, generally resulting into splitting of pollen into two almost equal halves. Exine incompletely tectate, foveolate, exine surface provided with club-shaped processes which are 4–6 μm long and 4–5 μm wide. Distal ends of processes having obtuse ends.

Occurrence Late Palaeocene, Akli Formation, Giral, Barmer, Rajasthan (Tripathi and Srivastava 2010).

Genus *Kielmeyerapollenites* Sah & Kar 1974

Type Species *Kielmeyerapollenites eocenicus* Sah & Kar 1974

Generic Description (Sah and Kar 1974, p. 173): Pollen grains mostly in tetrahedral tetrads, 3-colporate, colpi long, pore margin thickened, tetrads 60–77 μm in size. Exine thick, tegillate, retipilate.

Botanical Affinity *Kielmeyera* (Clusiaceae).

Species Recorded from India

Kielmeyerapollenites eocenicus Sah & Kar 1974 (Plate 14, Fig. 2)

Description (Sah and Kar 1974, p. 173): Pollen grains mostly in tetrahedral tetrads. 3-colporate, colpi distinct, pore margin thickened. Exine thick, retipilate.

Occurrence Early Eocene, Palana lignite, Bikaner, Rajasthan (Sah and Kar 1974); Palaeocene, Matanomadh Formation, Kutch, Gujarat (Kar 1985).

L

Genus *Lacrimapollis* Venkatachala & Rawat 1973

Type Species *Lacrimapollis pilosus* Venkatachala & Rawat 1973

Generic Description (Venkatachala and Rawat 1973, p. 245): Pollen grain spheroidal, tricolpate, 20.4–21.2 × 20.4–22.5 μm. Colpi short, teardrop-shaped, 4.7 μm deep in polar view, surrounded by up to 2 μm wide margo. Exine more or less 2 μm thick, sexine thicker than nexine, tectate, pilate, surface finely reticulate, reticulations less than 1 μm in diameter.

Emended Description (Venkatachala and Rawat in Thanikaimoni et al. 1984, p. 37): Pollen grain more or less circular in polar view, tricolporate, ectoaperture narrow and short, endoaperture 3–4 μm wide, sexine with perforate tectum and columellar infratectum, nexine thickened as teardrop-shaped costae around the endoaperture.

Botanical Affinity *Brownlowia* (Tiliaceae); mangrove.

Species Recorded from India

Lacrimapollis brownlowioides Ramanujam 1987 (Plate 15, Fig. 11)

Description (Ramanujam 1987, p. 33): Pollen grains isopolar, amb subcircular, 36–45 μm, suboblate to oblate equatorially, tricolporate, brevicolpate, colpi with rounded ends, margins and ends of colpi infolded, ora large, faint, rounded to lalongate with incrassate margins; exine up to 4.5 μm thick, crassisexinous, perfo-tectate, columella very prominent, up to 3 μm long, many of them forked, surface finely reticulate, brochi angular.

Occurrence Miocene, Warkalli Beds, Kerala (Ramanujam 1987).

Lacrimapollis pilosus Venkatachala & Rawat 1973 (Plate 15, Fig. 15)

Description (Venkatachala and Rawat 1973, p. 245): Pollen grain spheroidal, tricolporate, 20.4 × 20.4–22.4 μm deep in polar view, surrounded by up to 2 μm wide margo. Exine more or less 2 μm thick, sexine thicker than nexine, tectate, pilate, surface finely reticulate; reticulations less than 1 μm in diameter.

Occurrence Oligocene-Miocene, Cauvery Basin, Tamil Nadu (Venkatachala and Rawat 1973).

Genus *Ladakhipollenites* Mathur & Jain 1980

Type Species *Ladakhipollenites levis* (Sah and Dutta 1966) Mathur & Jain 1980

Generic Description (Mathur and Jain 1980, p. 64): Pollen grains oblate sub-spherical, generally preserved as polar compressions, oval, circular to subcircular; tricolpate, colpi long and not brevicolpate, fairly broad, ends generally pointed; exine laevigate to faintly and obscurely sculptured under high magnification (× 1000).

Remarks Mathur and Jain (1980) instituted *Ladakhipollenites* for psilate, tricolpate pollen and selected *Tricolpites levis* Sah and Dutta (1966) as its type species. The holotype of *Tricolpites levis* Sah and Dutta (1966) is not traceable. Mandal and Rao (2001) selected the specimen recorded by Sah and Kar (1974) as neotype. Mathur and Jain (1980) transferred *Tricolpites brevis* Sah & Kar 1970, *T. levis* Sah

& Dutta 1966, *T. longicolpus* Sah & Dutta 1966 and *T. minutes* Sah & Kar 1970 to *Ladakhipollenites*. Kar (1985) instituted *Intrareticulites* by selecting *Tricolpites brevis* Sah and Kar (1970) as holotype. Specimens referred to this species were therefore, placed accordingly by Mandal and Rao (2001).

Species Recorded from India

Ladakhipollenites constatus (Dutta and Sah 1970) Mandal & Rao 2001 (Plate 10, Fig. 3)

Description (Dutta and Sah 1970, p. 27): Size 32 μm, amb ovoid, oval elongate or subspheroidal; tricolpate, colpi long, tenuimarginate; sexine thicker than nexine, surface ornamentation psilate to faintly scabrate.

Occurrence Palaeocene, Cherra Formation, Meghalaya (Dutta and Sah 1970); Early Miocene, Bhuban Formation, Meghalaya and Assam (Rao et al. 1985).

Remarks Mandal and Rao (2001) transferred *Trifossapollenites constatus* Dutta and Sah (1970) to *Ladakhipollenites constatus*.

Ladakhipollenites elongatus Tripathi & Singh 1985 (Plate 10, Fig. 4)

Description (Tripathi and Singh 1985, p. 178, 179): Pollen grains prolate to perprolate in equatorial view; tricolporate, narrow, long; exine 1–1.5 μm thick, tectate, indistinctly structured, exine thickened at the two poles.

Occurrence Palaeocene, Therria Formation, Meghalaya (Tripathi and Singh 1985).

Ladakhipollenites levis (Sah and Dutta 1966) Mathur & Jain 1980 (Plate 10, Fig. 7)

Description (Sah and Dutta 1966, p.79): Size 24–30 μm, amb spherical to subspherical; tricolpate furrows thin, exine also thin, less than 1 μm, smooth, in some specimens a faint sculpture is discernible under oil immersion.

Description (Mandal and Rao 2001, p. 355): Pollen grains spheroidal in polar view, 30 × 32 μm. Tricolpate, longicolpate, colpi 7.5 μm deep, pointed at poles. Exine 2 μm thick and slightly thicker near colpi margins, sexine-nexine not separable, columellae indistinct. Surface smooth and intrastructured under high magnification.

Occurrence Palaeocene, Cherra Formation, Meghalaya (Sah and Dutta 1966); Early Eocene, Naredi Formation, Kutch, Gujarat (Sah and Kar 1970); Early Eocene,

Palana lignite, Rajasthan (Sah and Kar 1974); Palaeocene-Eocene, Akli lignite, Rajasthan (Naskar and Baksi 1978); Palaeocene, Dras Volcanics, Ladakh (Mathur and Jain 1980); Late Cretaceous-Palaeocene, subsurface sediments Bengal Basin, West Bengal (Baksi and Deb 1980); Palaeocene, Mikir Formation, Assam (Mehrotra 1983); Subsurface Early Miocene sediments, Meghalaya (Nandi and Sharma 1984); Oligocene, Barail Group, Meghalaya and Assam (Singh et al. 1985, 1986).

Genus *Lakiapollis* Venkatachala & Kar 1969

Type Species *Lakiapollis ovatus* Venkatachala & Kar 1969

Generic Description (Venkatachala and Kar 1969, p. 168): Pollen grains subcircular-circular in polar view. Tricolporate, brevicolpate, colpi narrow. Pores well-developed, oval elliptical in shape, broader than the colpus. Exine laevigate and infrastructured.

Remarks Different species of *Lakiapollis* are very frequent in Early Palaeogene of India, particularly in western and eastern regions. *Lakiapollis ovatus*, the most common species of the genus, shows affinity with the modern plant *Durio* which is a member of Indo-Malayan tropics (Thanikaimoni et al. 1984).

Botanical Affinity *Cullenia* and *Durio* (Bombacaceae); *Cullenia* is found in evergreen forests and *Durio* is found in evergreen as well as in swamp forests.

Species Recorded from India

Lakiapollis assamicus Tripathi & Singh 1985 (Plate 15, Fig. 13)

Description (Tripathi and Singh 1985, p. 180, 181): Pollen grains subspheroidal, rarely subtriangular in shape, 90–95 × 80–85 μm in size, tribrevicolporate, aperture irregularly placed, pore margin thickened, colpi thin and short, Exine thin, laevigate, folded.

Occurrence Palaeocene, Therria Formation, Jaintia Hills, Meghalaya (Tripathi and Singh 1985).

Lakiapollis matanomadhensis Venkatachala & Kar 1969

Now: *Tricolporopollis matanomadhensis* (Venkatachala and Kar 1969) Tripathi & Singh 1985

Lakiapollis ovatus Venkatachala & Kar 1969 (Plate 15, Fig. 14)

Description (Venkatachala and Kar 1969, p. 168): Pollen grains subcircular, 40–50 × 35–45 µm. Tribrevicolporate, colpi inconspicuous, hardly discernible in most specimens. Pores well-developed, oval-elliptical, outer margin thickened; pores mostly situated subequatorially. Exine up to 2.5 µm thick, generally psilate.

Occurrence Early Eocene, Laki Series, Matanomadh, Kutch, Gujarat (Venkatachala and Kar 1969); Early Eocene, Palana lignite, Bikaner, Rajasthan (Sah and Kar 1974); Palaeocene, Tura Formation, Nongwal Bibra, Garo Hills, Meghalaya (Singh 1977); Early Eocene, Naredi Formation, Kutch, Gujarat (Saxena 1979); Palaeocene, Matanomadh Formation, Kutch, Gujarat (Kar 1978, Saxena 1979); Palaeocene-Early Eocene, Mikir Formation, Garampani, North Cachar Hills, Assam (Mehrotra 1983).

Genus *Lanagiopollis* Morley 1982

Type Species *Lanagiopollis regularis* Morley 1982

Generic Description (Morley 1982, p. 68): Pollen grains large (60–105 µm), zonotricolporate or rarely ?di- or tetracolporate, radially symmetrical, oblate circular to subangular in polar view; ectoapertures colpi of various length, generally not markedly sunken at equator, more than 5 µm in width, borders discrete, with or without margins; endoapertures pori, with or without costae; nexine thinner or thicker than sexine, nexine often showing distinct thinning towards apertures; exine regularly or irregularly eureticulate, muri composed of distinct capita, each supported by a group of columella, muri simplibaculate, duplibaculate or pluribaculate.

Botanical Affinity *Alangium* (Alangiaceae).

Species Recorded from India

Lanagiopollis cambayensis Kumar 1996 (Plate 16, Fig. 1)

Description (Kumar 1996, p. 114): Pollen grains subcircular in shape, 60–70 µm in size. Tricolporate, colpi long, pore circular to oval, lalongate. Exine 4–5 µm thick, verrucate, verrucae 3–3.5 µm high, densely placed, verrucae heads variously shaped, free, sometimes fused, beset with microgranulate sculpture. Nexine up to 1 µm thick.

Occurrence Early Eocene, Tarkeshwar Formation, Rajpardi, Bharuch, Gujarat (Kumar 1996).

Lanagiopollis meghalayaensis Tripathi et al. 2000 (Plate 16, Fig. 2)

Description (Tripathi et al. 2000, p. 245): Pollen grains spheroidal to sub-spheroidal in shape. Tricolporate, colpi (ectoaperture) distinct, reaching almost up to the poles, ora (endoaperture) indistinct, lalongate. Exine irregularly reticulate, muri composed of distinct capita, each supported by group of columella, pluri-columellate, lumina 1.5–3 μm across, meshes smaller near colpi, free bacular heads seen in lumina. Exine fairly thick, sexine thicker than nexine.

Occurrence Early Eocene, Tura Formation, West Garo Hills, Meghalaya (Tripathi et al. 2000).

Lanagiopollis reitsmae Phadtare & Thakur in Saxena 2011 (Plate 16, Fig. 3)

Description (Phadtare and Thakur 1990, p. 289): Pollen large, subtriangular in polar view, oblate, isopolar, angulaperturate, 50–68 μm in equatorial diameter; apertures slightly asymmetric in distribution, tricolporate, minutely fossaperturate. Colpi about 12 μm long, brevicolpate, generally closed, crassimarginate, bordered by closely spaced, thick and small rugulae; ora comparatively small, 5–8 μm in diameter, lalongate, heavily costate. Nexine distinctly thicker than sexine, almost uniformly thick in mesocolpia and slightly thicker around endoaperture. Mesocolpial sexine more or less uniformly thick but around the endoaperture it is slightly thickened, verrucate along mesocolpia, otherwise clearly rugulate, dupli- to pluribaculate.

Occurrence Early-Middle Eocene, Tarkeshwar Formation, Rajpardi lignite, Bharuch, Gujarat (Phadtare and Thakur 1990).

Lanagiopollis rugulatus Phadtare & Thakur in Saxena 2011 (Plate 16, Fig. 4)

Description (Phadtare and Thakur 1990, p. 289): Pollen grains large, isopolar, triangular in polar view, oblate, angulaperturate, 62–70 μm in equatorial diameter, apertures slightly asymmetric in distribution, tricolporate. Colpi distinct, ca. 13 μm long, 8–10 μm wide at equator, almost brevicolpate; margin smooth, straight, thick, thicker around endoaperture, bordered by thick brochi, ends acute; ora distinct, circular to slightly elliptic, 10–12 μm in diameter, lalongate, heavily costate, margin incurved. Nexine is slightly thicker than sexine, generally thinning towards aperture, but thickening again around the endoaperture, forming conspicuously thick costae; sexine also is thicker in mesocolpia, gradually decreasing in thickness towards aperture and again increasing at the equatorial margin of colpi; apocolpial and mesocolpial sexine is prominently rugulate while around both sides of equatorial colpal margins it has a minute areolate appearance.

Occurrence Early-Middle Eocene, Tarkeshwar Formation, Rajpardi lignite, Bharuch, Gujarat (Phadtare and Thakur 1990); Early Eocene, Kharsalia Clay Formation, Bhavnagar, Cambay Basin, Gujarat (Samant 2000).

Lanagiopollisraguloreticulatus Phadtare & Thakur in Saxena 2011 (Plate 16, Fig. 5)

Description (Phadtare and Thakur 1990, p. 287, 289): Pollen grains large, isopolar, circular to subcircular in polar view, oblate, angulaperturate, 55–70 μm in equatorial diameter, tricolporate. Colpi about 16 μm long, 8–10 μm wide at equator, almost brevicolpate; margin smooth, straight, thin, bordered by randomly oriented columellae, ends acute; ora generally circular, 10–13 μm in diameter, costate and lalongate. Nexine 2.5 μm thick, almost double the thickness of sexine; sexine 1–1.5 μm thick, slightly thicker along the equatorial area of aperture, clearly regulate-reticulate.

Occurrence Early-Middle Eocene, Tarkeshwar Formation, Rajpardi lignite, Bharuch, Gujarat (Phadtare and Thakur 1990).

Lanagiopollis subglobosus Tripathi et al. 2000 (Plate 16, Fig. 6)

Description (Tripathi et al. 2000, p. 246): Pollen grains oval-spheroidal in shape. Tricolporate, colpi distinct, long, reaching almost up to the poles, ora distinct, subequatorial, lalongate, 8–10 μm across (transversely), ora margin thickened. Exine moderately thick, finely reticulate, muri formed by fusion of columellae, simplicolumellate, nexine slightly thicker than sexine.

Occurrence Early Eocene, Tura Formation, West Garo Hills, Meghalaya (Tripathi et al. 2000).

Lanagiopollis tetracolporatus Kumar 1996 (Plate 16, Fig. 7)

Description (Kumar 1996, p. 114): Pollen grains quadrangular in shape, 68–73 × 75–85 μm in size. Tetracolporate, colpi 20–25 μm long, open; pores circular, 10–12 μm in diameter, margin thickened, ora protruded. Exine 2–2.5 μm thick, pilate, pila heads fused to form microreticulate ornamentation. Lumina of various shapes, comparatively smaller near the poles and broader towards the equator; muri up to 1 μm thick. Nexine 0.5 μm thick.

Occurrence Early Eocene, Tarkeshwar Formation, Rajpardi, Bharuch, Gujarat (Kumar 1996).

Genus *Liliacidites* Couper 1953

Type Species *Liliacidites kaitangataensis* Couper 1953

Proposed synonymy:

1985—*Retimonosulcites* Kar, *The Palaeobotanist* 34: 70

1987—*Iridacidites* Ramanujam, *Journal of Palaeontological Society of India* 32: 29

Generic Description (Couper 1953, p. 56): Free anisopolar, bilateral, monosulcate, occasionally trichotomosulcate; sulcus long, broad; grain usually elongate; exine clearly reticulate, lumina of reticulum variable in size; clavate, baculate in optical section (muri composed of baculae or clavae).

Emended Description (Krutzsch 1970, p. 30): Monosulcate (abnormally also trichotomosulcate), pollen with coarsely reticulate exine, muri underlain by a distinct columellae or baculae layer, the components of which in part may protrude freely to the exterior as clavae. Medium sized to large forms.

Remarks *Liliacidites* was proposed to accommodate fossil pollen having liliaceous affinity. Kar (1985, p. 69) proposed *Retimonosulcites* mentioning that *Liliacidites* has broad and beaded reticulation with meshes larger in middle and smaller at ends. The diagnosis of *Liliacidites* does not mention that different size of lumina must be in the same specimen. Many species of this genus possess lumina of reticulation of different size whereas; in some others, these are of uniform size. In author's opinion, delineation of a new genus based on size of the lumina in a single specimen is not justified. *Retimonosulcites* Kar (1985) proposed for the reception of monocolpate pollen with uniform size of lumen throughout the grain is therefore, treated as junior synonym of *Liliacidites*.

Ramanujam (1987) proposed *Iridacidites* with following generic diagnosis: "Pollen grains heteropolar, amb elongate-ellipsoidal, monosulcate, longisulcate, surface foveoreticulate." *Iridacidites* comes under the generic circumscription of *Liliacidites* because *Iridacidites warkalliensis* Ramanujam (1987) and *Iridacidites plicatus* Ramanujam (1987) are distinctly reticulate. It is interesting to note that *Arecipites magnus* Ramanujam (1987, Plate 1, Fig. 13) and *Iridacidites warkalliensis* Ramanujam (1987, Plate 1, Fig. 14) are the two foci of a single specimen.

Botanical Affinity Liliaceae/Arecaceae.

Species Recorded from India

Liliacidites baculatus Venkatachala & Kar 1969 (Plate 2, Fig. 9)

Description (Venkatachala and Kar 1969, p. 160): Pollen grains oval-elliptical in shape, 45–65 × 30–45 μm. Colpus well developed, funnel-shaped. Exine intra-baculate, forming negative reticulum in surface view.

Occurrence Early Eocene, Naredi Formation, Matanomadh, Kutch, Gujarat (Venkatachala and Kar 1969).

Liliacidites circularis Dutta & Sah 1970 (Plate 2, Fig. 10)

Description (Dutta and Sah 1970, p. 31): Known size range 20–23 × 23–26 μm; amb spheroidal to subspheroidal; 1–sulcate, sulcus about 18 μm long, generally indistinct; lips wide open; sexine thicker than nexine, surface sculpture reticulate.

Affinity *Liliaceae*

Occurrence Palaeocene, Cherra Formation, Shillong Plateau, Meghalaya (Dutta and Sah 1970).

Liliacidites crassireticulatus Venkatachala & Rawat 1972 (Plate 2, Fig. 11)

Description (Venkatachala and Rawat 1972, p. 298): Pollen grains oval with slightly rounded lateral ends, monosulcate; 35.7 × 20.5 μm. Colpus long, broad and extending the whole length of the grain, more or less uniformly broad. Exine about 1.3–1.5 μm thick, tectate, sexine thicker than nexine, reticulate, muri irregular and projecting along the equatorial margin, lumina of irregular shape, ranging 1–1.3 μm in diameter.

Affinity *Liliaceae*

Occurrence Palaeocene-Eocene, Cauvery Basin, Tamil Nadu (Venkatachala and Rawat 1972).

Liliacidites ellipticus Venkatachala & Kar 1969 (Plate 2, Fig. 12)

Description (Venkatachala and Kar 1969, p. 159): Pollen grains oval-elliptical in shape, 30–42 × 25–28 μm. Sulcus broad, more or less boat-shaped. Exine finely intra-microreticulate.

Remarks Kar (1985, p. 69–70) transferred this species to *Retimonosulcites* (Kar 1985). Since *Retimonosulcites* is treated as synonym of *Liliacidites*, this species is retained in the latter genus.

Occurrence Early Eocene, Naredi Formation, Matanomadh, Kutch, Gujarat (Venkatachala and Kar 1969).

Liliacidites kutchensis Saxena 1979

Now: *Matanomadhiasulcites kutchensis* (Saxena 1979) Kar 1985

Liliacidites magnus Jain et al. 1973 (Plate 2, Fig. 13)

Description (Jain et al. 1973, p. 153): Pollen grains bilateral, monosulcate, oval-elliptical, 70–80 × 45 μm. Sulcus long, broad, elongate. Exine 1.5 μm thick, surface intra-microreticulate.

Occurrence Palaeocene, Barmer Sandstone, Barmer, Rajasthan (Jain et al. 1973).

Liliacidites major Singh in Sah & Singh 1974

Now: *Matanomadhiasulcites major* (Singh in Sah and Singh 1974) Saxena & Khare 2004

Liliacidites matanomadhensis Saxena 1979 (Plate 2, Fig. 14)

Description (Saxena 1979, p. 132): Pollen grains oval-elliptical, 86–150 × 58–94 μm. Monosulcate, sulcus broad, extends from one end to the other. Exine 2 3 μm thick, foveo-reticulate, foveola up to 2 μm wide, closely placed and evenly distributed.

Occurrence Palaeocene, Matanomadh Formation, Kutch, Gujarat (Saxena 1979).

Liliacidites microreticulatus Dutta & Sah 1970

Now: *Matanomadhiasulcites microreticulatus* (Dutta and Sah 1970) Kar & Kumar 1986

Liliacidites padappakkarensis Rao & Ramanujam 1978 (Plate 2, Fig. 15)

Description (Rao and Ramanujam 1978, p. 416): Pollen grains heteropolar, plano-convex to biconvex laterally, elliptical in polar view, 50–60 × 30–35 μm, monosulcate, sulcus long, often reaching both ends along long axis of pollen, uniformly narrow with tapering and blunt ends. Exine up to 4.5 μm thick, sexine thicker than nexine, surface retipilariate, pila heads prominent, rounded or angular, reticulum homobrochate, brochi large, up to 5 μm across, muri formed of groups of fused pila, duplipilate, up to 3.5 μm high, lumina large, angular to rounded, smooth.

Affinity Liliaceae

Occurrence Miocene, Quilon Beds, Padappakkara, Kerala (Rao and Ramanujam 1978).

Liliacidites pilosus Venkatachala & Rawat 1972 (Plate 3, Fig. 1)

Description (Venkatachala and Rawat 1972, p. 297): Pollen grains elongate, oval; monosulcate; 36.3 × 19.8 μm. Colpus well developed, extending along whole length of the body, broader at the lateral ends, narrows in the middle. Exine about 2 μm thick, sexine thicker than nexine, tectate, pila narrow, less than 1 μm, closely packed giving an appearance of reticulum on the surface.

Occurrence Palaeocene-Eocene, Cauvery Basin, Tamil Nadu (Venkatachala and Rawat 1972).

Liliacidites ramanujamii Saxena 1992 (Plate 3, Fig. 2)

= *Liliacidites magnus* Ramanujam 1987 (non Jain et al. 1973: 153)

Description (Ramanujam 1987, p. 29): Pollen grains heteropolar, amb ovoid-ellipsoidal, 58–75 × 42–55 μm, monosulcate, sulcus of medium size, tenuimarginate; exine up to 2.8 μm thick, subtectate, surface reticulate, heterobrochate, brochi polygonal, curvimurate, muri simplibaculate, locally with beaded look, lumina irregular, reticulum faint locally.

Occurrence Miocene, Warkalli Beds, Kerala (Ramanujam 1987).

Liliacidites ratnagiriensis Kulkarni et al. 1985 (Plate 3, Fig. 3)

Description (Kulkarni et al. 1985, p. 300): Pollen grains heteropolar, oval to elliptical in shape, 24 × 32.2 μm, monosulcate, sulcus closed, crassimarginate, fossaperturate, extended along the longer axis of pollen; sexine thicker than nexine, exine retipilate, muri homobrochate, simplibaculate, lumen more or less angular, smooth.

Occurrence Miocene, Ratnagiri Beds, subsurface sediments, Ratnagiri, Maharashtra (Kulkarni et al. 1985).

Liliacidites reticulatus Sah & Kar 1974 (Plate 3, Fig. 4)

Description (Sah and Kar 1974, p. 167): Pollen grains monosulcate, 45–60 μm, sulcus distinct, end-to-end. Exine reticulate, meshes broader in middle region and narrower at ends.

Occurrence Early Eocene, Palana lignite, Bikaner, Rajasthan (Sah and Kar 1974).

Genus *Longapertites* von Hoeken-Klinkenberg 1964

Type Species *Longapertites marginatus* von Hoeken-Klinkenberg 1964

Generic Description (von Hoeken-Klinkenberg 1964, p. 212): Monocolpate pollen grains with an aperture longer than half of the greatest circumference of the grain.

Proposed Emended Description Pollen grains subtriangular to oval-elliptical; size range 30–90 μm, monosulcate, sulcus long, extending up to more than half of the greatest circumference of pollen; exine reticulate, lumina 2–5 μm wide, muri 1–2 μm high.

Botanical Affinity and Ecology *Eugeissona* (Arecaceae); palms of lowland evergreen forests.

Species Recorded from India

Longapertites klinkenbergii Rao & Ramanujam 1978 (Plate 3, Fig. 5)

Description (Rao and Ramanujam 1978, p. 416): Pollen grains heteropolar, planoconvex laterally, 41.5–48 × 25–32 μm in size, monosulcate, sulcus very long, extending over 2/3rd of the circumference of grain, 5 μm wide in the middle, narrowing at ends, margins thickened and beset with clava. Exine up to 3 μm thick, sexine much thick than nexine, subtectate, columella distinct, surface reticulate, heterobrochate, brochi hexagonal to polygonal, curvimurate, muri up to 2 μm high, occasionally disjointed, simpliclavate, lumina angular to irregular with 1–3 free bacula in each lumen.

Occurrence Miocene, Quilon Beds, Padappakkara, Kerala (Rao and Ramanujam 1978); Miocene, Ratnagiri Beds, subsurface sediments, Ratnagiri, Maharashtra (Kulkarni et al. 1985); Miocene, Warkalli Beds, Kerala (Ramanujam 1987).

Longapertites ornatus Kulkarni et al. 1985 (Plate 3, Fig. 6)

Description (Kulkarni et al. 1985, p. 303): Pollen grains elliptical, heteropolar, planoconvex laterally, biconvex in polar view, monosulcate, 34 × 54 μm, sulcus more than half and less than two third the circumference, tenuimarginate, beset with pilae; sexine thicker than nexine, exine 2.7 μm thick, prominently ornate.

Occurrence Miocene, Ratnagiri Beds, subsurface sediments, Ratnagiri, Maharashtra (Kulkarni et al. 1985).

Longapertites psilatus Mathur & Chopra 1987 (Plate 3, Fig. 7)

Description (Mathur and Chopra 1987, p. 112): Shape in preserved condition generally planoconvex, convex side forming more than half circle, size 30–38 × 20–30 μm; longicolpate; tectum imperforate; more or less uniformly intra-granulate, surface psilate.

Occurrence Late Eocene, subsurface sediments, Bengal Basin, West Bengal (Mathur and Chopra 1987).

Longapertites retipilatus Kar 1985 (Plate 3, Fig. 8)

Description (Kar 1985 p. 120): Pollen grain bilaterally symmetrical, anisopolar, distally arched to provide a long colpus, covering more than 2/3rd of the greater circumference, 48–60 × 23–40 μm thick, exine thicker than nexine, retipilate.

Occurrence Eocene, subsurface sediments, Rataria, Kutch, Gujarat (Kar 1985).

Longapertites triangulatus Samant & Phadtare 1997 (Plate 3, Fig. 9)

Description (Samant and Phadtare 1997, p. 28): Pollen grains triangular in shape, 37–45 × 33–36 μm in size, monosulcate. Sulcus long, extending 2/3rd of the greatest circumference; exine microreticulate.

Occurrence Early Eocene, Tarkeshwar Formation, Rajpardi, Cambay Basin, Gujarat (Samant and Phadtare 1997).

Genus *Loranthipites* Rao & Ramanujam 1982

Type Species *Loranthipites elegans* Rao & Ramanujam 1982

Generic Description (Rao and Ramanujam 1982, p. 72): Pollen grains with tri-angular amb, distinctly 3-armed, sides concave, angles flatly arched or flared up, tricolpate, longicolpate, colpae almost extending up to the poles on one side; exine tectate, smoothly or finely patterned.

Botanical Affinity and Ecology Loranthaceae; parasites on shrubs and diverse forests.

Species Recorded from India

Loranthipites elegans Rao & Ramanujam 1982 (Plate 10, Fig. 10)

Description (Rao and Ramanujam 1982, p. 72): Pollen grains isopolar, amb triangular, distinctly 3-armed, prominently concave with slightly wavy margin, angles flatly arched, polar diameter 25.5–34 μm; tricolpate, longicolpate, becoming almost syncolpate on one side, colpal margins thin to slightly thickened, ends pointed. Exine 1.8 μm thick, tectate, sexine only slightly thicker or almost as thick as nexine, columella distinct, polar area thinner and smooth, rest of the surface psilate to finely patterned.

Occurrence Miocene, Quilon Beds, Padappakkara, Kerala (Rao and Ramanujam 1982); Miocene, Warkalli Beds, Kerala (Ramanujam 1987).

M

Genus *Maculoporites* Venkatachala & Rawat 1973

Now: *Coneopollis* Venkatachala & Rawat 1973

Genus *Malvacearumpollis* Nagy 1962

Type Species *Malvacearumpollis baconyensis* Nagy 1962

Generic Description (Nagy 1962, p. 159): Large (85–105 μm) globular pollen grains, the surface pilate and covered with spines in fairly regular, 8–14 μm large intervals. The spines are conical, tapering into acuminate tips, 6–12 μm large, rising on a bulging base formed of exine.

Botanical Affinity and Ecology Malvaceae; coastal palms.

Species Recorded from India

Malvacearumpollis paucibaculatus Venkatachala & Rawat 1973 (Plate 14, Fig. 3)

Description (Venkatachala and Rawat 1973, p. 255): Pollen grains spheroidal. Panaperturate, polyporate; 49 × 46.2 μm in size (excluding spines). Pores annulate, 8–10 in number, sparsely distributed, about 4.5 μm in diameter; sexine as thick as nexine or thinner, 1 μm or less, thickened at roots of spines up to 3.3 μm, tectate; spines few, sparsely distributed, about 6 μm long, about 2 μm thick at base, tapering distally, with conical to blunt tips. Exine between the spines prominently granulate, grana closely placed and 1 μm or more in diameter.

Occurrence Oligocene-Miocene, Cauvery Basin, Tamil Nadu (Venkatachala and Rawat 1973).

Malvacearumpollis rudis Kar 1979 (Plate 8, Fig. 12)

Description (Kar 1979, p. 31): Pollen grains subcircular, 52–76 μm, panporate, pore more or less elliptical. Exine spinose, spine with bulbous base and pointed tip; nexine thicker than sexine, latter with rod-like processes extending up to base of spines.

Occurrence Oligocene, Maniyara Fort Formation, Sarangwara, Kutch, Gujarat (Kar 1979).

Genus *Marginipollis* Clarke & Frederiksen 1968

Type Species *Marginipollis concinnus* Clarke & Frederiksen 1968

Generic Description (Clarke and Frederiksen 1968, p. 210): Pollen grain 3-colpate (syncolpate), prolate-spheroidal to prolate, colpi margins thick, generally smooth, typically becoming thicker and enlarged towards the poles.

Botanical Affinity and Ecology *Barringtonia* (Barringtoniaceae); trees of costal swamps.

Species Recorded from India

Marginipollis kutchensis (Venkatachala and Kar 1968) Kar 1978 (Plate 11, Fig. 8)

Description (Venkatachala and Kar 1968, p. 336): Pollen grains elliptical in equatorial view and circular to roundly triangular in polar view, 65 × 30 μm in equatorial view. This type of flattening is the most common condition observed. Tricolpate-colporoidate, syncolpate with crassimarginate colpi; colpi margins terminating near the poles in the form of extension appearing like beaked cuspidate projections. Sexine thicker than nexine. Mesocolpial sexine tegillate with irregular thickenings of the tegillum giving areoloidate appearance; in equatorial view these areoloidate areas border the colpi margins thus simulating a thick, broad spongiose zone along the colpi.

Occurrence Early Eocene, Naredi Formation, Kutch, Gujarat (Venkatachala and Kar 1968); Middle-Late Eocene, subsurface sediments, Rataria, Kutch, Gujarat (Kar and Saxena 1981).

Remarks *Rostriapollenites kutchensis* Venkatachala and Kar (1968) was transferred to *Marginipollis kutchensis* by Kar (1978).

Marginipollis psilatus Ramanujam et al. 1988 (Plate 11, Fig. 14)

Description (Ramanujam et al. 1988, p. 273): Pollen grains isopolar, radiosymmetric, prolate to subprolate equatorially, 32–42 × 26–30 μm, tricolpate, syncolpate, colpal margins incrassate (up to 2.5 μm thick), thickenings more prominent at polar regions forming beak like cushions (up to 3.5 μm thick), faint grooves separate colpal margins from mesocolpia, colpal membrane finely granular; exine up to 2 μm thick, tectate, columalla fine, indistinct, sexine thicker than nexine, surface psilate to locally finely flecked.

Occurrence Miocene, Neyveli Formation, Neyveli, Tamil Nadu (Ramanujam et al. 1988).

Marginipollis quilonensis Rao & Ramanujam 1982 (Plate 11, Fig. 15)

Description (Rao and Ramanujam 1982, p. 73): Pollen grains isopolar, ellipsoidal (prolate) equatorially, 36–43 × 23–29 μm; zonaperturate, tricolpate, syncolpate, colpae ending in beaked (knob-like) processes at poles, margins incrassate. Exine 1.8 μm thick, sexine thicker than nexine, colpae beaks provided with nexinous thickenings, surface of mesocolpia punctate to foveolate.

Occurrence Miocene, Quilon Beds, Padappakkara, Kerala (Rao and Ramanujam 1982); Miocene, Neyveli Formation, Neyveli, Tamil Nadu (Ramanujam et al. 1988).

Marginipollis verrucatus Ramanujam et al. 1988 (Plate 12, Fig. 1)

Description (Ramanujam et al. 1988, p. 272, 273): Pollen grains Isopolar, radiosymmetric, prolate to subprolate, 43–48 × 33–36 μm, tricolpate, syncolpate, colpal margin crassimarginate (up to 2 μm thick), thickenings more pronounced at polar regions forming beak-like cushions, up to 5 μm thick, colpal margins studded with verrucae and separated from mesocolpia by undulating grooves, colpal membranes provided with sparse low verrucate with polygonal heads, closely spaced, locally imparting a negatively reticulate look to mesocolpia.

Occurrence Miocene, Neyveli Formation, Neyveli, Tamil Nadu (Ramanujam et al. 1988).

Genus *Margocolporites* Ramanujam 1966 ex Srivastava 1969

Type Species *Margocolporites tsukadae* Ramanujam 1966 ex Srivastava 1969 (Designated by Srivastava 1969, p. 984)

Generic Description (Ramanujam 1966, p. 173): Pollen grains isopolar, oblate to suboblate, 3-zonimargocolporate, 37–56 μm, margocolpus smooth, granular or microreticulate. Exine reticulate or retipilate.

Botanical Affinity and Ecology Caesalpiniaceae; plants of coasts and dry forests.

Species Recorded from India

Margocolporites complexum Ramanujam 1966 (Plate 16, Fig. 8)

Description (Ramanujam 1966, p. 176): Pollen grains isopolar, suboblate, 3-zonimargocolporate, 33 × 40 μm. Amb sharply 3-lobed, lobes smoothly arched. Margo-colpi fusiform, margin thickened, surface finely granular, ora faint, narrow, lalongate. Margo-colpi 8–10 μm broad. Exine 2–8 μm thick, columellae indistinct, surface finely punctate-reticulate, meshes small, irregular, muri narrow, simplibaculate, locally curvimurate, punctae seen in lumina.

Occurrence Cuddalore Series, Neyveli, Tamil Nadu (Ramanujam 1966).

Margocolporites dubius Ramanujam 1966 (Plate 16, Fig. 9)

Description (Ramanujam 1966, p. 176): Pollen grains isopolar, suboblate, 3-zonimargocolporate, 50–56 μm in subequatorial view. Amb 3-lobed, lobes slightly bulging, poles depressed. Margocolpi long, just touching each other at poles to give an almost synmargocolporate look, margin considerably thickened, surface coarsely granular or microreticulate, ora faint, tenuimarginate, lolongate. Exine 2.5 μm thick, reticulate, meshes small, polygonal, muri narrow, simplibaculate, lumina small, angular, smooth.

Occurrence Cuddalore Series, Neyveli, Tamil Nadu (Ramanujam 1966).

Margocolporites eocenicus Samant & Phadtare 1997 (Plate 16, Fig. 10)

Description (Samant and Phadtare 1997, p. 31): Pollen grains isopolar, circular in polar view, distinctly trilobed, 40–45 μm in diameter; tricolporate; colpi long, characteristically funnel-shaped as seen in polar view; exine foveolate, dupli- to pluribaculate, 3–4 μm thick, sexine thicker than nexine.

Occurrence Early Eocene, Tarkeshwar Formation, Rajpardi, Cambay Basin, Gujarat (Samant and Phadtare 1997).

Margocolporites miocenicus Venkatachala & Rawat 1972 (Plate 16, Fig. 11)

Description (Venkatachala and Rawat 1972, p. 306, 307): Pollen grains with trilobed amb; 52.8 × 55.4 μm; tricolporate. Margocolpi considerably broad, about 14.5 μm wide at the equatorial region, deep in polar view, almost reaching the poles, ends tapering, margins heavily thickened. Pores large, lalongate. Exine 5–6 μm thick, tectate, sexine thicker than nexine; columellae distinct, reticulate, duplibaculate, lumina of different shape and size, rounded to angular.

Occurrence Palaeocene-Eocene, Cauvery Basin, Tamil Nadu (Venkatachala and Rawat 1972).

Margocolporites oligobrochatus Ramanujam 1966 (Plate 16, Fig. 12)

Description (Ramanujam 1966, p. 177): Pollen grains isopolar, suboblate, 3-zonimargocolporate, 48 × 53 μm. Amb distinctly 3-lobed, margocolpi considerably broad, 12–18 μm wide at equatorial region, almost reaching poles, ends tapering, margin heavily thickened, ora tenuimarginate, lolongate, surface coarsely granular. Exine 4 μm thick, crassisexinous, heavily and ruggedly reticulate, columella distinct. Surface oligobrochate, brochi large, hexa- to polygonal, crassimurate, duplibaculate, lumina large with one or two free bacula. Basal diameter of bacula in lumina smaller than or almost equal to mural bacula.

Occurrence Cuddalore Series, Neyveli, Tamil Nadu (Ramanujam 1966).

Margocolporites prominatus Mathur & Chopra 1987 (Plate 16, Fig. 13)

Description (Mathur and Chopra 1987, p. 130): Pollen grains isopolar, 38–45 μm in diameter; tricolporate, margocolpate, margo 3–5 μm thick; exine 4–6 μm thick, reticulate, muri sausage-shaped, thicker than lumina.

Occurrence Palaeocene, subsurface sediments, Bengal Basin, West Bengal (Mathur and Chopra 1987).

Margocolporites sahnii Ramanujam 1966 (Plate 16, Fig. 14)

Description (Ramanujam 1966, p. 174): Pollen grains isopolar, suboblate, 3-zonimargocolporate, equatorial diameter 56 μm. Amb subtriangular, distinctly 3-lobed. Margocolpi large, margin considerably thickened, surface finely granular locally coarsely so, ora faint, lolongate. Exine 3.3 μm thick, columellae faint, reticulate, homobrochate, brochi hexa- to polygonal, simplibaculate, each lumen with single baculum basal diameter of which is larger than mural bacula.

Occurrence Cuddalore Series, Neyveli, Tamil Nadu (Ramanujam 1966).

Margocolporites sitholeyi Ramanujam 1966 (Plate 16, Fig. 15)

Description (Ramanujam 1966, p. 174): Pollen grains isopolar, oblate spheroidal, 3-zonimargocolporate, equatorial diameter 38 µm, Amb distinctly 3-lobed, lobes slightly bulging. Margocolpus long, but not reaching poles, straight, margin thickened, ends pointed, surface granular, ora lolongate, pouting as seen in polar view. Exine 3.3 µm thick, sexine thicker than nexine, surface prominently reticulate, columellae distinct, reticulum polygonal, homobrochate, muri simplibaculate, lumina smooth or with free bacula.

Occurrence Cuddalore Series, Neyveli, Tamil Nadu (Ramanujam 1966).

Margocolporites tsukadae Ramanujam 1966 (Plate 17, Fig. 1)

Description (Ramanujam 1966, p. 173, 174): Pollen grains isopolar, suboblate, 3-zonimargocolporate, 48 × 54 µm. Amb prominently 3-lobbed, lobes widely spaced because of large margocolpae. Margocolpus broad, 12–16 µm, gradually tapering towards poles (fusiform), microreticulate, rimmed by a thickened wall (margo), ora distinct, small, lolongate. Exine 3.5 µm thick, sexine thicker than nexine, surface baculate-reticulate. Reticulum homobrochate, polygonal, muri simplibaculate, fairly thick, lumina rounded or angular, smooth.

Occurrence Cuddalore Series, Neyveli, Tamil Nadu (Ramanujam 1966).

Genus *Matanomadhiasulcites* Kar 1985

Type Species *Matanomadhiasulcites maximus* (Saxena 1979) Kar 1985

Generic Description (Kar 1985, p. 34): Pollen grains oval-elliptical in shape. Monosulcate, sulcus distinct to indistinct, almost extending end to end. Exine retipilate-retibaculate.

Botanical Affinity *Annona* (Annonaceae).

Species Recorded from India

Matanomadhiasulcites kutchensis (Saxena 1979) Kar 1985 (Plate 3, Fig. 10)

Description (Saxena 1979, p. 132): Pollen grains oval-elliptical, 63–100 × 38–58 µm. Monosulcate, sulcus mostly indistinct, wide, generally extend from pole to pole. Exine 1.5–2.5 μm thick, retipilate, sculptural elements 3.4 μm long, closely placed.

Occurrence Palaeocene, Matanomadh Formation, Matanomadh, Kutch, Gujarat (Saxena 1979; Kar 1985); Palaeocene, Lakadong Sandstone, Khasi Hills, Meghalaya (Kar and Kumar 1986).

Remarks *Liliacidites kutchensis* Saxena (1979) was transferred to *Matanomadhiasulcites kutchensis* by Kar (1985).

Matanomadhiasulcites major (Singh in Sah and Singh 1974) Saxena & Khare 2004 (Plate 3, Fig. 11)

Description (Singh in Sah and Singh, p. 96): Pollen grains oval, 78–100 μm. Monosulcate, sulcus distinct to indistinct. Exine pitted to retibaculate.

Occurrence Palaeocene, Tura Formation, Nongwal Bibra, Garo Hills, Meghalaya (Singh in Sah and Singh 1974; Singh 1977); Palaeocene-Early Eocene, Mikir Formation, Garampani, North Cachar Hills (Mehrotra 1983); Palaeocene Therria Formation, Jowai-Badarpur Road, Jaintia Hills, Meghalaya (Tripathi and Singh 1985); Late Palaeocene, subsurface sediments, Tiruchirapalli, Tamil Nadu (Saxena and Khare 2004).

Remarks *Liliacidites major* Singh in Sah and Singh (1974) was transferred to the genus *Matanomadhiasulcites* by Saxena and Khare (2004).

Matanomadhiasulcites maximus (Saxena 1979) Kar 1985 (Plate 3, Fig. 12)

Description (Saxena 1979, p. 132): Pollen grains oval-elliptical, 135–205 × 90–145 μm. Monosulcate, sulcus may be distinct or indistinct, wide, extending from pole to pole. Exine 2–3 μm thick, generally retipilate to retibaculate, pila/bacula 3–8 μm long, muri thick, raised, lumina 3–8 μm wide.

Occurrence Palaeocene, Matanomadh Formation, Matanomadh, Kutch, Gujarat (Saxena 1979); Palaeocene, Lakadong Sandstone, Khasi Hills, Meghalaya (Kar and Kumar 1986).

Remarks *Liliacidites maximus* Saxena (1979) was transferred to *Matanomadhiasulcites* by Kar (1985).

Matanomadhiasulcites microreticulatus (Dutta and Sah 1970) Kar & Kumar 1986 (Plate 3, Fig. 13)

Description (Dutta and Sah 1970, p. 30): Size range 72–89 μm; holotype 87 μm; amb more or less spherical to oval elongate; 1-sulcate; exine thin, sexine slightly thicker than nexine, surface sculpture finely reticulate.

Occurrence Palaeocene, Cherra Formation, Shillong Plateau, Meghalaya (Dutta and Sah 1970); Palaeocene, Lakadong Sandstone, Khasi Hills, Meghalaya (Kar and Kumar 1986); Palaeocene-Eocene, subsurface sediments, Kapurdi, Barmer, Rajasthan (Tripathi 1995); Late Palaeocene, Akli Formation, Barmer, Rajasthan (Tripathi et al. 2003).

Remarks *Liliacidites microreticulatus* Dutta and Sah (1970) was transferred to *Matanomadhiasulcites* by Kar and Kumar (1986).

Genus *Meliapollis* Sah & Kar 1970

Type Species *Meliapollis ramanujamii* Sah & Kar 1970

Generic Description (Sah and Kar 1970, p. 133): Pollen grains circular-subcircular, 3–5 colporate, brevicolpate. Pore distinct with thickened margin, circular-lalongate. Exine laevigate to weakly intrastructured.

Emended Description (Thanikaimoni et al. 1984, p. 54): Pollen isopolar, triangular, pentagonal to subcircular in polar view while suboblate to subprolate in equatorial view. 36–140 × 27–140 μm, tri-, tetra- and pentacolporate, but tetracolporate condition is most common. Longicolpate to brevicolpate, colpi tenuimarginate, generally faintly visible, pore mostly well-defined, 5–10 μm in diameter, more or less circular, lalongate, rarely lolongate, surrounded by thickened rim. Mostly in case of grains with thin exine, colpi membrane dissolves rendering deepening of colpi and pore untraceable. Such grains look like cross. Exine thin to very thick (2–14 μm). Sexine is generally thinner (1–2.5 times) than nexine. Nexine is continuous throughout the pollen via thickening at the pore. In case of pollen with thick nexine, the nexine makes 1–2 bulges inwards. This is probably to accommodate thickened nexinal part of the pores and mesocolpi. These bulges are mostly seen clearly in tetra- and pentacolporate grains. Surface laevigate, sometimes weakly intrastructured, at least in this assemblage most of the grains bear scars of biological degradation.

Botanical Affinity Meliaceae.

Species Recorded from India

Meliapollis iratus (Sah and Kar 1974) Navale & Misra 1979 (Plate 20, Fig. 9)

Description (Navale and Misra 1979, p. 233): Isopolar, quadrangular pollen, 66–90 × 64–90 μm in size. Tetracolporate, colpi (membrane dissolved) appearing long, rendering the grain cross-like appearance, pores distinct (but all pores are not intact), 9–11 μm in diameter, margins thickened. Exine 3–4 μm thick, sexine thinner than or equal to nexine, surface laevigate.

Occurrence Early Eocene, Palana lignite, Bikaner, Rajasthan (Sah and Kar 1974); Miocene, Neyveli Lignitefield, Neyveli (Navale and Misra 1979).

Remarks *Platoniapollenites iratus* Sah and Kar (1974) was transferred to *Meliapollis iratus* by Navale and Misra (1979).

Meliapollis melioides (Ramanujam 1966) Sah & Kar 1970 (Plate 20, Fig. 11)

Description (Ramanujam 1966, p. 180): Pollen grains isopolar, prolate spheroidal, tetrazonicolporate, 36 × 27 μm. Amb four-sided, quadrangular. Colpi narrow, tenuimarginate, almost straight, long but not reaching the poles, ends blunt, ora conspicuous with thickened rims, lalongate. Exine 1.8 μm thick, columellae not found, surface giving a flecked appearance.

Occurrence Miocene-Pliocene, Cuddalore, Series, Neyveli, Tamil Nadu (Ramanujam 1966); Early Eocene, Naredi Formation, Kutch, Gujarat (Kar 1978); Palaeocene, Matanomadh Formation, Matanomadh, Kutch, Gujarat (Saxena 1979).

Remarks *Tetracolporites melioides* Ramanujam (1966) was transferred to *Meliapollis* by Sah and Kar (1970).

Meliapollis navalei Sah & Kar 1970 (Plate 20, Fig. 12)

Description (Sah and Kar 1970, p. 134): Pollen grains subcircular-circular, 60–75 × 57–70 μm. 5-colporate, brevicolpate, pore distinct with thickened margin. Exine 1.5–3.5 μm thick, laevigate or weakly intrastructured.

Occurrence Early Eocene, Laki Series, Kutch, Gujarat (Sah and Kar 1970); Early Eocene, Naredi Formation, Kutch, Gujarat (Kar 1978).

Meliapollis pachydermis Navale & Misra 1979 (Plate 20, Fig. 13)

Description (Navale and Misra 1979, p. 233): Isopolar, subtriangular pollen grain, 95 × 95 μm in size. Tricolporate, brevicolpate, colpi faintly visible, grandiporate, crassimarginate, pores 17 μm in diameter. Exine 10 μm thick, sexine thinner than nexine, surface laevigate.

Occurrence Miocene, Neyveli Lignitefield, Neyveli (Navale and Misra 1979).

Meliapollis quadrangularis (Ramanujam 1966) Sah & Kar 1970 (Plate 20, Fig. 14)

Description (Ramanujam 1966, p. 178, 180): Pollen grains isopolar, prolate to subprolate, tetrazonicolporate, large, 54 × 33 μm. Amb four-sided, quadrangular, colpi tenuimarginate, end portions rather faint, pointed, long but not reaching the poles, ora prominent, slightly lalongate with thickened rims. Exine 2.5 μm thick, columellae not distinct, surface psilate.

Occurrence Miocene-Pliocene, Cuddalore, Series, Neyveli, Tamil Nadu (Ramanujam 1966); Early Eocene, Laki Series, Kutch, Gujarat (Sah and Kar 1970); Early Eocene, Naredi Formation, Kutch, Gujarat (Kar 1978); Palaeocene, Matanomadh Formation, Matanomadh, Kutch, Gujarat (Saxena 1979).

Remarks *Tetracolporites quadrangularis* Ramanujam (1966) was transferred to *Meliapollis* by Sah and Kar (1970).

Meliapollis quilonensis Rao & Ramanujam 1982 (Plate 20, Fig. 15)

Description (Rao and Ramanujam 1982, p. 81): Pollen grains isopolar, amb rounded to more or less squarish, polar diameter 12–17 μm, zonaperturate, tetracolporate, colpae considerably short with thin margins, ends pointed to blunt; ora faint, rounded. Exine 1.5 μm thick, sexine as thick as nexine, columella distinct, surface smooth to flecked.

Occurrence Miocene, Quilon Beds, Kerala (Rao and Ramanujam 1982).

Meliapollis ramanujamii Sah & Kar 1970 (Plate 21, Fig. 1)

Description (Sah and Kar 1970, p. 134): Pollen grains circular-subcircular, 50–65 × 48–62 μm. 4-colporate, brevicolpate, pore circular-lalongate, margin thickened. Exine 1.5–2.5 μm, laevigate.

Occurrence Early Eocene, Laki Series, Kutch, Gujarat (Sah and Kar 1970); Early Eocene, Palana lignite, Bikaner, Rajasthan (Sah and Kar 1974); Early Eocene, Naredi Formation, Kutch, Gujarat (Kar 1978); Palaeocene, Matanomadh Formation, Matanomadh, Kutch, Gujarat (Saxena 1979); Miocene, Neyveli Lignitefield, Neyveli, Tamil Nadu (Navale and Misra 1979); Miocene, Warkalli Beds, Kerala (Ramanujam 1987).

Meliapollis raoi Sah & Kar 1970 (Plate 21, Fig. 2)

Description (Sah and Kar 1970, p. 134): Pollen grains circular-subcircular, 50–60 × 50–58 μm. 3-colporate, brevicolpate, pore distinct with thickened margin, lalongate to circular. Exine 1.5–2.5 μm thick, laevigate.

Occurrence Early Eocene, Laki Series, Kutch, Gujarat (Sah and Kar 1970); Early Eocene, Naredi Formation, Kutch, Gujarat (Kar 1978); Palaeocene, Matanomadh Formation, Matanomadh, Kutch, Gujarat (Saxena 1979); Miocene, Neyveli Lignitefield, Neyveli, Tamil Nadu (Navale and Misra 1979); Early Eocene, Naredi Formation, subsurface sediments, Kutch, Gujarat (Kar 1985).

Meliapollis simplex Navale & Misra 1979 (Plate 21, Fig. 3)

Description (Navale and Misra 1979, p. 234): Isopolar, subcircular to pentagonal pollen grain, 53 × 52 μm in size. Pentacolporate, brevicolpate, colpi faint to indistinct, grandiporate, lalongate, 7–8 μm in diameter with appreciably thickened margins. Exine 7–8 μm thick, sexine (1.5 μm) 3–4 times thinner than nexine (6–6.5 μm). Internal margin of the mesocolpial nexine bulged out at places due to accommodation of much thickened pore margins, surface laevigate.

Occurrence Miocene, Neyveli Lignitefield, Neyveli (Navale and Misra 1979)

Meliapollis tamilii Navale & Misra 1979 (Plate 21, Fig. 4)

Description (Navale and Misra 1979, p. 234): Isopolar, subcircular to pentagonal pollen, 85–100 × 85–95 μm in size. Pentacolporate, brevicolpate, colpi faint to distinct, some colpi margins dissolved appearing deeper than others. Pores distinct, 11–14 μm in diameter, lalongate with thickened margins. Exine 3–5.5 μm thick, sexine equal to or slightly thinner than nexine, surface laevigate.

Occurrence Miocene, Neyveli Lignitefield, Neyveli (Navale and Misra 1979)

Meliapollis triangulus Saxena 1979

Now: *Dermatobrevicolporites triangulus* Kar 1985

Genus *Meyeripollis* Baksi & Venkatachala 1970

Type Species *Meyeripollis naharkotensis* Baksi & Venkatachala 1970

Generic Description (Baksi and Venkatachala 1970, p. 83): Triangular, 26–40 μm; with rounded angles and prominent convex sides, angles distinguished by the presence of two large tubercles or gemmae on either side; polar compression more than equatorial ones. Trisyncolpate, colpi often obscured by the ornamentation, colpi meeting at distal pole giving the appearance of a trilete mark. Exine ornamented with gemmae, tubercles and verrucae of different sizes; broad prominent gemmae present at angles of the pollen at distal pole.

Emended Description (Baksi and Venkatachala in Thanikaimoni et al. 1984, p. 57): Pollen grain tricolporate; triangular in polar view; ectoaperture more or less united at the poles; exine with striate warts which are bigger at the equatorial angles as well as at the poles.

Species Recorded from India

Meyeripollis baksii Rao & Ramanujam 1982 (Plate 17, Fig. 2)

Description (Rao and Ramanujam 1982, p. 73): Pollen grains isopolar, amb rounded triangular to subspherical, polar diameter 25–32 μm; zonaperturate, tricolpate, syncolpate, colpae usually obscured by heavy sculpturing, narrow at equator but widening towards poles where they meet, colpal margins densely beset with gemmae. Exine up to 4.5 μm thick, sexine much thicker than nexine, tectate, densely studded with supratectal gemmate-tuberculate processes; gemmae of different sizes, up to 3.5 μm high, rounded in surface view; apices surrounded by 2–4 gemmae only, slightly larger or almost of the same size as other processes.

Occurrence Miocene, Quilon Beds, Padappakkara, Kerala (Rao and Ramanujam 1982).

Meyeripollis laudabilis Salujha et al. 1972 (Plate 17, Fig. 3)

Description (Salujha et al. 1972, pp. 280, 281): Brown, roundly triangular with straight to slightly convex sides, measuring 30.4–33.6 μm; trisyncolpate, colpi faintly to clearly discernible; exine bearing tubercles, two large tubercles 6.4–8 μm wide at each of the angles, smaller tubercles, 3–4 μm wide, present throughout, occasionally coalescing to give a broadly reticulate appearance; in case of gap in between the tubercles exine faintly structured.

Occurrence Palaeocene, Garo Hills, Meghalaya (Salujha et al. 1972).

Meyeripollis naharkotensis Baksi & Venkatachala 1970 (Plate 17, Fig. 4)

Description (Baksi and Venkatachala 1970, p. 83): Same as generic description.

Occurrence Upper Eocene-Late Miocene, Assam (Baksi and Venkatachala 1970).

Genus *Monocolpopollenites* Pflug & Thomson in Thomson and Pflug 1953

Type Species: *Monocolpopollenites tranquillus* (Potoniè 1934) Jansonius & Hills 1976

Basionym: *Pollenites tranquillus* Potoniè 1934

Generic Description (Pflug & Thomson in Thomson and Pflug 1953, p. 62): Pollen grains with only one plane of symmetry. Monocolpate. Colpus in the plain of symmetry, i.e. located on the line bisecting the proximal face, but asymmetric with regard to the equator. Overall shape not bean-shaped. Colpus commonly with marginal folds are thickenings.

Emended diagnosis (Saxena 2010, p. 106): Pollen grains oval-elliptical in shape. Small to medium sized. Monosulcate, sulcus as long as the longer axis or may be shorter, may be associated with marginal folds or thickenings. Exine generally thin, psilate (or more or less psilate), may be faintly punctate, granulate, scabrate or baculate.

Remarks Forms assigned to *Monocolpopollenites* Pflug & Thomson in Thomson and Pflug (1953) represent monosulcate fossil pollen having affinity with the family Arecaceae. These pollen frequently occur in Palaeogene and Neogene sediments of India. Several monocolpate pollen assignable to this genus have been described under *Palmaepollenites* Potoniè ex Potoniè 1958 and *Palmaepites* Biswas 1962. Diagnosis of *Monocolpopollenites* has been emended by various palynologists resulting into placement of diverse monocolpate pollen under it. Saxena (2010) discussed taxonomic and nomenclatural implications of Monocolpate genera frequently found in Indian Tertiary sediments and considered *Palmaepollenites* Potoniè ex Potoniè (1958) and *Palmaepites* Biswas (1962) as junior synonyms of *Monocolpopollenites*.

Species Recorded from India

Monocolpopollenites andamanicus (Mathur and Mathur 1980) Saxena 2010 (Figure not given)

= *Palmaepollenites andamanicus* Mathur & Mathur 1980

Description (Mathur and Mathur 1980, p. 57): Pollen grains monocolpate, oval-elliptical in shape with more or less narrow lateral ends, 30–35 × 15–17 μm in size; colpi almost extending from one end to the other end, lips thickened, 2–3 μm broad. Exine less than 1 μm thick, scabrate.

Occurrence Early Oligocene, Baratang Formation, Middle Andaman, Andaman and Nicobar Islands (Mathur and Mathur 1980).

Monocolpopollenites baculatus (Venkatachala and Rawat 1972) Saxena 2010 (Plate 4, Fig. 3)

= *Palmaepollenites baculatus* Venkatachala & Rawat 1972

Description (Venkatachala and Rawat 1972, p. 297): Pollen grain oval with rounded lateral ends; 29.7 × 19.8 μm; monocolpate, colpus distinct, narrow in the middle and broader at the ends, funnel shaped, running along the whole length of the pollen and reaching up to the margins. Exine about 1.3 μm thick, tectate, sexine thicker than nexine, baculate, bacula fine.

Occurrence Palaeocene-Eocene, Cauvery Basin, Tamil Nadu (Venkatachala and Rawat 1972); Eocene, Neyveli lignite, Cuddalore District, Tamil Nadu (Venkatachala 1973).

Monocolpopollenites cauveriensis (Venkatachala and Rawat 1972) Saxena 2010 (Plate 4, Fig. 4)

= *Palmaepollenites cauveriensis* Venkatachala & Rawat 1972

Description (Venkatachala and Rawat 1972, pp. 296–297): Pollen grain oval with rounded lateral ends, 39.6 × 23.1 μm, monocolpate; colpus broad, more or less extending along the length of the grain, more or less reaching up to the margin. Exine about 2.6 μm thick, psilate, sexine thicker than nexme.

Occurrence Palaeocene-Eocene Cauvery Basin, Tamil Nadu (Venkatachala and Rawat 1972); Early-Middle Miocene Warkalli Beds, Kerala (Ramanujam and Rao 1977).

Monocolpopollenites communis (Sah and Dutta 1966) Kumar & Takahashi 1991 (Plate 4, Fig. 5)

= *Palmaepollenites communis* Sah & Dutta 1966

Description (Sah and Dutta 1966, p. 77): Size range 30 × 15 to 34 × 20 μm; bilateral, amb ellipsoidal. Monosulcate, sulcus straight, usually extending from pole to pole, sometimes slightly shorter; lips slightly raised. Exine thin, less than 1 μm; generally smooth, but sometimes appears to be faintly sculptured.

Occurrences Palaeocene, Cherra Formation, Shillong Plateau, Meghalaya (Sah and Dutta 1966; Dutta and Sah 1970); Palaeocene, Nongwal Bibra, Garo Hills, Meghalaya (Singh et al. 1976); Palaeocene, Tura formation, Nongwal Bibra, Garo Hills, Meghalaya (Singh 1977).

Monocolpopollenites cymbatus (Rawat et al. 1977) Saxena 2010 (Figure not given)

= *Palmaepollenites cymbatus* Rawat et al. 1977

Description (Rawat et al. 1977, pp. 180–181): Pollen elongated; oval with slightly rounded lateral ends; monocolpate, 25 × 55–68 μm.Colpus distinct, broad, running along the length of the pollen, more or less extending up to lateral ends. Exine up to 2 μm thick, sexine as thick as nexine, granulate; grana closely placed, up to 0.5 μm in diameter.

Occurrence Early Eocene, Kadi Formation Cambay Basin, Gujarat (Rawat et al. 1977).

Monocolpopollenites elongatus (Kulkarni et al. 1985) Saxena 2010 (Figure not given)

= *Palmaepollenites elongatus* Kulkarni et al. 1985

Description (Kulkarni et al. 1985, pp. 297–298): Pollen grains peroblate, heteropolar, biconvex laterally, 16.6 × 50.0 μm; monosulcate, sulcus slightly opened, longisulcate, crassimarginate; exine 1.6 μm thick, psilate.

Occurrence Miocene, subsurface sediments, Ratnagiri Beds, Maharashtra (Kulkarni et al. 1985).

Monocolpopollenites eocenicus (Biswas 1962) Saxena 2010 (Plate 4, Fig. 6)

= *Palmaepites eocenicus* Biswas 1962

= *Palmaepollenites eocenicus* (Biswas) Sah & Dutta 1966

Description (Biswas 1962, p. 40): Equatorial compression; exine smooth and rather thin; grain measurements 36 × 24; sulcus nearly as long as the long diameter of the grain, measuring 28 × 8 μm.

Description (Sah and Dutta 1966, p. 77): Size range 35 × 24–44 × 40 μm; amb oval to elliptical. Monosulcate, sulcus long but not reaching the equator, lips thin and usually gaping, exine up to 2 μm thick, smooth but at places appears to be faintly sculptured.

Occurrence Early Eocene, Tura Formation, Meghalaya (Biswas 1962); Palaeocene, Cherra Formation, Meghalaya (Sah and Dutta 1966).

Monocolpopollenites indicus (Ramanujam1966) Saxena 2010 (Plate 4, Fig. 8)

= *Palmaepollenites indicus* Ramanujam1966

Description (Ramanujam 1966, p. 157): Pollen bilateral, oval to elliptical, poles tapering, monosulcate. 38 × 22 µm, sulcus very narrow, long, ends blunt. Margin, tenuimarginate, exine 1.5 µm thick, surface smooth.

Occurrence Miocene-Pliocene, Cuddalore Series, Neyveli, Tamil Nadu (Ramanujam 1966).

Monocolpopollenites keralensis (Rao and Ramanujam 1978) Saxena 2010 (Plate 4, Fig. 9)

= *Palmaepollenites keralensis* Rao & Ramanujam 1978

Description (Rao and Ramanujam 1978, p. 411): Pollen grains heteropolar, amb roughly rhombic, narrow towards one end and broader towards other end, 37.5–42 × 26–28 µm, monosulcate, sulcus long but not reaching the equatorial margin, club-shaped, narrow at one end and gradually widening at other end, margins thickened at mid region, ends blunt. Exine 1.5 µm thick, sexine as thick as nexine, surface smooth.

Occurrence Miocene, Quilon Beds, Padappakkara, Kerala (Rao and Ramanujam 1978).

Monocolpopollenites kutchensis (Venkatachala and Kar 1969) Saxena & Trivedi 2009 (Plate 4, Fig. 10)

= *Palmaepollenites kutchensis* Venkatachala & Kar 1969

Description (Venkatachala and Kar 1969, p. 159): Pollen grains oval with more or less equally broad lateral ends, 25–30 × 10–15 µm. Monocolpate, colpus broad, boat-shaped, not reaching up to margins. Exine more or less laevigate.

Occurrence Early Eocene, Laki Series, Matanomadh, Kutch, Gujarat (Venkatachala and Kar 1969).

Monocolpopollenites magnus (Sah and Kar 1970) Saxena 2010 (Plate 4, Fig. 11)

= *Palmaepollenites magnus* Sah & Kar 1970

Description (Sah and Kar 1970, p. 129): Pollen grains oval, 130–160 × 50–70 μm. Monocolpate, colpus closed, extending from one end to other end. Exine more or less laevigate.

Occurrence Early Eocene, Laki Series, Kutch, Gujarat (Sah and Kar 1970).

Monocolpopollenites medicolpus (Mathur and Mathur 1969) Saxena 2010 (Figure not given)

= *Palmaepollenites medicolpus* Mathur & Mathur 1969

Description (Mathur and Mathur 1969, p. 7): Apolar, bilaterally symmetrical, 29 μm long and 24 μm broad, oval, monocolpate. Colpi not extending from one margin to another, narrow. Exine thin, intragranulose, brownish yellow.

Occurrence Pliocene, Naera Formation, Baraia and Naera, Kutch District, Gujarat (Mathur and Mathur 1969); Pliocene, Baraia and Naera, Kutch District, Gujarat (Kar 1985); Early Miocene, Bhuban Formation, Chawngte, Lawngtlai District, Mizoram (Mandaokar 2007).

Monocolpopollenites nadhamunii (Venkatachala and Kar 1969) Saxena & Trivedi 2009 (Plate 4, Fig. 12)

= *Palmaepollenites nadhamunii* Venkatachala and Kar 1969

Description (Venkatachala and Kar 1969, p. 159): Pollen grains elliptical, spindle-shaped, 25–30 × 15–22 μm. Monocolpate, colpi narrow, extending from one margin to other. Exine laevigate.

Occurrence Early Eocene, Laki Series, Matanomadh, Kutch, Gujarat (Venkatachala and Kar 1969); Early Eocene, Laki Series, Kutch, Gujarat (Sah and Kar 1970).

Monocolpopollenites neyvelii (Ramanujam 1966) Saxena 2010 (Plate 4, Fig. 13)

= *Palmaepollenites neyvelii* Ramanujam 1966

Description (Ramanujam 1966, p. 157): Pollen bilateral, elliptical, often sharply tapering at both poles, monosulcate, 35 × 20 μm. Sulcus long, narrow, ends blunt, margin crassimarginate, 2.5 μm thick. Exine 2.5 μm thick, surface smooth.

Occurrence Miocene-Pliocene, Cuddalore Series, Neyveli, Tamil Nadu (Ramanujam 1966).

Monocolpopollenites ovatus (Sah and Kar 1970) Saxena 2010 (Plate 4, Fig. 14)

= *Palmaepollenites ovatus* Sah & Kar 1970

Description (Sah and Kar 1970, p. 129): Pollen grains roundly oval, 38–45 × 35–40 μm. Monocolpate, colpus restricted in middle region. Exine intragranular, about 2 μm thick.

Occurrence Early Eocene, Laki Series, Kutch, Gujarat (Sah and Kar 1970).

Monocolpopollenites plicatus (Sah and Kar 1970) Saxena 2010 (Plate 4, Fig. 15)

= *Palmaepollenites plicatus* Sah & Kar 1970

Description (Sah and Kar 1970, p. 129): Pollen grain elliptical, 48–60 × 20–40 μm. Monocolpate, colpus extending from end to end, associated with folds; exine more or less laevigate.

Occurrence Early Eocene, Laki Series, Kutch, Gujarat (Sah and Kar 1970).

Monocolpopollenites rajpardiensis Nom. Nov. (Plate 4, Fig. 7)

Homonym: *Palmaepollenites elongatus* Samant & Phadtare 1997, *Palaeontographica*, p. 38, Plate 4, Fig. 5.

Description (Samant and Phadtare 1997, p. 38): Pollen grains heteropolar, prolate, appears like pea pod, 40–45 × 26–30 μm, monosulcate, sulcus extending from end to end, obliquely placed, narrow, crassimarginate, margin smooth; exine 1–2 μm thick, nexine thicker than sexine, psilate to micropunctate.

Occurrence Early Eocene, Tarkeshwar Formation, Rajpardi, Cambay Basin (Samant and Phadtare 1997).

Monocolpopollenites ratnagiriensis (Kulkarni et al. 1985) Saxena 2010 (Plate 5, Fig. 1)

= *Palmaepollenites ratnagiriensis* Kulkarni et al. 1985

Description (Kulkarni et al. 1985, p. 297): Pollen grains heteropolar, biconvex in equatorial view, monosulcate, 35 × 49 μm, sulcus prominently wide, 16.6 × 45.7 μm, ends pointed, crassimarginate, exine about 2.5 μm thick, sexine thicker than nexine, psilate.

Occurrence Miocene, subsurface sediments, Ratnagiri Beds, Maharashtra (Kulkarni et al. 1985).

Genus *Myricipites* Wodehouse 1933

Type Species *Myricipites dubius* Wodehouse 1933

Generic Description (Wodehouse 1933, p. 505): Pollen approximately spheroidal. Pores three, greatly protruding, of the 'tarsus' pattern. Texture smooth.

Species Recorded from India

Myricipites harrisii (Couper 1953) Dutta & Sah 1970 (Plate 25, Fig. 6)

Restated Description (Dutta and Sah, p. 47): Size range 18–35 μm; amb triangular to roundly triangular; 3-aperturate, angulaperturate, distinctly porate, pores more or less circular to sometimes lalongate, crassimarginate; exine fairly thick, surface sculpture psilate to minutely scabrate.

Occurrence Palaeocene-Early Eocene, Shillong Plateau, Meghalaya (Dutta and Sah 1970).

Myricipites singhii Rao 1995 (Plate 25, Fig. 7)

Description (Rao 1995, p. 331): Pollen grains subtriangular-subcircular associated with folds. Size range 75–10 × 67–82 μm. Triporate, pore distinct, 12–14 μm in diameter. Exine 3–4 μm thick. Ornamentation psilate to scabrate.

Occurrence Palaeocene-Early Miocene, subsurface sediments, Alappuzha (Alleppey) District, Kerala (Rao 1995).

Myricipites vulgaris Dutta & Sah 1970 (Plate 25, Fig. 8)

Description (Dutta and Sah 1970, p. 48): Size range 31–36 μm; amb triangular to roundly triangular; 3-aperturate, angulaperturate, distinctly porate, pores generally lolongate, crassimarginate; exine thin, undifferentiated, surface sculpture minutely granulate.

Occurrence Palaeocene-Early Eocene, Shillong Plateau, Meghalaya (Dutta and Sah 1970); Middle-Late Eocene, subsurface sediments, Rataria, Kutch, Gujarat (Kar and Saxena 1979).

N

Genus *Neocouperipollis* (Venkatachala and Kar 1969) Kar & Kumar 1987

Remarks Venkatachala and Kar (1969) proposed *Couperipollis* for monosulcate pollen with spinose/baculate ornamentation and designated *Monosulcites perspinosus* Couper (1953) as type species of this genus. Morphological features of *Monosulcites perspinosus* cannot be affirmed as, its holotype is lost and the original drawing (Couper 1953, Plate 8, Fig. 133) does not provide enough information. Considering *Couperipollis* Venkatachala and Kar (1969) a nomen ambiguoum, Kar and Kumar (1987) proposed the genus *Neocouperipollis* and designated *Couperipollis kutchensis* its type species.

Singh and Misra (1991c) re-examined the specimens of *Couperipollis kutchensis* Venkatachala and Kar (1969) and found that all specimens resembling this species were ill preserved, folded or broken. These authors also observed that none of the studied specimens showed presence of sulcus. Probably, the exinal folds were considered as sulcus. Singh and Misra (1991c) studied some other specimens of *Couperipollis* such as, *C. rarispinosus* (Sah and Kar 1974, Plate 1, Fig. 26), *C. brevispinosus* (Sah and Kar 1974, Plate 1, Fig. 28) and *C. wodehousei* (Kar and Kumar 1986, Plate 5, Fig. 8) and found that these show at least two distinct pores which were obscured by heavy ornamentation.

Singh and Misra (1991c) re-examined the holotype of *Acanthotricolpites bulbospinosus* (Kar 1985, Plate 11, Figs. 4–6) and found that it is a spinose porate form and not tricolpate as described by Kar (1985). Singh and Misra (1991c) recovered large number of tri- to tetraporate pollen having spines from Neyveli Lignitefield, Tamil Nadu. These workers observed wide range of variation in the spine morphology overlapping with the spine structure of *Neocouperipollis*. In order to resolve the generic confusion and incongruous speciation among spinose genera *Neocouperipollis* and *Acanthotricolpites* large number of spinose pollen described by Venkatachala and Kar (1969), Sah and Kar (1970, 1974), Saxena (1979), Kar (1985), Kar and Kumar (1986, 1987) and those recorded from Neyveli Lignite, Tamil Nadu were studied by Singh and Misra (1991c). These authors observed the assemblage is characterized by two types of pollen, monosulcate and tri- to tetraporate. Exine ornamentation in these forms was in general spinose. However, shape of pollen, fine structure of spines and inter-spinal areas in monosulcate and porate forms are different.

Singh and Misra (1991c, p. 221) observed that holotype of *Neocouperipollis kutchensis* (Venkatachala & Kar) Kar and Kumar (1987) was porate instead of monosulcate hence, these authors transferred it to *Acanthotricolpites* and *Neocouperipollis* became obligate junior synonym. As, there was no suitable genus to accommodate spinose monosulcate pollen described under *Neocouperipollis*, a new genus *Spinomonosulcites* Singh and Misra (1991c) was proposed.

Species Recorded from India

Neocouperipollis ankleshwarensis Kar & Bhattacharya 1992

Now: *Spinomonosulcites ankleshwarensis* (Kar and Bhattacharya 1992) Samant & Phadtare 1997

Neocouperipollis achinatus (Sah and Kar 1970) Kar & Kumar 1987

Now: *Spinomonosulcites achinatus* (Sah and Kar 1970) Singh & Misra 1991c

Neocouperipollis cymbatus (Venkatachala and Rawat 1972) Saxena & Khare 2004 (Plate3, Fig. 15)

Description (Venkatachala and Rawat 1972, p. 296): Pollen grain elongate-oval, monosulcate; 23–25.8 × 42–43 µm. Sulcus well-developed, running the whole length of the pollen, broader at the ends and slightly constricted in the middle. Exine 1.5–2 µm thick, spinose, spines with bulbous base, 3.3 µm wide, with tapering ends, 5.3 µm long, placed 3.5–5.5 µm apart, uniformly distributed, surface in between spines granulose.

Occurrence Palaeocene-Eocene, Cauvery Basin, Tamil Nadu (Venkatachala and Rawat 1972); Late Palaeocene-Middle Eocene, subsurface sediments, Tiruchirapalli, Tamil Nadu (Saxena and Khare 2004).

Remarks *Couperipollis cymbatus* Venkatachala and Rawat (1972) was transferred to *Neocouperipollis* by Saxena and Khare (2004).

Neocouperipollis rarispinosus (Sah and Dutta 1966) Singh 1990 (Plate 4, Fig. 1)

Description (Sah and Dutta 1966, p. 76): Size range 44–56 µm; amb spheroidal to subspheroidal. Monosulcate, sulcus usually distinct, long (extending almost up to the poles) and usually broad. Exine thin, less than 2 µm, undifferentiated; spinose, spines sparse, small (less than 2 µm) and with pointed tips.

Occurrence Early Eocene, Cherra Formation, South Shillong Plateau, Meghalaya (Sah and Dutta 1966).

Remarks *Monosulcites rarispinosus* Sah and Dutta (1966) and *Couperipollis rarispinosus* Venkatachala and Kar (1969) were transferred to *Neocouperipollis* by Singh (1990).

Genus *Neotrichotomosulcites* Samant & Phadtare 1997

Type Species *Neotrichotomosulcites foveolatus* Samant & Phadtare 1997

Generic Description (Samant and Phadtare 1997, p. 33): Pollen grains triangular to subtriangular in polar view; trichotomosulcate; foveolate to reticulate.

Botanical Affinity *Elaeis guineensis* (Arecaceae).

Species Recorded from India

Neotrichotomosulcites foveolatus Samant & Phadtare 1997 (Plate 12, Fig. 2)

Description (Samant and Phadtare 1997, p. 33, 34): Pollen grains with triangular amb, heteropolar, 32–38 μm in diameter, trichotomosulcate, sulcus reaching up to the margin which is psilate, exine 0.5–1.5 μm thick, semitectate, sexine and nexine undifferentiated.

Occurrence Early Eocene, Tarkeshwar Formation, Rajpardi, Cambay Basin, Gujarat (Samant and Phadtare 1997).

Genus *Nymphaeacidites* Sah 1967

Type Species *Nymphaeacidites typicus* Sah 1967

Generic Description (Sah 1967, p. 108): Pollen grains usually large, amb spheroidal to prolate-spheroidal; usually 1–porate, aperture sometimes operculate; ornamentation reticulate or occasionally spiniferous or baculate.

Botanical Affinity Nymphaeaceae.

Species Recorded from India

Nymphaeacidites clarus Dutta & Sah 1970 (Plate 22, Fig. 15)

Description (Dutta and Sah 1970, p. 45): Size range 39–47 μm; amb sub-prolate; 1–porate or distinctly equatorial, operculate, operculum 32 μm in diameter. Exine thick, sexine as thick as nexine or slightly thicker, surface finely structured.

Occurrence Palaeocene, Cherra Formation, Shillong Plateau, Meghalaya (Dutta and Sah 1970).

Nymphaeacidites giganteus Mehrotra 1983 (Plate 23, Fig. 1)

Description (Mehrotra 1983, p. 16): Pollen grains spheroidal; size range 58–92 μm; monosulcate, sulcus large running parallel to the margin; exine undifferentiated, 1–2 μm thick, punctate to scabrate.

Occurrence Palaeocene-Eocene, Mikir Formation, North Cachar Hills, Assam (Mehrotra 1983).

Nymphaeacidites sphericus Dutta & Sah 1970 (Plate 23, Fig. 2)

Description (Dutta and Sah 1970, p. 45): Size range 48–60 μm; amb spheroidal to sometimes prolate-spheroidal; 1–sulcate, sulcus equatorial, exine very thin, stratification obscure, surface ornamentation scabrate.

Occurrence Palaeocene, Cherra Formation, Shillong Plateau, Meghalaya (Dutta and Sah 1970).

Genus *Nyssapollenites* Thiergart 1937 ex Potonié 1960

Type Species *Nyssapollenites pseudocruciatus* Thiergart 1937 ex Potonié 1960

Generic Description (Potonié 1960, p. 103): Pollen with suboblate shape, rounded, oval or almost spherical; meridian often rhombic; equator more or less rounded triangular; tricolporate, equatorial rugae more or less distinct, meridional rugae (= colpi) with more or less thickened border; exine infrapunctate.

Remarks *Pseudonyssapollenites* Kar (1985) is morphologically similar to *Nyssapollenites* hence should be treated as synonymous to the latter one.

Species Recorded from India

Nyssapollenites barooahii Sah & Dutta 1968 (Plate 17, Fig. 5)

Description (Sah and Dutta 1968, p. 189): Size range 21–26 μm; amb roundly triangular in polar view, ovoid in equatorial view, 3-zonaperturate, colporate, pores rounded, colpi wide at the equator, narrow towards the poles; exine two layered, sexine thickened at the aperture region, ornamentation psilate to scabrate.

Occurrence Miocene, Tipam Group, Upper Assam (Sah and Dutta 1968).

Nyssapollenites incertus Dutta & Sah 1970 (Plate 17, Fig. 6)

Description (Dutta and Sah 1970, p. 39): Size range 22–28 μm; amb oblate-spheroidal to subspheroidal in polar and oval-elliptical in equatorial view; 3-zonaperturate; colporate, colpi long, extending more or less to the poles, crassimarginate, ora lalongate; exine thick, sexine as thick as nexine, or slightly thinner, surface sculpture appears to be scabrate.

Occurrence Palaeocene, Cherra Formation, Shillong Plateau, Meghalaya (Dutta and Sah 1970).

Nyssapollenites kutchensis Venkatachala & Kar 1969 (Plate 17, Fig. 7)

Description (Venkatachala and Kar 1969, p. 167): Pollen grains roundly triangular, 20–35 × 18–32 μm. Tricolporate, colpi short, narrow, inconspicuous, pores well developed with rounded margin. Exine thick, laevigate and intragranulate.

Occurrence Early Eocene, Laki Series, Matanomadh, Kutch, Gujarat (Venkatachala and Kar 1969).

O

Genus *Ocimumpollenites* Kar 1996

Type Species *Ocimumpollenites indicus* Kar 1996

Generic Description (Kar 1996, p. 46): Pollen grains subcircular in polar view, size range 56–72 μm, 6-colpate, brevicolpate, colpi slit-funnel shaped in polar view. Exine 2–4 μm thick, sexine thicker than nexine, tectate, columellate, some columellae fused at top but free at base, reticulation broad, muri pluricolumellate, columella also present in lumina.

Botanical Affinity *Ocimum* (Lamiaceae).

Species Recorded from India

Ocimumpollenites indicus Kar 1996 (Plate 12, Fig. 14)

Description (Kar 1996, p. 46): Pollen grains mostly found in polar view, subcircular-circular, 56–68 × 54–65 μm, hexacolpate, brevicolpate, colpi generally slit-like, distinct. Exine 2–4 μm thick, sexine thicker than nexine; tectate, columellate, columella closely placed, some columellae fused at the top but free at base. Ornamentation reticulate, reticulum broad, square to rhomboidal in shape, muri pluricolumellate, few columellae also present in lumina.

Occurrence Early Eocene, Palana Formation, subsurface sediments, Bikaner, Rajasthan (Kar 1996); Late Palaeocene, subsurface sediments, Bikaner, Rajasthan (Kar and Sharma 2001); Early Eocene, Koti, Solan, Himachal Pradesh (Sarkar and Prasad 2002); Late Palaeocene, Akli Formation, Barmer, Rajasthan (Tripathi et al. 2003).

Genus *Ornatetradites* Rao & Ramanujam 1982

Type Species *Ornatetradites droseroides* Rao & Ramanujam 1982

Generic Description (Rao and Ramanujam 1982, p. 87): Pollen grains in tetrahedral tetrads, individual grains rounded to suboblate, porate; pores 4–8 or more, usually indistinct and masked by heavy ornamentation, confined more towards proximal facets. Exine tectate, clavate, gemmate or verrucate, surface between sculptural elements punctate.

Species Recorded from India

Ornatetradites chandae Rao and Ramanujam 1982 (Plate 25, Fig. 11)

Description (Rao and Ramanujam 1982, pp. 87, 88): Pollen grains in tetrahedral tetrads, occasionally with a tendency towards rhomboidal nature, 35–43 μm; individual grain rounded, 16–23.5 × 18.5–25 μm; porate, pores 4–6 in number, rounded to ellipsoidal, confined more towards proximal facets, tenuimarginate, obscured by heavy sculptural elements. Exine up to 3 μm thick, tectate, columella distinct; surface verrucate, verrucae up to 2 μm high, surface between verrucae psilate to finely punctate.

Occurrence Miocene, Quilon Beds, Padappakkara and Edavai, Kerala (Rao and Ramanujam 1982); Miocene, Warkalli Beds, Kerala (Ramanujam 1987).

Ornatetradites droseroides Rao & Ramanujam 1982 (Plate 25, Fig. 14)

Description (Rao and Ramanujam 1982, p. 87): Pollen grains in tetrahedral tetrads, 34–45 μm, individual grains rounded suboblate equatorially, 15–25 × 20–27 μm; porate, pores up to 8 in number, somewhat obscured by closely placed sculptural processes, confined more towards proximal facets, circular to flattened, 2.5–3 μm across, with thin margins. Exine up to 3.5 μm thick, sexine thicker than nexine, tectate, columella distinct, appearing as radial striae at the periphery of the pollen; supratectal ornamentation gemmate-tuberculate; gemmae up to 2.5 μm high, rather densely placed all over, surface in between gemmae regulate-punctate.

Occurrence Miocene, Quilon Beds, Padappakkara and Edavai, Kerala (Rao and Ramanujam 1982); Miocene, Warkalli Beds, Kerala (Ramanujam 1987).

Ornatetradites keralaensis Rao 1995 (Plate 25, Fig. 15)

Description (Rao 1995, p. 333): Pollen grains in tetrahedral tetrads. Size range 53–70 × 52–60 μm. Individual grain 32–43 × 22–38 μm. Porate, 4–6 pores, obscured by heavy sculptural elements. Exine 2–2.5 μm thick, sexine and nexine not differentiated, tectate, columella distinct, verrucate, verrucae 3–4 μm high, sparsely placed, surface between verrucae finely reticulate.

Occurrence Eocene-Early Miocene, subsurface sediments, Alappuzha (Alleppey), Kerala (Rao 1995).

Genus *Ornatriporites* Rao & Ramanujam 1982

Type Species *Ornatriporites elegans* Rao & Ramanujam 1982

Generic Description (Rao and Ramanujam 1982, p. 84): Pollen grains isopolar, subtriangular to rounded triangular, triporate, pore margin ornate, beset with gemmae or tubercles, surface gemmate-tuberculate; locally confluent to form vermiform to cribellate structures.

Species Recorded from India

Ornatriporites elegans Rao & Ramanujam 1982 (Plate 25, Fig. 9)

Description (Rao and Ramanujam 1982, p. 84): Pollen grains isopolar, amb subtriangular, sides convex, polar diameter 18–24 μm; zonaperturate, triporate, pores up to 3 μm, pore margin incrassate, ornate, studded with 4–6 prominent gemmae or tubercles. Exine up to 3.5 μm thick, intectate, surface gemmate-tuberculate, processes up to 2.5 μm high, densely placed all over, locally anastomosing to form vermiform to cribellate structures.

Occurrence Miocene, Quilon Beds, Kerala (Rao and Ramanujam 1982).

P

Genus *Padappakkarapollis* Rao & Ramanujam 1982

Type Species *Padappakkarapollis venkatachalae* Rao & Ramanujam 1982

Generic Description (Rao and Ramanujam 1982, p. 82): Pollen grains isopolar, amb spheroidal to stellate spheroidal; stephanocolpate, brevicolpate, colpae 6–8 in number, ora lalongate. Exine differentially sculptured nexine with undulating or vermiform thickenings, sexine finely pitted.

Botanical Affinity Ctenolophonaceae.

Species Recorded from India

Padappakkarapollis venkatachalae Rao & Ramanujam 1982 (Plate 21, Fig. 5)

Description (Rao and Ramanujam 1982, p. 82): Pollen grains isopolar, amb spheroidal to stellate spheroidal; polar diameter 38–50 μm; zonaperturate, stephanocolporate, brevicolpate, colpal margins thin, ends blunt. Ora lalongate with thickened margins. Exine up to 3 μm thick, sexine thicker than nexine, surface differentially sculptured, nexine showing characteristic vermiculate thickenings and sexine finely pitted, pits coalescing at many places to form short canals.

Occurrence Miocene, Quilon Beds, Padappakkara, Kerala (Rao and Ramanujam 1982).

Genus *Palaeocoprosmadites* Ramanujam 1966

Type Species *Palaeocoprosmadites arcotense* Ramanujam 1966

Generic Description (Ramanujam 1966, p. 172): Pollen grains isopolar, oblate, 3-zonicolporate, colpi typically brevicolpate, extremely short, ora prominently lalongate and equal to or slightly longer than colpi, surface smooth.

Species Recorded from India

Palaeocoprosmadites arcotense Ramanujam 1966 (Plate 17, Fig. 8)

Description (Ramanujam 1966, p. 172): Pollen grain isopolar, oblate, 25 × 31 μm. Amb subtriangular, colpi three, pronouncedly brevicolpate, ends blunt or pointed. Exine 1.5 μm thick, columellae indistinct, surface smooth.

Occurrence Miocene-Pliocene, Cuddalore Series, Neyveli, Tamil Nadu (Ramanujam 1966); Early Eocene, Laki Series, Kutch, Gujarat (Venkatachala and Kar 1969; Sah and Kar 1970).

Genus *Paleoaraliaceaepites* Biswas 1962

Type Species *Paleoaraliaceaepites indica* Biswas 1962

Description of Genus (Biswas 1962, p. 39): Equatorial compression; exine thick, reticulate, thicker and minutely plicated in the polar region; tricolporate with prominent longitudinal furrows, apparently functional; transverse furrow 3 μm wide, continuous across the equatorial region; 29 × 24 μm.

Botanical Affinity Araliaceae.

Species Recorded from India

Paleoaraliaceaepites Indica Biswas 1962 (Plate 17, Fig. 9)

Description (Biswas 1962, p. 39): Same as that of the genus

Occurrence Early-Middle Eocene, Tura Formation, Meghalaya (Biswas 1962).

Paleoaraliaceaepites indistinctus Samant & Phadtare 1997 (Plate 17, Fig. 10)

Description (Samant and Phadtare 1997, pp. 35, 36): Pollen grains prolate spheroidal to subprolate, 40–43 × 39–42 μm in size in equatorial view; tricolporate; longicolpate, colpi not reaching up to the poles; ora endexinous; exine reticulate, heterobrochate, about 2 μm thick, slightly thickened at equatorial region, tectate, columellate, sexine and nexine undifferentiable.

Occurrence Early Eocene, Tarkeshwar Formation, Rajpardi, Cambay Basin, Gujarat (Samant and Phadtare 1997); Early Eocene, Kharsalia Clay Formation, Bhavnagar, Cambay Basin, Gujarat (Samant 2000).

Genus *Paleocaesalpiniaceaepites* Biswas 1962

Type Species *Paleocaesalpiniaceaepites eocenica* Biswas 1962

Generic Description (Biswas 1962, p. 43): Polar compression; diameter 40 μm; exine is differentiated into three lobes and three inter-lobes; lobes thickened, coarsely reticulate, 24 μm wide at the equator; inter-lobe areas nearly half as wide as the lobes.

Botanical Affinity *Caesalpinia* (Caesalpiniaceae).

Species Recorded from India

Paleocaesalpiniaceaepites eocenica Biswas 1962 (Plate 17, Fig. 11)

Description (Biswas 1962, p. 43): Same as generic description.

Occurrence Early-Middle Eocene, Tura Formation, Meghalaya (Biswas 1962).

Paleocaesalpiniaceaepites ratnamii Ramanujam 1966 (Plate 17, Fig. 12)

Description (Ramanujam 1966, p. 170): Pollen grains isopolar, prolate, 3-zonicolporate, 45 × 32 μm. Amb 3-lobed, lobes slightly arched but not bulging.

Colpi broad, gaping at equator, longicolpate, ora distinct. Exine 4.2 μm thick, crassisexinous, surface retipilate, pila heads rounded or slightly angular in surface view, meshes hexa-polygonal, lumina rounded, or angular, smooth.

Occurrence Miocene-Pliocene, Cuddalore Series, Neyveli, Tamil Nadu (Ramanujam 1966).

Genus *Paleosantalaceaepites* Biswas 1962 ex Dutta & Sah 1970

Type Species *Paleosantalaceaepites dinoflagellatus* Biswas 1962 ex Dutta & Sah 1970

Generic Description (Biswas 1962, p. 40): Equatorial compression; exine moderately thick, smooth; grain measurement 42 × 34 μm; longitudinal colpae conspicuous, apparently functional, but do not have pole to pole extension, transverse furrow less prominent but well-defined, 3 μm across, having superficial resemblance with the equatorial "band" of the dinoflagellates.

Emended Description (Dutta and Sah 1970, p. 40): Pollen grains small to medium in size; rounded triangular to subspheroidal in polar view, oblate spheroidal to more or less spheroidal in equatorial view; 3-colporate, brevicolpate to longicolpate, ora distinct, large, generally zonorate, sometimes lalongate-grandorate; exine thickness variable, sexine smooth or with columella, surface ornamentation smooth, to minutely granulate, to finely reticulate.

Botanical Affinity Rhizophoraceae.

Species Recorded from India

Paleosantalaceaepites claricolpus Venkatachala & Rawat 1972 (Plate 17, Fig. 13)

Description (Venkatachala and Rawat 1972, p. 306): Pollen grain broadly oval in equatorial view; 33 × 25.7 μm; tricolporate. Colpi long but not reaching up to the poles and tapering; ora lalongate. Exine about 1.5 μm thick, sexine thicker than nexine; surface smooth to faintly sculptured.

Occurrence Palaeocene-Eocene, Cauvery Basin, Tamil Nadu (Venkatachala and Rawat 1972).

Paleosantalaceaepites dinoflagellatus Biswas 1962 Ex Dutta & Sah 1970 (Plate 17, Fig. 14)

Description (Dutta and Sah 1970, p. 41): Pollen grains small, size range 32–42 μm along the polar and 24–34 μm along the equatorial axis; amb roundly triangular in equatorial axis; amb roundly triangular in polar view and oblate spheroidal to more or less spheroidal in equatorial view; 3-zonaperturate, distinctly colporate, colpi long, pores lalongate, distinctly or faintly zonorate; exine up to 2 µm thick at the apertures and up to 1 µm at the inter-apertural sides, sexine as thick as nexine, surface ornamentation usually obscure, in some specimens appear to be psilate, in others minutely granulate.

Occurrence Early-Middle Eocene, Tura Formation, Meghalaya (Biswas 1962); Palaeocene, Cherra Formation, Shillong Plateau, Meghalaya (Dutta and Sah 1970).

Paleosantalaceaepites ellipticus Sah & Kar 1970 (Plate 17, Fig. 15)

Description (Sah and Kar 1970, p. 136): Pollen grains elliptical-oval in equatorial view, 46–55 × 25–40 µm. 3-colporate, colpi well developed, pore lalongate. Exine thick, more or less laevigate to finely intrastructured.

Occurrence Early Eocene, Laki Series, Kutch, Gujarat (Sah and Kar 1970).

Paleosantalaceaepites miocenicus (Ramanujam 1966) Ramanujam et al. 1985 (Plate 18, Fig. 1)

Description (Ramanujam 1966, p. 169): Pollen grains isopolar, subprolate, poles smoothly rounded, 3-zonicolporate, 30–34 × 22–26 µm. Colpi short, slender, streak-like, faint, often vestigial, ora prominent. Ora of adjacent colpi extend along equator and fuse with each other forming a conspicuous zonorate band around equator. Exine 3.3 µm thick, columellae distinct, surface smooth.

Occurrence Miocene-Pliocene, Cuddalore Series, Neyveli, Tamil Nadu (Ramanujam 1966).

Remarks Ramanujam et al. (1985) tansferred *Baksipollis miocenicus* Ramanujam (1966) to *Paleosantalaceaepites*.

Paleosantalaceaepites reticulatus Samant & Phadtare 1997 (Plate 18, Fig. 2)

Description (Samant and Phadtare 1997, p. 36): Pollen grains prolate-spheroidal to subprolate, 24–25 × 21–22 µm in equatorial view, tricolporate, colpi short, almost half of the longer axis; ora endexinous, distinctly zonicostate, exine about 1.5 µm thick, semitectate, sexine slightly thicker than nexine; reticulate, muri almost polygonal in shape, heterobrochate, reticulum increasing in size towards pore.

Occurrence Early Eocene, Tarkeshwar Formation, Rajpardi, Cambay Basin, Gujarat (Samant and Phadtare 1997).

Genus *Palmidites* Couper 1953

Type Species *Palmidites maximus* Couper 1953

Generic Description (Couper 1953, p. 57): Free anisopolar, monosulcate grains large, more than 50 μm long. Exine flecked to psilate.

Remarks Singh et al. (1985) proposed *Jaintiapollenites* with following diagnosis: "Pollen grains subtriangular to subcircular, 50–60 μm, monocolpate, colpi distinct or indistinct, associated with one or two folds, folds well-developed, extending throughout the colpus. Exine 2–3 μm thick, more thickened at fold margins. Sexine as thick as nexine. Ornamentation psilate to scabrate." This diagnosis is very similar to that of *Palmidites plicatus*. Therefore, *Jaintiapollenites* has been merged with *Palmidites plicatus*.

Botanical Affinity Arecaceae.

Species Recorded from India

Palmidites aplicatus Kar & Kumar 1986 (Plate 5, Fig. 2)

Description (Kar and Kumar 1986, p. 192): Pollen grains oval, 48–79 μm. Monocolpate, colpus distinct, stretching from one end to other. Exine up to 2 μm thick, laevigate, sometimes slightly intrastructured.

Occurrence Palaeocene, Lakadong Sandstone, Khasi Hills, Meghalaya (Kar and Kumar 1986).

Palmidites excellensus Kar & Kumar 1986 (Plate 5, Fig. 3)

Description (Kar and Kumar 1986, p. 192): Pollen grains oval-elliptical, 108–120 × 68–114 μm in size, monocolpate, colpus distinct, extending from one end to the other. Exine up to 3 μm thick, laevigate.

Occurrence Palaeocene, Lakadong Sandstone, Khasi Hills, Meghalaya (Kar and Kumar 1986).

Palmidites granulatus Mehrotra 1983 (Plate 5, Fig. 4)

Description (Mehrotra 1983, p. 13): Pollen grains oval-elliptical, bilateral, size range 108–125 μm, monosulcate, sulcus distinct or indistinct, running parallel to the longer axis, exine thick, granulose.

Occurrence Palaeocene-Early Eocene, Mikir Formation, Garampani, North Cachar Hills, Assam (Mehrotra 1983).

Palmidites naviculus Kar & Saxena 1981 (Plate 5, Fig. 5)

Description (Kar and Saxena 1981, pp. 111, 112): Pollen grains monocolpate, one end broader than other, 80–110 × 40–56 μm; colpus well developed, extending from one end to the other, broader at one side. Exine 1–2 μm thick, laevigate.

Occurrence Middle-Late Eocene, subsurface sediments, Rataria, Kutch, Gujarat (Kar and Saxena 1981; Kar 1985).

Palmidites noviculatus Sarkar & Singh 1988 (Plate 5, Fig. 6)

Description (Sarkar and Singh 1988, p. 75): Pollen grains oval to elliptical in shape, mostly with rounded and narrower ends. Monosulcate, sulcus long, running from pole to pole, middle part measuring 12–15 μm in width, lips thickened and associated with the folds on either side, usually flapping over, exine up to 1.5 μm thick, scabrate to intrafoveolate, lumina more or less 2 μm in diameter.

Occurrence Eocene, Subathu Formation, Sirmaur, Himachal Pradesh (Sarkar and Singh 1988).

Palmidites obtusus Tripathi & Singh 1985 (Plate 5, Fig. 7)

Description (Tripathi and Singh 1985, p. 177): Pollen grains ovoid, 115–130 × 75–95 μm in size; bilateral, lateral ends obtusely pointed; monocolpate, colpus distinct, long, narrowly open; exine 1–2 μm thick, laevigate to granulose.

Occurrence Palaeocene, Therria Formation, Jowai-Badarpur Road, Jaintia Hills, Meghalaya (Tripathi and Singh 1985).

Palmidites plicatus Singh in Sah & Singh 1974 (Plate 5, Fig. 8)

Synonym: *Jaintiapollenites* Singh et al. 1985, *Journal of Palynology,* 21: 28–55.

Description (Singh in Sah and Singh 1974, p. 96): Pollen grains oval to elliptical, 98–120 μm. Monocolpate, colpus generally extending from end to end, associated with folds on each side. Exine laevigate.

Occurrence Palaeocene, Tura Formation, Nongwal Bibra, Garo Hills, Meghalaya; Palaeocene, Lakadong Sandstone, Khasi Hills, Meghalaya (Kar and Kumar 1986).

Palmidites punctatus Mehrotra 1983 (Plate 5, Fig. 9)

Description (Mehrotra 1983, pp. 12, 13): Pollen grains oval to elliptical, bilateral; size range 105–140 μm. Monosulcate, sulcus running parallel to the longer axis, extending three fourth or more of its length, lips thickened, exine thin, punctate.

Occurrence Palaeocene-Early Eocene, Mikir Formation, Garampani, North Cachar Hills, Assam (Mehrotra 1983).

Genus *Paravuripollis* Rao & Ramanujam 1978

Type Species *Paravuripollis mulleri* Rao & Ramanujam 1978

Generic Description (Rao and Ramanujam 1978, p. 415): Pollen grains zonosulcate, ellipsoidal or ovoid, sulcus running all around equatorial zone splitting the pollen body into two more or less equal halves, sulcus margins beset with pila or clava. Exine densely pilate-clavate.

Botanical Affinity and Ecology *Salacca* (Arecaceae); palms of evergreen forests and swamp forests.

Species Recorded from India

Paravuripollis mulleri Rao & Ramanujam 1978 (Plate 5, Fig. 10)

Description (Rao and Ramanujam 1978, p. 415): Pollen grains ellipsoidal or ovoid, with or without truncate ends, 22.5–29 × 19–23 μm, zonosulcate, sulcus running all along equatorial zone splitting the pollen into two more or less equal halves, sulcus margins slightly incrassate, studded with pila or clava. Exine up to 4.5 μm thick, sexine much thicker than nexine, intectate, surface densely beset with pilate-clavate processes up to 3.5 μm high, heads of processes angular to locally rounded.

Occurrence Miocene, Quilon Beds, Paravur, Kerala (Rao and Ramanujam 1978).

Genus *Pellicieroipollis* Sah & Kar 1970

Type Species *Pellicieroipollis langenheimii* Sah & Kar 1970

Generic Description (Sah and Kar 1970, p. 133): Pollen grains triangular-subtriangular in polar view, 3-colporate, brevicolpate. Pore distinct, lalongate, generally with thickened margin, Exine thick, tegillate, bacula forming negative reticulum in surface view.

Restated generic description (Phadtare and Thakur 1990, p. 284): Pollen grains large, oblate, triangular to subtriangular in polar view, tricolporate, rarely di- or tetracolporate, generally brevicolpate; ora costate, distinctly lalongate; exine thick, mesocolpial nexine slightly thicker than sexine; ornamentation reticulate, generally homobrochate, often heterobrochate, muri simple to duplibaculate.

Botanical Affinity and Ecology *Alangium* (Alangiaceae); lowland humid deciduous forests.

Species Recorded from India

Pellicieroipollis langenheimii Sah & Kar 1970 (Plate 18, Fig. 3)

Description (Sah and Kar 1970, p. 133): Pollen grains triangular-subtriangular, 58–75 × 55–70 μm. 3-colporate, brevicolpate, pore lalongate with thickened margin. Exine thick, tegillate, bacula forming negative reticulum in surface view.

Revised description (Phadtare and Thakur 1990, p. 284): Pollen grains large (59 μm in equatorial diameter), oblate, isopolar, rounded triangular in polar view, tricolporate, brevicolpate; colpi tenuimarginate, margin straight, bordered with more or less elongated and slightly radiating brochi, ends acute; ora clearly costate, about 10 μm wide, lalongate, mesocolpial nexine slightly thicker than sexine, ornamentation clearly reticulate.

Occurrence Early Eocene, Laki Series, Kutch, Gujarat (Sah and Kar 1970); Early Eocene, Naredi Formation, Kutch, Gujarat (Kar 1978); Middle-Late Eocene, subsurface sediments, Rataria, Kutch, Gujarat (Kar and Saxena 1981; Kar 1985).

Genus *Piladiporocolpites* Kar 1995

Type Species *Piladiporocolpites caratinii* Kar 1995

Generic Description (Kar 1995, p. 383): Pollen grains elliptical to oval, 49–60 × 35–45 μm. Diporocolpate, pores big, distinct, sometimes looking like ora, placed opposite to each other on equator, pore margin spinose, pores connected by a colpus, colpus margin spinose. Exine 2–4 μm thick, pilate, pila 3–5 μm long, with narrow base and swollen tips, sparsely placed, inter-pilate exine laevigate.

Botanical Affinity Arecaceae.

Species Recorded from India

Piladiporocolpites caratinii Kar 1995 (Plate 23, Fig. 5)

Description (Kar 1995, p. 383): Proximo-distally flattened specimens oval-elliptical, meridionally preserved specimens mostly subcircular in outline, 48–55 × 37–48 μm. Diporocolpate aperture distinct in proximo-distally flattened specimens, obscure in others, pores and colpus placed in a line in meridionally preserved specimens, margin of pores spinose, spines 2–4 μm long, delicate.

Colpus distinct, uniformly broad, connects each pore at equator, colpus margin also spinose, spines 2–4 μm long, slender. Exine strongly built, 2–4 μm thick, pilate, pila well-developed, sparsely placed, 2–5 μm long, narrow at base but swollen at tip, mostly appressed to provide circular wart-like structures on the spine, inter-pilate exine laevigate.

Occurrence Early Eocene, Palana Formation, subsurface sediments, Bikaner, Rajasthan (Kar 1995; Kar and Sharma 2001).

Genus *Pilatrisyncolpites* Kar et al. 1994

Type Species *Pilatrisyncolpites triangulatus* Kar et al. 1994

Generic Description (Kar et al. 1994, p. 287): Pollen triangular in polar view; trisyncolpate, colpi long, distinct. Exine up to 2 μm thick, sculpture psilate, interpilar exine ornamentation pseudomicroreticulate.

Emended generic description (Mandal and Rao 2001, p. 357): Pollen triangular-subtriangular in polar view; trichotomosulcate, colpi long and join at pole. Exine moderately thick, sculpture pilate, interpilar exine ornamentation pseudo-microreticulate.

Species Recorded from India

Pilatrisyncolpites triangulatus Kar et al. 1994 (Plate 12, Fig. 3)

Description (Kar et al. 1994, p. 287, 288): Pollen grains triangular in polar view, size range 36–52 × 34–50 μm, apices rounded, interapical margin convex. Trisyncolpate, colpi distinct, broader at equator and tapering at pole. Exine pilate, pila 4–8 μm long, 2–5 μm broad, narrow at base and swollen at tip, sparsely placed, interpilar exine granulose, very closely placed, forming pseudomicroreticulate ornamentation in surface view.

Emended Description (Mandal and Rao 2001, p. 357): Pollen grains triangular in polar view, size range 36–52 × 34–40 μm, apices rounded, interapical margin convex. Trichotomosulcate, colpi distinct, broader at equator and tapering at pole. Exine pilate, pila 4–8 μm long, 2–5 μm broad, narrow at base and swollen at tip, sparsely placed, interpilar exine granulose, very closely placed forming pseudoreticulate ornamentation in surface view.

Occurrence Oligocene, Barail Group, Assam (Kar et al. 1994); Early Eocene, Naredi Formation, Kutch, Gujarat (Mandal 2000).

Genus *Pinjoriapollis* Saxena & Singh 1981

Type Species *Pinjoriapollis magnus* Saxena & Singh 1981

Generic Description (Saxena and Singh 1981, p. 418): Pollen grains elliptical or lanceolate in shape (length two and half times of breadth), heteropolar, large in size, 120–170 × 51–70 μm. Monosulcate, sulcus wide throughout its length, extending from one end to the other. Exine up to 3.5 μm thick, laevigate, faintly intrapunctate, tegillate.

Botanical Affinity Magnoliaceae.

Species Recorded from India

Pinjoriapollis lanceolatus Saxena & Singh 1981 (Plate 5, Fig. 11)

Description (Saxena and Singh 1981, p. 418): Pollen grains lanceolate, 123–170 × 51–62 μm. Monosulcate, sulcus generally reaching from one end to the other, occasionally shorter. Exine up to 1 μm thick, laevigate.

Occurrence Pliocene, Pinjor Formation, Chandigarh (Saxena and Singh 1981); Palaeocene Therria Formation, Jowai-Badarpur Road, Jaintia Hills, Meghalaya (Tripathi and Singh 1985).

Pinjoriapollis magnus Saxena & Singh 1981 (Plate 5, Fig. 12)

Description (Saxena and Singh 1981, p. 418): Pollen grains oval, elliptical, 120–148 × 62–70 μm. Monosulcate, sulcus wide, extending from one end to the other, sulcus widening more towards one end. Exine 2.5–3.5 μm thick, laevigate, occasionally finely intrapunctate.

Occurrence Pliocene, Pinjor Formation, Chandigarh (Saxena and Singh 1981).

Genus *Platoniapollenites iratus* Sah and Kar 1974

Now: *Meliapollis iratus* Navale and Misra 1979.

Genus *Plumbaginacipites* Navale & Misra 1979

Type Species *Plumbaginacipites neyvelii* Navale & Misra 1979

Generic Description (Navale and Misra 1979, p. 227): Pollen grain isopolar, oblate, suboblate to spheroidal. Size ranges from 60–95 × 60–70 μm. Tricolpate, colpi simple, long and deep, do not reach the poles, mesocolpi rounded. Exine always thick (5–7 μm), sexine thicker than nexine, sexine with distinct rod layer, rods are long, papillate to clavate showing a granular to microreticulate surface pattern.

Botanical Affinity Plumbaginaceae.

Species Recorded from India

Plumbaginacipites neyvelii Navale & Misra 1979 (Plate 10, Fig. 11)

Description (Navale and Misra 1979, p. 227): Isopolar pollen, oblate to spheroidal, 60–95 × 60–70 μm in size. Tricolpate, longicolpate, deep but do not reach the poles, mesocolpi are rounded at the periphery. Exine 5–7 μm thick, sexine thicker (2–3 times) than nexine, gradually thinning towards the colpi margins. Distinct rod layer is present in the sexine, rods are long, papillate to clavate forming granulate to finely microreticulate surface structure.

Occurrence Miocene, Neyveli Lignitefield, Tamil Nadu (Navale and Misra 1979).

Genus *Polybrevicolporites* Venkatachala & Kar 1969

Type Species *Polybrevicolporites cephalus* Venkatachala & Kar 1969

Generic Description (Venkatachala and Kar 1969, p. 171): Pollen grains subcircular-circular, polycolporate, brevicolpate. Exine 3–8 μm thick, intrabaculate.

Species Recorded from India

Polybrevicolporites antiquum (Ramanujam 1966) Venkatachala & Kar 1969 (Plate 21, Fig. 6)

Description (Venkatachala and Kar 1969, p. 171): Pollen grains subcircular in equatorial view, 24–34 μm. Polybrevicolporate with six apertures. Exine thick, generally finely microreticulate, sometimes seems to be granulose.

Occurrence Miocene-Pliocene, Cuddalore Series, Neyveli, Tamil Nadu (Ramanujam 1966); Early Eocene, Laki Series, Matanomadh, Kutch, Gujarat (Venkatachala and Kar 1969).

Remarks Venkatachala and Kar (1969) transferred *Nothofagidites antiquum* Ramanujam (1966) to *Polybrevicolporites*.

Polybrevicolporites cephalus Venkatachala & Kar 1969 (Plate 21, Fig. 7)

Description (Venkatachala and Kar 1969, p. 171): Pollen grains subcircular in equatorial view, 20–35 × 17–30 μm. Polycolporate, usually pentacolporate, brevicolpate, pores well developed. Exine 3–6 μm thick, intrabaculate.

Occurrence Early Eocene, Laki Series, Matanomadh, Kutch, Gujarat (Venkatachala and Kar 1969).

Polybrevicolporites karii Rao & Ramanujam 1982 (Plate 21, Fig. 8)

Description (Rao and Ramanujam 1982, p. 82): Pollen grains isopolar, spheroidal and distinctly lobed in polar view, more or less rhomboidal in equatorial view, 21–27 × 23–29 μm, zonaperturate, pentacolporate, colpae constricted and arching at equator, tenuimarginate, ends blunt, ora prominent, lalongate, 7.5 μm across, ora of adjacent colpae almost touching each other. Exine up to 3.5 μm thick, thicker at poles than elsewhere, sexine thicker than nexine, tectate, columella distinct, surface intrareticulate with columella below tectum aligned in a reticuloid manner. Mesocolpia with 2–4 prominent thickenings.

Occurrence Miocene, Quilon Beds, Edavai, Kerala (Rao and Ramanujam 1982).

Polybrevicolporites reticulata Samant 2000 (Plate 21, Fig. 9)

Description (Samant 2000, pp. 104/106): Pollen grains isopolar; circular in polar view; 40–41 μm in size; 7–8 colporate; colpi short, open; ora almost indistinct; exine 5 μm thick, nexine thicker than sexine, nexine forms cresent-shaped thickening in intercolpal region, thin at aperture margins, columellate, semitectate, reticulate, simplibaculate, homobrochate.

Occurrence Early Eocene, Kharsalia, Clay Formation, Bhavnagar, Cambay Basin, Gujarat (Samant 2000).

Genus *Polycolporites* Mehrotra 1983

Type Species *Polycolporites indicus* Mehrotra 1983

Generic Description (Mehrotra 1983, p. 15): Pollen grains circular-subcircular, polycolporate, colpi long, reaching up to radius; exine ornamentation scabrate.

Species Recorded from India

Polycolporites indicus Mehrotra 1983 (Plate 21, Fig. 10)

Description (Mehrotra 1983, p. 15): Pollen grains circular; size range 44–52 μm, hexacolporate, colpi reaching up to the radius, pore distinct, 5–6 μm wide; elongate; exine 2 μm thick, scabrate.

Occurrence Palaeocene-Early Eocene, Mikir Formation, Garampani, North Cachar Hills, Assam (Mehrotra 1983); Late Palaeocene-Early Eocene, Palana Formation, Bikaner, Rajasthan (Kar and Sharma 2001).

Genus *Polygalacidites* Sah & Dutta 1966

Type Species *Polygalacidites clarus* Sah & Dutta 1966

Generic Description (Sah and Dutta 1966, p. 81): Pollen grains small; amb spheroidal to subspheroidal; polycolporate, generally synorate, colpi long or short; exine smooth to finely sculptured.

Botanical Affinity and Ecology Polygalaceae; herbs and shrubs of variable ecology.

Species Recorded from India

Polygalacidites clarus Sah & Dutta 1966 (Plate 21, Fig. 11)

Description (Sah and Dutta 1966, p. 81): Size range 22–27 μm; grains spheroidal to subspheroidal in polar view, more or less prolate-spheroidal in equatorial view; 5–6 colporate, colpi long, ora faint, in some specimens distinctly synorate; exine up to 1 μm thick, sculptureless.

Occurrence Palaeocene, Cherra Formation, Meghalaya (Sah and Dutta 1966)

Polygalacidites gujratensis Kar 1985 (Plate 21, Fig. 12)

Description (Kar 1985, p. 87): Pollen grains 38–46 × 36–44 μm in equatorial view, 5–6 colporate, colpi long, narrow, pore distinct or indistinct. Exine 2–3 μm thick, sexine thicker than nexine, weakly intrastructured.

Occurrence Eocene, Naredi Formation, subsurface sediments, Bandra, Kutch, Gujarat (Kar 1985).

Polygalacidites magnus Samant & Phadtare 1997 (Plate 21, Fig. 13)

Description (Samant and Phadtare 1997, p. 42): Pollen grains circular to subcircular in polar view, 51–60 × 50–55 μm in size, polycolporate; generally 10–11 apertures, longicolpate; colpi about 25 μm long in polar view, narrow, tenuimarginate, ora endexinous with nexine crassimarginate, synorate, forming characteristic equatorial girdle; exine about 2 μm thick, sexine thicker than nexine; almost psilate to micropunctate.

Occurrence Early Eocene, Tarkeshwar, Rajpardi, Cambay Basin, Gujarat (Samant and Phadtare 1997); Early Eocene, Kharsalia clay Formation, Bhavnagar, Cambay Basin, Gujarat (Samant 2000).

Polygalacidites rhomboidus Kar & Bhattacharya 1992 (Plate 21, Fig. 14)

Description (Kar and Bhattacharya 1992, p. 256): Pollen grains more or less rhomboidal in equatorial view, 42–62 × 40–55 μm, 5–7 colporate, pores large, joining together to form synorate condition, colpi long, distinct, extending almost from one end to other in equatorial view. Exine 1.5–2.5 μm thick, retipilate.

Occurrence Early Eocene, Rajpardi lignite mine, Cambay Basin, Gujarat (Kar and Bhattacharya 1992).

Genus *Polygonaceaepites* Baksi 1962

Type Species *Polygonaceaepites zonoides* Baksi 1962

Generic Description (Baksi 1962, p. 20): Equatorial compression; longitudinal diameter 34 μm; body zoned, outward portion appears like a flange which becomes appreciably wider near poles, a few discrete radiating striae mark the flange, otherwise more granular-reticulate than main body; tricolpate, colpi extend from pole to pole (out limit the inner margin), adjacent.

Botanical Affinity Polygonaceae.

Species Recorded from India

Polygonaceaepites zonoides Baksi 1962 (Plate 10, Fig. 12)

Description (Baksi 1962, p. 20): As generic description

Occurrence Miocene, Surma Group, Meghalaya (Baksi 1962).

Genus *Polygonacidites* Sah & Dutta 1968

Type Species *Polygonacidites frequens* Sah & Dutta 1968

Generic Description (Sah and Dutta 1968, p. 192, 193): Pollen grains usually more or less spheroidal; panporate; apertures distinct or indistinct, usually not well defined, exine ornamentation variable, usually reticulate.

Species Recorded from India

Polygonacidites frequens Sah & Dutta 1968 (Plate 14, Fig. 7)

Description (Sah and Dutta 1968, p. 193): Size range 41–61 μm; amb globular, polyforate, apertures very small, inconspicuous; exine thick, sexine slightly thicker than nexine, ornamentation reticulate, duplibaculate, tegillate.

Occurrence Miocene-Pliocene, Tipam Series, Assam (Sah and Dutta 1968).

Genus *Polymorphocolporites* Kar & Kumar 1986

Remarks Merged with *Polycolporites* Mehrotra (1983) by Venkatachala et al. (1996, pp. 130, 131).

Genus *Polyporina* Naumova 1939 ex Potonié 1960

Type Species *Polyporina multistigmosa* (Potonié 1931) ex Potonié 1960

Generic Description (Potonié 1960, p. 134): With numerous foveae, more or less closely placed, distinctly recessed, with double contours; exine between the circular foveae more distinctly granulate than in foveae.

Botanical Affinity Amaranthaceae-Chenopodiaceae.

Species Recorded from India

Polyporina excavatus Venkatachala & Rawat 1973 (Plate 18, Fig. 6)

Description (Venkatachala and Rawat 1973, p. 256): Pollen grains spheroidal, amb more or less circular. Panporate, polyporate; 39.6–44 × 39.6–47.5 μm. Pores large, circular to ellipsoidal, 3.5–6.6 μm in diameter 25–30 in numbers, 4–5 μm apart. Exine about 4 μm thick, sexine thicker than nexine, tegillate, pilate, pila distinct, less than 1 μm thick, small coni form the suprategillar ornamentation, coni very small, about 1 μm high, sparsely placed.

Occurrence Oligocene-Miocene, Cauvery Basin, Tamil Nadu (Venkatachala and Rawat 1973).

Polyporina excellens Sah & Dutta 1968 (Plate 19, Fig. 7)

Description (Sah and Dutta 1968, p. 192): Size range 22–34 μm; amb globular, panaperturate, polyporate, pores about 50 in number; exine fairly thick, sexine thicker than nexine, ornamentation finely punctate.

Occurrence Miocene-Pliocene, Tipam Series, Assam (Sah and Dutta 1968); Palaeocene, Cherra Formation, Shillong Plateau, Meghalaya (Dutta and Sah 1970).

Genus *Potamogetonaceaepites* Biswas 1962

Type Species *Potamogetonaceaepites pramathii* Biswas 1962

Now: *Proxapertites pramathii* Singh 1975

Genus *Prectenolophonidites* Samant & Phadtare 1997

Type Species *Prectenolophonidites primitiva* Samant & Phadtare 1997

Generic Description (Samant and Phadtare 1997, p. 44): Pollen grains circular in polar view (generally found in polar view), polycolporate; longi- to brevicolpate; ora endexinous, generally indistinct under LM. However, SEM shows distinct endexinous ora; exine thick, along mesocolpial area, sexine and nexine of same thickness or sexine is thicker than nexine, while apocolpial area shows primitive to distinct nexinal thickening; perforate.

Botanical Affinity Ctenolophonaceae.

Species Recorded from India

Prectenolophonidites primitiva Samant & Phadtare 1997 (Plate 21, Fig. 15)

Description (Samant and Phadtare 1997, p. 44): Pollen grains circular in polar view, 26–42 μm in diameter, 6–8 colporate; generally brevicolpate, rarely longicolpate, colpi 5–10 μm long in polar view, commonly tenuimarginate; ora almost indistinct in LM observations, however SEM illustrations show endexinous circular to lalongate ora in widely open colpi. Exine 0.5–3 μm thick, tectate, more or less columellate, nexine thicker at apocolpial region resulting in the formation of undulations or warts at the apocolpia, these undulations in LM appear as rugulae or vermiculae; exine perforate, homobrochate, lumina < 1 μm in width, circular to slightly angular, muri about 1 μm thick.

Occurrence Early Eocene, Tarkeshwar Formation, Rajpardi, Cambay Basin, Gujarat (Samant and Phadtare 1997).

Genus *Proteacidites* Cookson 1950 ex Couper 1953

Type Species *Proteacidites adenanthoides* Cookson 1950

Generic Description (Cookson 1950, p. 172): Free Isopolar or subisopolar, triorate, occasionally diorate; grain triangular to subtriangular, sides concave to convex between ora in polar view; exine clearly differentiated into nexinous and sexinous layers. Sexine baculate, clavate or tuberculate, forming a very variable pitted-reticulate, reticulate or pseudoreticulate sculpture in surface view.

Species Recorded from India

Proteacidites excertus Dutta & Sah 1970 (Plate 25, Fig. 10)

Description (Dutta and Sah 1970, p. 49): Known size range 42–67 μm; amb triangular in polar view, interapertural sides convex, generally one side concave; angulaperturate, 3-aperturate, porate, pores conspicuously protruding, ora distinct, lalongate; exine thick, sexine as thick as nexine, tegillate, surface sculpture coarsely reticulate.

Occurrence Palaeocene, Cherra Formation, Shillong Plateau, Meghalaya (Dutta and Sah 1970).

Proteacidites protrudus Sah & Kar 1970 (Plate 25, Fig. 12)

Description (Sah and Kar 1970, p. 139): Pollen grains triangular, 50–60 × 48–55 μm. 3-porate, pores distinct, slightly protruding. Exine up to 3 μm thick, sexine thicker than nexine, scrobiculate.

Occurrence Early Eocene, Laki Series, Kutch, Gujarat (Sah and Kar 1970).

Proteacidites triangulus Kar & Jain 1981 (Plate 25, Fig. 13)

Description (Kar and Jain 1981, pp. 123, 124): Pollen grains triangular in polar view, 49–60 μm; 3-porate, pores distinct, not protruding, margin not thickened. Exine psilate, tegillate, sexine thicker than nexine, retipilate.

Occurrence Neogene, around Quilon and Varkala, Kerala (Kar and Jain 1981).

Genus *Proxapertites* van der Hammen 1956

Type Species *Proxapertites operculatus* (van der Hammen 1956) van der Hammen 1956

Generic Description (van der Hammen 1956, p. 113): Pollen grains with a big and wide aperture at the proximal side of the grain (remnants of an extremely thin exine layer can be found sometimes in the aperture), grains of this type are still found in dyads (or sometimes in tetrads, etc.).

Emended Description (Singh 1975, p. 101): Pollen grains circular to subcircular in shape, zonisulcate, sulcus more or less parallel to margin, resulting into breaking of the pollen into two more or less equal halves. Size range 20–70 × 25–50 μm. Exine tectate, variable in ornamentation and sculpture, generally ranging from punctate, psilate, granulose to microfoveolate and reticulate.

Remarks The genus *Proxapertites* was considered by van der Hammen (1956, Plate 1, Fig. 5) to possess a big aperture on proximal side. He also observed that sometimes these grains are found in dyads or tetrads. Muller (1968) regarded dyads of this genus as zonisulcate pollen having tendency to split into two slightly unequal parts. Potonié (1970) stated that these pollen consist of a fine groove running along the periphery. Singh (1975) while emending diagnosis of this genus also regarded it as zonisulcate. Singh (1975) transferred *Schizosporis assamicus* Sah & Dutta (1966), *Schizosporis crassimurus* Sah & Dutta (1966),

Potamogetonaceaepites Biswas (1962), *Microreticulatipites* Baksi (1962) and *Nymphaeoipollis* Venkatachala & Kar (1969) to *Proxapertites*.

The genus *Assamialetes*, diagnosed by Singh (1975) as non-aperturate is a zonisulcate pollen with reticulate ornamentation (Singh and Tripathi 1986). However, because of nature of the sulcus, *Assamialetes* is now considered as junior synonym of *Proxapertites* (Kar & Kumar 1986).

Botanical Affinity and Ecology *Raphidophora aurea* (Arecaceae); climbers in evergreen forests.

Species Recorded from India

Proxapertites assamicus (Sah and Dutta 1966) Singh 1975 (Plate 5, Fig. 13)

Description (Sah and Dutta 1966, p. 79): Size range 41–50 µm; amb circular to slightly elliptical, usually found to be splitting subequatorially into two equal disc-like halves. Exine up to 1.5 µm thick, unsculptured, but a faint LO pattern seen under oil immersion.

Occurrence Palaeocene, Cherra Formation, Shillong Plateau, Meghalaya (Sah and Dutta 1966); Palaeocene-Early Eocene, Mikir Formation, Garampani, North Cachar Hills, Assam (Mehrotra 1983); Palaeocene, Therria Formation, Jaintia Hills, Meghalaya (Tripathi and Singh 1985); Palaeocene, Matanomadh Formation, Matanomadh, Kutch, Gujarat (Kar 1985).

Remarks Singh (1975) transferred *Schizosporis assamica* Sah & Dutta 1966 to *Proxapertites*.

Proxapertites crassimurus (Sah and Dutta 1966) Singh 1975 (Plate 5, Fig. 14)

Description (Sah and Dutta 1966, p. 78): Size range 36–53 µm, amb oval, more or less flattened at the poles, dividing subequatorially into two equal, boat-shaped halves. Exine reticulate, crassimurate, muri as much as 3 µm broad, lumina variously shaped, usually of the same width as muri or sometimes slightly wider.

Occurrence Palaeocene, Cherra Formation, Shillong Plateau, Meghalaya (Sah and Dutta 1966); Palaeocene, Tura Formation, Garo Hills, Meghalaya (Singh 1975); Palaeocene-Early Eocene, Mikir Formation, Garampani, North Cachar Hills, Assam (Mehrotra 1983).

Remarks Singh (1975) transferred *Schizosporis crassimurus* Sah and Dutta (1966) to *Proxapertites*.

Proxapertites emendatus (Sah and Dutta 1966) Kar & Kumar 1986 (Plate 5, Fig. 15)

Emended Description (Kar and Kumar 1986, p. 196): Equatorial outline subcircular, 58 × 50 μm, pollen grains generally split into two halves. Exine 3–5 μm thick, distinctly tegillate, coarsely reticulate, sculptured with pila/muri, muri more or less uniform, 2.5 μm thick and fairly high, lumina of different shapes and sizes, 4–8 μm in diameter.

Occurrence Palaeocene, Cherra Formation, Shillong Plateau, Meghalaya (Sah and Dutta 1966); Palaeocene Lakadong Sandstone, Jaintia Hills, Meghalaya (Kar and Kumar 1986).

Remarks Kar and Kumar (1986) transferred *Assamialetes emendatus* (Sah and Dutta 1966) to *Proxapertites.*

Proxapertites granulatus Singh 1977 (Plate 6, Fig. 1)

Description (Singh 1977, p. 197): Pollen grains subcircular to oval. Size 46–50 μm, zonisulcate, sulcus mostly distinct, sometimes indistinct, parallel to margin. Exine up to 2 μm thick, granulose, grana up to 1 μm high, closely placed and uniformly distributed.

Occurrence Palaeocene, Tura Formation, Garo Hills, Meghalaya (Singh 1975).

Proxapertites hammenii Venkatachala & Rawat 1972 (Plate 6, Fig. 2)

Description (Venkatachala and Rawat 1972, p. 295): Pollen grain rounded, angular to oval in outline, proxaperturate (36.5–40 × 40–51 μm). Pollen generally separated by a continuous equatorial aperture into two slightly unequal parts. Exine about 1.3 μm thick, sexine slightly thicker than nexine, tectate, columellae distinct, less than 1 μm thick, finely reticulate-punctate, punctae less than 1 μm in diameter, sometimes few punctae fused giving an irregular pattern.

Occurrence Palaeocene-Eocene, Cauvery Basin, Tamil Nadu (Venkatachala and Rawat 1972).

Proxapertites intecta (Baksi 1962) Singh 1975 (Plate 6, Fig. 3)

Description (Baksi 1962, p. 17): Long, diameter 50 μm, intectate; surface punctate-reticulate, lumina larger than muri, ?acolporate or monocolpate (?colpus peripheral); may be related to Nelumbiaceae.

Occurrence Tertiary, Simsang River Section, Shillong Front, Meghalaya (Baksi 1962).

Remarks *Microreticulatipites intecta* Baksi (1962) was transferred to *Proxapertites* by Singh (1975).

Proxapertites marginatus (Venkatachala and Kar 1969) Singh 1975 (Plate 6, Fig. 4)

Description (Venkatachala and Kar 1969, p. 163): Pollen grains subcircular-circular, zonisulcate. Sulcus more or less circular along the equatorial margin, 35–45 × 30–43 μm. Exine tectate, scrobiculate. Pollen grains in many specimens divided into two more or less equal halves due to equatorial breaking.

Occurrence Early Eocene, Naredi Formation, Matanomadh, Kutch, Gujarat (Venkatachala and Kar 1969).

Remarks *Nymphaeoipollis marginatus* Venkatachala & Kar (1969) was transferred to *Proxapertites* by Singh (1975).

Proxapertites microreticulatus Jain et al. 1973 (Plate 6, Fig. 5)

Description (Jain et al. 1973, p. 154): Spores medium-sized, equatorial diameter 40–60 μm, amb circular to subcircular or oval splitting into saucer-shaped halves. Exine 1–1.5 μm thick, intramicroreticulate.

Occurrence Palaeocene, Barmer Sandstone, Rajasthan (Jain et al. 1973); Early-Middle Eocene, Naredi and Harudi Formations, Kutch, Gujarat (Kar 1978); Palaeocene, Matanomadh Formation, Kutch, Gujarat (Saxena 1979).

Proxapertites pramathii (Biswas 1962) Singh 1975 (Plate 6, Fig. 6)

Description (Biswas 1962, p. 48): Exine moderately thick, uniformly reticulate, though comparatively finely; non-aperturate; size 44 μm.

Occurrence Early-Middle Eocene, Sylhet Limestone Formation, Therriaghat, Shillong Plateau, Meghalaya (Biswas 1962).

Remarks *Potamogetonaceaepites pramathii* Biswas (1962) was transferred to *Proxapertites* by Singh (1975).

Proxapertites reticulatus (Kar and Saxena 1981) Kar 1985 (Plate 6, Fig. 7)

Description (Kar 1985, pp. 120, 121): Pollen grains subcircular-oval, 40–52 × 36–50 μm, zonisulcate, pollen often split into two slightly unequal halves along aperture. Exine up to 2 μm thick, reticulate, meshes of two types, larger one found at outer side and smaller one inside.

Occurrence Middle-Late Eocene, subsurface sediments, Rataria, Kutch, Gujarat (Kar and Saxena 1981).

Remarks *Assamialetes reticulatus* Kar & Saxena (1981) was transferred to *Proxapertites* by Kar (1985).

Proxapertites rugulatus Samant & Phadtare 1997 (Plate 6, Fig. 8)

Description (Samant and Phadtare 1997, p. 47): Pollen grains circular to subcircular, isopolar, radially symmetrical; 30–33 μm in diameter, zonisulcate; sulcus closed, almost encircling the grain and divides it into two more or less unequal halves, margins thick, regular, continuous; exine about 0.5 μm thick, columellate, columellae distinct, free; sexine thicker than nexine, distinctly rugulate.

Occurrence Early Eocene, Tarkeshwar Formation, Rajpardi, Cambay Basin, Gujarat (Samant and Phadtare 1997).

Proxapertites scabratus Jain et al. 1973 (Plate 6, Fig. 9)

Description (Jain et al. 1973, p. 154): Size 60 × 30 μm, amb oval to elliptical, flattened at poles dividing subequatorially into two equal boat-shaped halves. Exine 1 μm thick, scabrate.

Occurrence Palaeocene, Barmer Sandstone, Rajasthan (Jain et al. 1973).

Genus *Pseudonothofagidites* Venkatachala & Kar 1969

Type Species *Pseudonothofagidites kutchensis* Venkatachala & Kar 1969

Generic Description (Venkatachala and Kar 1969, p. 173): Pollen grains more or less subcircular with regular lobed margins. Polyporate, pores rimmed, invaginated inwards. Exine up to 3 μm thick, finely granulose.

Remarks Samant and Phadtare (1997) transferred the genus *Gangapollenites* Mathur and Chopra (1987) to *Pseudonothofagidites* Venkatachala and Kar (1969).

Species Recorded from India

Pseudonothofagidites cerebrus Venkatachala & Kar 1969 (Plate 20, Fig. 3)

Description (Venkatachala and Kar 1969, p. 174): Pollen grains lobed with broadly lobed margins. Pores 8 in number, situated at inter-marginal area of the lobe, rimmed. Sexine roughly granulose.

Occurrence Early Eocene, Laki Series, Matanomadh, Kutch, Gujarat (Venkatachala and Kar 1969); Early Eocene, Naredi Formation, subsurface sediments, Kutch, Gujarat (Kar 1985).

Pseudonothofagidites kutchensis Venkatachala & Kar 1969 (Plate 20, Fig. 10)

Description (Venkatachala and Kar 1969, p. 173): Pollen grains more or less subcircular with regular lobed margins. Polyporate, pores rimmed, invaginated inwards. Exine up to 3 μm thick, finely granulose.

Occurrence Early Eocene, Laki Series, Matanomadh, Kutch, Gujarat (Venkatachala and Kar 1969); Early Eocene, Palana lignite, Bikaner, Rajasthan (Sah and Kar 1974); Palaeocene, Tura Formation, Nongwal Bibra, Garo Hills, Meghalaya (Singh 1977).

Pseudonothofagidites microreticulatus Samant & Tapaswi 2001 (Plate 25, Fig. 3)

Description (Samant and Tapaswi 2001, p. 126): Pollen grains spherical to subspherical in polar view, with lobed margins and distinctly raised apocolpial region; measuring 27–35 μm in diameter; polyporate, 5–7 pores; pores circular and are situated in depression between lobes, pores 4–6 μm in diameter and crassimarginate; exine about 1.0–1.5 μm thick and slightly thicker at poles, sexine and nexine undifferentiable; sculpture microreticulate to almost psilate.

Occurrence Early Eocene, Cambay Shale Formation, Tarkeshwar, Surat, Gujarat (Samant and Tapaswi 2001).

Genus *Psiladiporocolpites* Kar 1995

Type Species *Psiladiporocolpites pachyexinus* Kar 1995

Generic Description (Kar 1995, p. 383): Pollen grains generally oval-elliptical, 42–55–33–47 μm. Diporocolpate, pores equatorial, distinct, big, margin spinose, spines delicate. Colpus distinct, connecting the two pores, margin spinose, occasionally folded. Exine 2–5 μm thick, laevigate.

Species Recorded from India

Psiladiporocolpites pachyexinus Kar 1995 (Plate 23, Fig. 6)

Description (Kar 1995, p. 383): Proximo-distally flattened specimens oval elliptical, meridionally preserved specimens subcircular, 45–62 × 35–45 μm. Diporocolpate, pores at equatorial margin, big, distinct, margin spinose. Exine 2–5 μm thick, laevigate.

Occurrence Early Eocene, Palana Formation, subsurface sediments, Bikaner, Rajasthan (Kar 1995).

Genus *Psilodiporites* Varma & Rawat 1963

Type Species *Psilodiporites hammenii* Varma & Rawat 1963

Generic Description (Varma and Rawat 1963, p. 131): Diporate pollen grains with psilate exine (which may sometimes appear finely scabrate under high power).

Emended generic description (Venkatachala and Rawat 1972, p. 300): Barrel-shaped; diporate, exine smooth or variously pitted or ornamented. Pores distinct, terminal, without structural differentiation.

Species Recorded from India

Psilodiporites bengalensis (Varma and Rawat 1963) Venkatachala & Rawat 1972 (Plate 23, Fig. 7)

Description (Venkatachala and Rawat 1972, p. 301): Pollen grains 2-porate, bilateral, isopolar, barrel-shaped, laterally elongated, 45.6 × 27 μm. Pores circular, about 5.7–8.6 μm in diameter, slightly sunken, not covering the entire width of the lateral ends. Exine evenly thick, about 1.5 μm. Nexine about twice as thick as sexine.

Occurrence Palaeocene-Eocene, Cauvery Basin, Tamil Nadu (Venkatachala and Rawat 1972).

Remarks *Retidiporites bengalensis* Varma & Rawat (1963) was transferred to *Psilodiporites* by Venkatachala and Rawat (1972).

Psilodiporites erdtmanii Venkatachala & Rawat 1972 (Plate 23, Fig. 8)

Description (Venkatachala and Rawat 1972, p. 301): Pollen grain 2-porate, bilateral, isopolar, elongated and barrel-shaped, 42.5 × 30 μm. Pores simple, sunken, presumably circular (14.3 μm in diameter), occupying the entire lateral ends. Exine granulose-scabrate, 1.4 μm thick, sexine thicker than nexine.

Occurrence Palaeocene-Eocene, Cauvery Basin, Tamil Nadu (Venkatachala and Rawat 1972).

Remarks *Granodiporites erdtmanii* Varma & Rawat (1963) was transferred to *Psilodiporites* by Venkatachala and Rawat (1972).

Psilodiporites hammenii Varma & Rawat 1963 (Plate 23, Fig. 9)

Description (Varma and Rawat 1963, p. 131): Pollen grains 2-porate, isopolar-paraisopolar, bilateral (somewhat barrel-shaped), usually 29–34 × 43–51 μm in size. Pores equatorial, one at each end of the grains, simple, slightly sunken, presumably circular (diameter about 11.4 μm). Exine psilate, about 1–1.4 μm thick, sexine and nexine of same thickness.

Occurrence Early-Middle Eocene (locality not reported).

Q

Genus *Quilonipollenites* Rao & Ramanujam 1978

Type Species *Quilonipollenites sahnii* Rao & Ramanujam 1978

Generic Description (Rao and Ramanujam 1978, p. 417): Pollen grains heteropolar, broadly oval to subspherical in polar view, somewhat biconvex laterally, monosulcate, sulcus long and extended type, extending over a smaller or greater part of the proximal facet, extended part of the sulcus usually broader than the rest, sulcus margins thin or thickened, even or uneven. Exine tectate, columella distinct, surface prominently and coarsely reticulate, retipilariate or retipilate-clavate.

Botanical Affinity and Ecology Arecaceae; palms of lowland evergreen forests.

Species Recorded from India

Quilonipollenites crassimarginatus Phadtare & Kulkarni 1984 (Plate 6, Fig. 10)

Description (Phadtare and Kulkarni 1984, p. 220): Pollen grain heteropolar, biconvex in polar view, planoconvex laterally, 31–34.5 × 55 μm; monosulcate, longisulcate, sulcus of extended type, distinctly crassimarginate, closed in middle and opened on either ends; sexine thicker than nexine, exine 2.5–3 μm thick, distinctly retipilate to microreticulate, reticulum heterobrochate, decreasing in size towards the proximal side, muri simplibaculate, lumina small, smooth.

Occurrence Miocene, Ratnagiri Beds, Ratnagiri, Maharashtra (Phadtare and Kulkarni 1984).

Quilonipollenites minutus Samant & Phadtare 1997 (Plate 6, Fig. 11)

Description (Samant and Phadtare 1997, p. 51): Pollen grains heteropolar, laterally biconvex, 29–35 × 16–28 μm in size, monosulcate, sulcus long, extended; exine about 1.5 μm thick, semitectate sexine and nexine of more or less same thickness, coarsely reticulate, homobrochate, simply to duplibaculate, lumina 1–1.5 μm wide and muri 0.5–1 μm thick.

Occurrence Early Eocene, Tarkeshwar Formation, Rajpardi, Cambay Basin, Gujarat (Samant and Phadtare 1997).

Quilonipollenites ornatus Rao & Ramanujam 1978 (Plate 6, Fig. 12)

Description (Rao and Ramanujam 1978, p. 418): Pollen grains subspheroidal in polar view, 40–48 × 35–39 μm, monosulcate, sulcus of extended type, extending to a short distance on proximal side, 15–18 μm broad, margins uneven and ends rounded. Exine up to 2.5 μm thick, sexine thicker than nexine, tectate, columella distinct, surface reticulate (supra-reticulate), reticulum heterobrochate, meshes 1.5 μm in diameter, decreasing in size towards proximal side, muri thick, simplibaculate, lumina small, angular, smooth.

Occurrence Miocene, Quilon Beds, Padappakkara, Kerala (Rao and Ramanujam 1978).

Quilonipollenites reticulatus Samant and Phadtare 1997 (Plate 6, Fig. 13)

Description (Samant and Phadtare 1997, p. 50): Pollen grains planoconvex laterally, 26–30 × 40–43 μm in size, monosulcate, sulcus long, extended, margins thin, closed, rarely open; exine 1–2 μm thick, tectate, tectum about 0.6 μm thick, columellate, columellae distinct, reticulate, heterobrochate, lumina slightly angular, about 1 μm in width, muri < 1 μm thick, simplibaculate.

Occurrence Early Eocene, Tarkeshwar Formation, Rajpardi, Cambay Basin, Gujarat (Samant and Phadtare 1997).

Quilonipollenites sahnii Rao & Ramanujam 1978 (Plate 6, Fig. 14)

Description (Rao and Ramanujam 1978, p. 417): Pollen grains biconvex laterally, subspheroidal in polar view, 50–55 × 40–52.5 μm, monosulcate, sulcus long, of extended type, considerably wide, 20–29 μm wide, margins thin. Exine up to 4 μm thick, thinner towards proximal facet, sexine much thicker than nexine, tectate, surface conspicuously and coarsely reticulate (supra-reticulate), reticulum with large hexa- to polygonal meshes, meshes up to 5.5 μm broad, muri thick, formed of fused clava, simpliclavate, rarely dupliclavate, up to 3 μm high, beaded in appearance, occasionally incomplete, lumina angular with 1–5 free baculoid processes.

Occurrence Miocene, Quilon Beds, Padappakkara, Kerala (Rao and Ramanujam 1978).

Quilonipollenites trianguliformis Phadtare & Kulkarni 1984 (Plate 6, Fig. 15)

Description (Phadtare and Kulkarni 1984, p. 221): Pollen grains heteropolar, distally aperturate, triangular with rounded ends in lateral view, proximally plane, distally convex; 33–35 × 50–50 μm; monosulcate, sulcus equal to 2/3 of the circumference along the convexity in lateral view, of extended type, more or less tenuimarginate; sexine about 3.2 μm thick, coarsely reticulate, reticulum heterobrochate, decreasing in size towards proximal side, muri simplibaculate, straight, lumina angular to frequently circular, smooth.

Occurrence Miocene, Ratnagiri Beds, Ratnagiri, Maharashtra (Phadtare and Kulkarni 1984).

R

Genus *Ratariacolporites* Kar 1985

Type Species *Ratariacolporites plicatus* Kar 1985

Generic Description (Kar 1985, p. 123): Pollen grains triangular-subtriangular in polar view, tribrevicolporate, apertures situated at angles, generally associated with folds, exine more or less laevigate to intrastructured.

Species Recorded from India

Ratariacolporites plicatus Kar 1985 (Plate 18, Fig. 4)

Description (Kar 1985, p. 123): Pollen grains triangular-subtriangular in polar view, 50–60 × 48–58 μm. Tricolporate, brevicolpate, colpi 10–18 μm, colpi generally indistinct, pore circular-oval. Exine 1–2.5 μm thick, more or less laevigate, sometimes weakly granulose, feebly intrastructured. Exine folded regularly in inter-apertural region, folds originate in one apertural region and end in other, run more or less parallel to margin.

Occurrence Eocene, subsurface sediments, Rataria, Kutch, Gujarat (Kar 1985).

Genus *Retidiporocolpites* Kar 1995

Type Species *Retidiporocolpites exellensus* Kar 1995

Generic Description (Kar 1995, p. 383): Pollen grains oval-elliptical in shape, 35–60 × 30–42 μm in size. Diporocolpate, pores equatorially placed, big, prominent, colpus distinct, sometimes obscured by ornamentation, extending from one margin to other and connecting two pores. Exine 1–2.5 μm thick, tectate, columellate, sexine slightly thicker than nexine, reticulum well-developed, muri thick, lumina of different sizes and shapes, smaller sizes near pore margin.

Species Recorded from India

Retidiporocolpites exellensus Kar 1995 (Plate 23, Fig. 10)

Description (Kar 1995, p. 384): Pollen grains oval-elliptical in shape, 33 42 × 30–35 μm in size. Diporocolpate, pores at equator, very big, margin thin and smooth. Colpus not very distinct, obscured by reticulation, extending from one end to other and connecting the two pores. Exine 2.5 μm thick, tectate columellate, sexine slightly thicker than nexine, reticulation broad, muri thick, lumina of different sizes.

Occurrence Early Eocene, Palana Formation, Bikaner, Rajasthan (Kar 1995; Kar and Sharma 2001).

Genus *Retimonosulcites* Kar 1985

Remarks *Retimonosulcites* Kar (1985) is treated as synonym of *Liliacidites* for the reasons explained under *Liliacidites*.

Genus *Retipilonapites* Ramanujam 1966 ex Potonié 1970

Type Species *Retipilonapites arcotense* Ramanujam 1966 ex Potonié 1970

Generic Description (Ramanujam 1966, p. 153): Pollen grains spheroidal, non-aperturate but provided with a central depression on one side. Exine retipilate.

Botanical Affinity *Potamogeton* (Potamogetonaceae).

Species Recorded from India

Retipilonapites arcotense Ramanujam 1966 (Plate 1, Fig. 4)

Description (Ramanujam 1966, p. 154): Pollen spheroidal or ovoid. Non-aperturate but with a central depression which is 30.6 μm in diameter. Exine 1.5 μm thick, retipilate, pila 1.5–2.5 μm high, locally lobed. Reticulum coarse.

Occurrence Miocene-Pliocene, Cuddalore Series, Neyveli, Tamil Nadu (Ramanujam 1966); Early Eocene, Palana lignite, Palana, Bikaner (Sah and Kar 1974); Palaeocene, Tura Formation, Nongwal Bibra, Garo Hills, Meghalaya (Singh 1977).

Retipilonapites conspicua Rao & Ramanujam 1978 (Plate 1, Fig. 5)

Description (Rao and Ramanujam 1978, p. 410): Pollen grains irregularly spheroidal, polar diameter 25–31 μm, inaperturate but provided with a large central depression on one side. Exine 3–4 μm thick, much thicker than nexine, intectate, surface prominently reticlavate-retipilate, clava/pila up to 3.5 μm high, heads rounded, locally clava/pila processes fused, meshes rounded to angular, formed of reticulate alignment of clava or pila.

Occurrence Miocene, Quilon Beds, Padappakkara, Kerala (Rao and Ramanujam 1978).

Retipilonapites delicatissimus Ramanujam 1966 (Plate 1, Fig. 6)

Description (Ramanujam 1966, p. 154): Pollen grains broadly oval or spheroidal, non-aperturate but provided with a faint central depression of irregular contour on one side. 30–40 μm in diameter. Exine 2.2 μm thick, surface beset with extremely fine pila or coni, locally forming a delicate reticulum.

Occurrence Miocene-Pliocene, Cuddalore Series, Neyveli, Tamil Nadu (Ramanujam 1966).

Retipilonapites tertiaris Rao & Ramanujam 1978 (Plate 1, Fig. 7)

Description (Rao and Ramanujam 1978, p. 410): Pollen grains spheroidal to subspheroidal, polar diameter 22–40 μm; inaperturate, but sometimes provided with a large and irregular central depression on one side. Exine 2.5–3.5 μm thick, sexine much thicker than nexine, intectate, retipilate; muri discontinuous, formed of prominent pila, up to 2.5 μm high, oligobrochate, brochi hexagonal to polygonal, lumina angular, often with a few free bacula.

Occurrence Miocene, Quilon Beds, Padappakkara, Kerala (Rao and Ramanujam 1978).

Genus *Retipollenites* Gonzalez Guzman 1967

Type Species *Retipollenites confuses* Gonzalez Guzman 1967

Generic Description (Gonzalez Guzman 1967, p. 61): Pollen grains with heavy reticulum that is apparently not placed on a basal layer. Apparently, there are no apertures. There seems no clear similarities with other genera.

Botanical Affinity *Borassodendron* (Arecaceae).

Species Recorded from India

Retipollenites borassodendrii Samant & Phadtare 1997 (Plate 7, Fig. 1)

Description (Samant and Phadtare 1997, p. 53): Pollen grains generally spheroidal in shape, monosulcate, 45–55 μm in diameter; characteristic ciliate projections radially arranged in ectoaperture, exine distinctly columellate and coarsely reticulate, inner body distinctly psilate.

Occurrence Early Eocene, Tarkeshwar Formation, Rajpardi, Cambay Basin, Gujarat (Samant and Phadtare 1997).

Retipollenites echinatus Samant & Phadtare 1997 (Plate 7, Fig. 2)

Description (Samant and Phadtare 1997, p. 54): Pollen grains spheroidal, monosulcate; 55–60 μm in diameter; inner body about 44–49 μm in size; sulcus generally distinct to often indistinct; aperture generally subcircular to slightly elliptic; exine coarsely reticulate; lumina about 4–10 μm wide, muri about 2–3 μm thick; inner body characteristically echinate.

Occurrence Early Eocene, Tarkeshwar Formation, Rajpardi, Cambay Basin, Gujarat (Samant and Phadtare 1997).

Retipollenites monosulcites Samant & Phadtare 1997 (Plate 7, Fig. 3)

Description (Samant and Phadtare 1997, p. 55): Pollen grains oval to ellipsoidal, 40–50 μm along the longer axis; inner body 32–43 μm long, distinctly monosulcate; sulcus closed, long, extending end to end; sexine coarsely reticulate with elliptic aperture, lumina 8–10 μm in diameter, variously shaped, muri 2–3 μm thick.

Occurrence: Early Eocene, Tarkeshwar Formation, Rajpardi, Cambay Basin, Gujarat (Samant and Phadtare 1997).

Genus *Retitrescolpites* Sah 1967

Type Species *Retitrescolpites typicus* Sah 1967

Generic Description (Sah 1967, p. 56): Grains tricolpate or colporoidate, spheroidal to subspheroidal to rounded triangular. Exine retipila(ria)te to sometimes reticulate.

Species Recorded from India

Retitrescolpites assamicus Dutta & Sah 1970 (Plate 18, Fig. 5)

Description (Dutta and Sah 1970, p. 34): Size range 41–73 μm; amb spheroidal to subspheroidal, sometimes roundly triangular; 3-colp(oroid)ate, colpi long, ora indistinct; exine thick, surface ornamentation coarsely reticulate, crassimurate, lumina as wide as muri or even less.

Occurrence Palaeocene, Cherra Formation, Meghalaya (Dutta and Sah 1970).

Retitrescolpites minor Dutta & Sah 1970 (Plate 18, Fig. 7)

Description (Dutta and Sah 1970, p. 34): Size range 22–28 μm; amb more or less spheroidal to subprolate in polar view; tricolpate, colpi long; exine thick, sexine thicker than nexine, pilate, non-tegillate, surface sculpture retipilariate, reticulum formed of thick, irregular muri and small lumina.

Occurrence Palaeocene, Cherra Formation, Meghalaya (Dutta and Sah 1970).

Retitrescolpites robustus Kar & Saxena 1981

Now: *Tricolporopilites robustus* (Kar and Saxena 1981) Kar 1985

Retitrescolpites singularis Rao & Ramanujam 1982 (Plate 18, Fig. 8)

Description (Rao and Ramanujam 1982, p. 70): Pollen grain isopolar, amb subspheroidal to rounded triangular, lobes prominent, polar diameter 20–28 µm, zonaperturate, tricolpate, longicolpate, colpi reaching poles but not uniting, colpal surface finely granular. Exine up to 4 µm thick, sexine much thicker than nexine, surface irregularly retipilate, heterobrochate, brochi hexa- to polygonal, 3–5 µm in diameter, curvimuroid, lumina irregular with free baculoid processes.

Occurrence Miocene, Quilon Formation, Kerala (Rao and Ramanujam 1982).

Genus *Retitrisyncolpites* Mandal et al. 1994

Type Species *Retitrisyncolpites reimannii* Mandal et al. 1994

Generic Description (Mandal et al. 1994, p. 360): Pollen grains trisyncolpate, sometimes colporoidate, triangular-subtriangular in shape. Colpi distinct, long, occasionally funnel-shaped or bifurcated at tips. Exine thick, sexine thicker than nexine; tectate, tectum perforated; infratectum columellar, columellae either of same size or coarser at polar and inter-apertural areas. Surface reticulate.

Emended Description (Mandal and Rao 2001, p. 360): Pollen trichotomosulcate, rarely colporoidate, colpi join together at poles; other characters as above.

Species Recorded from India

Retitrisyncolpites reimannii Mandal et al. 1994 (Plate 12, Fig. 4)

Description (Mandal and Rao 2001, p. 360): Triangular-subtriangular pollen, size range 41–47 µm. Trichotomosulcate, sometimes open, funnel-shaped, rarely seems to be colporoidate. Exine tectate, perforate, size of perforation variable, infratectum columellar, columellae stronger at polar and apertural region. Columellae 1–2 µm long and 1–1.5 µm broad, 0.5–0.8 µm long at corners. Exine 1–2.5 µm thick, sexine thicker than nexine, nexine very thin, 0.5 µm, always not distinct. Surface reticulate, meshes coarse and irregular, muri more or less 1 µm high.

Occurrence Early Eocene, Middle Andaman (Mandal et al. 1994).

Retitrisyncolpites tha-ungii Mandal et al. 1994 (Plate 12, Fig. 5)

Emended Description (Mandal and Rao 2001, p. 360): Trichotomosulcate pollen with size range of 31.5–45.5 µm, triangular in shape. Colpi meet at poles, sometimes open at apertural end. Exine 0.7–1 µm thick, tectate, perforate, perforation more or less of same size, more or less 0.7 µm high. Sexine and nexine not separable. Surface reticulate, lumina circular, more or less 0.5 μm, more or less equal in size.

Occurrence Early Eocene, Middle Andaman (Mandal et al. 1994).

Genus *Retiverrumonosulcites* Tripathi 1994

Type Species *Retiverrumonosulcites barmerensis* Tripathi 1994

Generic Description (Tripathi 1994, p. 62): Pollen grains subspherical to oval; monosulcate, sulcus not extending up to the poles, with thickened lips. Exine microreticulate, possessing verrucae or clavae of varying sizes over the microreticulate surface.

Botanical Affinity Arecaceae.

Species Recorded from India

Retiverrumonosulcites barmerensis Tripathi 1994 (Plate 7, Fig. 4)

Description (Tripathi 1994, p. 64): Pollen grains ovoid in shape, 43–48 × 58 μm in size (excluding processes). Monosulcate, sulcus not reaching up to the poles, possessing thickened lips. Exine 2–4 μm thick, microreticulate, muri about 1 μm thick, lumina about 1 μm across. Verrucae (2.5 μm in diameter) and clavae (7–23 μm long, 5–10 μm in diameter at distal part) present over the microreticulate surface.

Occurrence Late Palaeocene-Early Eocene, subsurface sediments, Barmer, Rajasthan (Tripathi 1994; Tripathi et al. 2003; Tripathi and Srivastava 2010).

Genus *Rhoipites* Wodehouse 1933

Type Species *Rhoipites bradleyi* Wodehouse 1933

Generic Description (Wodehouse 1933, p. 513): Ellipsoidal, tricolpate pollen with furrows long and pointed; furrow and pore thickenings conspicuous, projecting deeply inwards; exine rather finely reticulate-pitted.

Species Recorded from India

Rhoipites anacardioides Ramanujam 1987 (Plate 18, Fig. 9)

Description (Ramanujam 1987, p. 32): Pollen grains essentially isopolar, amb subtriangular, equatorial view prolate to subprolate, 30–36 × 22–26 μm, tricolporate, longicolpate, colpi extending up to polar region, constricted at equator, bracket-shaped, ora distinct, lalongate, exine up to 2.5 μm thick, sexine thicker than nexine, tectate, columella fine, surface finely reticulate to locally striate-reticulate

(reticulate - rigged), brochi very fine, narrow, irregular, more or less meridionally aligned and locally parallely aligned.

Occurrence Miocene, Warkalli Beds, Kerala (Ramanujam 1987).

Rhoipites baculiferus Venkatachala & Rawat 1972 (Plate 18, Fig. 10)

Description (Venkatachala and Rawat 1972, p. 308): Pollen grains broadly oval in shape (equatorial view); 28.4 × 23.7 µm; tricolporate. Colpi long, reaching up to the poles; ora distinct, lalongate. Exine up to 2 µm thick, sexine thicker than nexine, baculate-intrabaculate; surface finely reticulate.

Occurrence Palaeocene-Eocene, Cauvery Basin, Tamil Nadu (Venkatachala and Rawat 1972).

Rhoipites conatus Venkatachala & Rawat 1972 (Plate 18, Fig. 11)

Description (Venkatachala and Rawat 1972, p. 308): Pollen grains oval in equatorial view (23.76 × 18.5 µm) and roundly triangular in polar view; tricolporate. Colpi long, reaching almost to the poles; ora lalongate. Exine about 1.3 µm, thickening at polar region about 2.5 µm; tegillate, sexine thicker than nexine, clearly psilate, surface punctitegillate forming close reticulations. Suprategillar surface is formed by fine coni; coni about 1.3 µm in height.

Occurrence Palaeocene-Eocene, Cauvery Basin, Tamil Nadu (Venkatachala and Rawat 1972).

Rhoipites kutchensis Venkatachala & Kar 1969 (Plate 18, Fig. 12)

Description (Venkatachala and Kar 1969, p. 166): Pollen grains oval in polar view, 15–25 × 10 18 µm. Colpi narrow, long, extending almost from one end to other, pores easily recognizable. Exine thin, finely inframicroreticulate.

Occurrence Early Eocene, Laki Series, Matanomadh, Kutch, Gujarat (Venkatachala and Kar 1969); Early Eocene, Naredi Formation, Kutch, Gujarat (Kar 1978).

Genus *Rostriapollenites* Venkatachala & Kar 1968

Now: *Marginipollis* Clarke & Frederiksen 1968

Genus *Rousea* Srivastava 1969

Type Species *Rousea subtilis* Srivastava 1969

Generic Description (Srivastava 1969, p. 52): Tricolpate, angulaperturate, colpi long, reaching polar area; amb subtriangular or rounded, sides convex; sexine thick, reticulate, lumina larger in mesocolpia, becoming smaller at colpus margins and apocolpia.

Species Recorded from India

Rousea matanomadhensis (Saxena 1979) Mandal & Rao 2001 (Plate 10, Fig. 13)

Description (Saxena 1979, p. 134): Pollen grain more or less circular in polar view, 72–90 μm. Tricolpate, colpi well-developed, long. Exine 2–4 μm thick, retibaculate to retipilate, duplibaculate/pilate, tegillate, bacula closely placed and almost joined with each other forming a perfect reticulum, meshes bigger at mesocolpi region.

Description (Mandal and Rao 2001, p. 360): Pollen grain subspherical in polar view, 93 × 95 μm. Tricolpate, colpi long, 40 μm deep, widely gaping at equator. Exine 4.5 μm, sexine as thick as nexine, thins out towards colpi, tectate, pilate, columellae 2.5 μm long, 1 μm broad, 1–3 μm apart, closer near colpi, capita up to 2.5 μm in diameter. Surface retipilate, reticulum bigger at mesocolpia (4.5 μm), narrower towards colpi and apocolpia (1 μm), lumina of irregular shape, muri simplicolumellate, 2 μm broad.

Occurrence Palaeocene, Matanomadh Formation, Kutch, Gujarat (Saxena 1979).

Remarks *Tricolpites matanomadhensis* Saxena (1979) was transferred to *Rousea* by Mandal and Rao (2001).

Rousea meghalayensis Mandal & Rao 2001 (Plate 10, Fig. 14)

Description (Mandal and Rao 2001, p. 361): Pollen grain spheroidal, 56.5 × 60 μm in equatorial view, mesocolpia convex. Tricolpate, colpi long, 50–52 μm, narrow and gaping at equator, margin uniformly 2 μm thick and smooth. Exine 5–6 μm thick, tectum at mesocolpia 2–2.5 μm; columellae robust, 2–2.5 μm long, 1 μm broad, 1–1.5 μm apart, capita 1.5–2.5 μm broad. Surface crassireticulate, reticulum larger at equator and gradually narrows towards pole; muri simplicolumellate, 1.5–2.5 μm wide, narrower near pole; lumina oval to elongated, 1–2 μm long, 1 μm wide at mesocolpia, nearly circular (less than 1 μm) at poles.

Occurrence Late Palaeocene, Lakadong Sandstone Member, Meghalaya (Kar and Kumar 1986).

Remarks *Tricolpites crassireticulatus* Dutta & Sah 1970 in Kar & Kumar 1986 (Non: *Tricolpites crassireticulatus* Dutta & Sah 1970) was transferred to *Rousea meghalayensis* by Mandal and Rao (2001).

Rousea saxenae Mandal & Rao 2001 (Plate 10, Fig. 15)

Description (Mandal and Rao 2001, p. 361): Pollen grains spheroidal, 68–70 μm in polar view, mesocolpia convex. Tricolpate, colpi long, 24 μm, gaping at equator, margin minutely wrinkled. Exine 4.5 μm; sexine 3 μm, nexine uniformly 1.5 μm; semitectate, tectum 1.5 μm at equator, less thick at colpus margins; columellae baculate to club-shaped, 1.5 μm long, 1 μm broad and about 1 μm apart, capita 1.5–2.5 μm wide. Surface reticulate, reticulum larger at equator and colpi margins, smaller at poles, larger reticula often broken at mesocolpia, simplicolumellate; lumina irregular, narrow and elongated at poles, larger lumina 5 μm long.

Occurrence Late Palaeocene, Matanomadh Formation, Kutch Basin, Gujarat (Saxena 1979).

Remarks *Tricolpites crassireticulatus* Dutta & Sah 1970 in Saxena 1979 (Non: *Tricolpites crassireticulatus* Dutta & Sah 1970) was transferred to *Rousea saxenae* by Mandal and Rao (2001).

Rousea sitholeyi (Ramanujam 1966) Mandal & Rao 2001 (Plate 11, Fig. 1)

Description (Ramanujam 1966, p. 163): Pollen grains isopolar, subprolate, tri-zonicolpate, equatorial diameter 26–30 μm. Amb rounded, 3-lobed, lobes not bulging. Colpi broadly gaping at equator, ends sharply pointed, longicolpate. Exine 2.2 μm thick, sexine thicker than nexine, columellae faint, surface reticulate, meshes polygonal, larger at poles, smaller at mesocolpia, muri simplibaculate, look finely beaded under high magnification.

Occurrence Miocene, Neyveli lignite, Tamil Nadu (Ramanujam 1966).

Remarks Mandal and Rao (2001) transferred *Retitricolpites sitholeyi* Ramanujam (1966) to *Rousea*.

S

Genus *Sastriipollenites* Venkatachala & Kar 1969

Type Species *Sastriipollenites trilobatus* Venkatachala & Kar 1969

Generic Description (Venkatachala and Kar 1969, p. 169): Pollen grains tricolporate, margocolporate, three-lobed; colpi long, funnel-shaped, ora distinct to indistinct. Exine granulose to intrabaculate.

Botanical Affinity and Ecology *Callophyllum polyanthum* (Cluciaceae); tropical forests.

Species Recorded from India

Sastriipollenites trilobatus Venkatachala & Kar 1969 (Plate 18, Fig. 13)

Description (Venkatachala and Kar 1969, p. 169): Pollen grains isopolar, trilobed in equatorial view. Tricolporate, margocolpate, colpi well-developed, invaginated inwards, ora distinct to indistinct. Exine up to 2 μm thick, intrabaculate.

Occurrence Early Eocene, Laki Series, Matanomadh, Kutch, Gujarat (Venkatachala and Kar 1969; Sah and Kar 1970); Early Eocene, Naredi Formation, Kutch, Gujarat (Kar 1978); Palaeocene, Lakadong Sandstone, Khasi Hills, Meghalaya (Kar and Kumar 1986).

Genus *Sonneratioipollis* Venkatachala & Kar 1969

Type Species *Sonneratioipollis bellus* Venkatachala & Kar 1969

Generic Description (Venkatachala and Kar 1969, p. 172): Pollen grains oval-spindle-shaped in equatorial view, without any constriction in middle. Triporate, zoniporate, pores oval, lolongate. Exine psilate.

Species Recorded from India

Sonneratioipollis bellus Venkatachala & Kar 1969 (Plate 26, Fig. 1)

Description (Venkatachala and Kar 1969, p. 173): Pollen grains oval to spindle-oval in shape, 18–30 × 10–15 μm. Triporate, zoniporate, pore margins do not show any marked thickening, lalongate. Exine more or less tectate, smooth.

Occurrence Early Eocene, Laki Series, Matanomadh, Kutch, Gujarat (Venkatachala and Kar 1969); Early Eocene, Laki Series, Kutch, Gujarat (Sah and Kar 1970).

Genus *Spinizonocolpites* Muller 1968

Type Species *Spinizonocolpites echinatus* Muller 1968

Generic Description (Muller 1968, p. 11): Pollen grains with one encircling colpus of the type occurring in the recent genus *Nypa* (Palmae) and with finely reticulate wall which is covered with scattered baculate or echinate spines, when dispersed grains mostly split into two halves.

Botanical Affinity Arecaceae.

Species Recorded from India

Spinizonocolpites bulbospinosus Singh 1990 (Plate 7, Fig. 5)

Description (Singh 1990, p. 226): Pollen subcircular, annulocolpate; exine sculpture bulbospinosus, spine with small, acuneate, protruding tips, tectum microreticulate, infratectum columellate.

Occurrence Palaeocene, Tura Formation, Langrin coalfield, Garo Hills, Meghalaya (Singh 1990).

Spinizonocolpites indicus Singh 1990 (Plate 7, Fig. 6)

Description (Singh 1990, p. 226): Pollen subcircular, annulocolpate; exine sculpture spinose, spine base swollen with gradually tapering apex, tectum microreticulate, infratectum columellate.

Occurrence Palaeocene, Tura Formation, Langrin coalfield, Garo Hills, Meghalaya (Singh 1990).

Spinizonocolpites quilonensis Rao & Ramanujam 1978 (Plate 7, Fig. 7)

Description (Rao and Ramanujam 1978, p. 415): Amb ovoid to subspheroidal, 35–42 × 26–30 µm, zonisulcate, monosulcate, sulcus extending all around equator and splitting the grain into two more or less equal halves, sulcus margin thin. Exine up to 3.5 µm thick, sexine thicker than nexine, echinate, spines widely placed, up to 3 µm high, slender, tips pointed, straight or flexuous.

Occurrence Miocene, Quilon Beds, Padappakkara, Kerala (Rao and Ramanujam 1978).

Spinizonocolpites spinulosus Ramanujam 1987 (Plate 7, Fig. 8)

Description (Ramanujam 1987, p. 30): Pollen grains ellipsoidal, 23–28.5 × 14.5–18 µm, zonosulcate, sulcus extending all around the grain dividing it into two halves; exine up to 1.8 μm thick, structure not distinct, surface spinulate, surface between spines microreticulate.

Occurrence Miocene, Warkalli Beds, Kerala (Ramanujam 1987).

Spinizonocolpites venkatachalae Saxena & Khare 2004 (Plate 7, Fig. 9)

Description (Saxena and Khare 2004, pp. 78, 80): Pollen grains subcircular to oval, size range 43–50 × 32–45 μm; zonisulcate, exine 1–1.5 μm thick, spinose, spines 5–7.5 μm long with pointed tips, inter-spinal area laevigate.

Occurrence Late Palaeocene-Middle Eocene, subsurface sediments, Tiruchirapalli, Tamil Nadu (Saxena and Khare 2004).

Spinizonocolpites wodehousei Singh 1990 (Plate 7, Fig. 10)

Description (Singh 1990, p. 225): Pollen circular, medium-sized, annulocolpate; exine sculpture spinose, spines long, beaked to flask-shaped with swollen base and rounded apex; tectum microreticulate, infratectum columellate.

Occurrence Palaeocene, Tura Formation, Langrin coalfield, Garo Hills, Meghalaya (Singh 1990).

Genus *Spinomonosulcites* Singh & Misra 1991c

Type Species *Spinomonosulcites varispinosus* Singh & Misra 1991c

Generic Description (Singh and Misra 1991c, p. 223): Pollen grains oval, elliptical, monosulcate; sulcus long, margin smooth or ornamented; exine dominantly spinose, occasionally spinules, baculae, verrucae are also present, spines deeply rooted, broad based, with pointed or globose apex, interspinal area psilate.

Botanical Affinity Arecaceae.

Species Recorded from India

Spinomonosulcites achinatus (Sah and Kar 1970) Singh & Misra 1991c (Plate 7, Fig. 11)

Description (Sah and Kar 1970, p. 180): Pollen grains generally oval with equally broad lateral ends. Colpus distinct, more or less uniformly broad. Exine 1.5–2 μm thick, spinose, spines 5–8 μm long, generally mixed with more or less pinheaded bacula, spines not closely placed, uniformly distributed, spaces between spines laevigate.

Occurrence Early Eocene, Laki Series, Kutch, Gujarat (Sah and Kar 1970); Early Eocene, Naredi Formation, Kutch, Gujarat (Kar 1978); Palaeocene, Matanomadh Formation, Matanomadh, Kutch, Gujarat (Saxena 1979; Kar 1985); Palaeocene Lakadong Sandstone, Khasi Hills, Meghalaya (Kar and Kumar 1986); Late Palaeocene-Early Eocene, subsurface sediments, Kapurdi, Barmer, Rajasthan (Tripathi 1995); Late Palaeocene-Middle Eocene, subsurface sediments, Tiruchirapalli, Tamil Nadu (Saxena and Khare 2004).

Remarks *Couperipollis achinatus* Sah and Kar (1970) and *Neocouperipollis achinatus* Sah and Kar (1970) Kar and Kumar (1987) were transferred to *Spinomonosulcites* by Singh and Misra (1991c).

Spinomonosulcites ankleshwarensis (Kar and Bhattacharya 1992) Samant & Phadtare 1997 (Plate 3, Fig. 14)

Description (Kar and Bhattacharya 1992, p. 252): Pollen grains oval, 58–84 × 44–62 μm in size. Monocolpate, colpus long, distinct, stretching almost from end to end. Spinose, spines well-built with bulbous base and pointed tips, 4–8 μm long, 2–4 μm broad at base, interspinal exine more or less laevigate.

Occurrence Early Eocene, Rajpardi lignite mine, Cambay Basin, Gujarat (Kar and Bhattacharya 1992); Early Eocene, Tarkeshwar Formation, Rajpardi, Cambay Basin, Gujarat (Samant and Phadtare 1997).

Remarks *Neocouperipollis ankleshwarensis* Kar and Bhattacharya (1992) was transferred to *Spinomonosulcites* by Samant and Phadtare (1997).

Spinomonosulcites magnus (Dutta and Sah 1970) Singh & Misra 1991c (Plate 7, Fig. 12)

Description (Dutta and Sah 1970, p. 28): Size range 70–95 μm; holotype 90 μm; amb elliptical to oval-elongate; monosulcate, sulcus distinct, extending from pole to pole; exine moderately thick, undifferentiated, surface densely covered with long spinose processes generally with a bulbous base.

Occurrence Palaeocene, Cherra Formation, Shillong Plateau, Meghalaya (Dutta and Sah 1970); Late Palaeocene-Early Eocene, subsurface sediments, Kapurdi, Barmer, Rajasthan (Tripathi 1995).

Remarks *Monosulcites magnus* Dutta and Sah (1970) and *Neocouperipollis magnus* Kar and Kumar (1987) were transferred to *Spinomonosulcites* by Singh and Misra (1991c).

Spinomonosulcites spinorobustus (Kar and Kumar 1986) Singh & Misra 1991c (Plate 7, Fig. 13)

Description (Kar and Kumar 1986, p. 195): Pollen grains oval, monocolpate, 65 × 48 μm, colpus generally discernible, sometimes obscured, extending from one end to the other. Exine spinose, spines robustly built, uniformly broad except at tip, sparsely placed, interspinal space more or less granulose.

Occurrence Palaeocene, Lakadong Sandstone, Khasi Hills, Meghalaya (Kar and Kumar 1986).

Remarks *Neocouperipollis spinorobustus* Kar and Kumar (1986) was transferred to *Spinomonosulcites spinorobustus* by Singh and Misra (1991c).

Spinomonosulcites varispinosus Singh & Misra 1991c (Plate 7, Fig. 14)

Description (Singh and Misra 1991c, pp. 223, 224): Pollen grains elliptical, monosulcate, length 70–85 μm × breadth 49–66 μm (including spines); sulcus long (almost reaching ends), margin uneven, beset with spines, spinules and verrucae etc.; exine spinose, 2–3 μm (excluding spines), spines 1–7 μm long, varying in size, deeply rooted, base broad, apex rounded or pointed or pinheaded, when broken spines without globose head; nexine thin (1–1.5 μm), sunken below spines; interspinal area psilate.

Occurrence Miocene, Cuddalore Formation, Neyveli Lignitefield, Neyveli, Tamil Nadu (Singh and Misra 1991c; Singh et al. 1992).

Genus *Spinulotricolporites* Saxena & Misra 1987

Type Species *Spinulotricolporites spinosus* (Saxena 1979) Saxena & Misra 1987

Generic Description (Saxena and Misra 1987, p. 20): Pollen grains circular to subcircular in shape, small to medium-sized. Tribrevicolporate, colpi elliptical to spindle-shaped, pore subcircular to oval with thickened and rugged margin, Exine thin to moderately thick, crassisexinous, spinose, spinules short, delicate and sparsely placed.

Species Recorded from India

Spinulotricolporites spinosus (Saxena 1979) Saxena & Misra 1987 (Plate 18, Fig. 14)

Description (Saxena and Misra 1987, p. 20): Pollen grains circular to subcircular in shape, size range 30–52 × 29–50 microns. Tribrevicolporate; colpi spindle-shaped, 9–13 microns long, colpi margin not thickened but beset with sparsely placed spinules, pores circular or oval, 6–9 microns in diameter, pore margin thickened (costate) and rugged. Exine 2–3 microns thick (excluding supratectal spinules), sexine appreciably thicker than nexine (up to 0.5 microns), tectate, ornamentation spinulose, spinules 0.5–1.25 microns long (rarely up to 1.5 microns), sparsely placed (2–6 microns apart) and more or less evenly distributed all over the exine. Surface finely scabrate or microreticulate.

Occurrence Palaeocene, Matanomadh Formation, Matanomadh, Kutch, Gujarat (Saxena 1979).

Remarks *Lakiapollis spinosus* Saxena (1979) was transferred to a new genus *Spinulotricolporites* by Saxena and Misra (1987).

Genus *Striacolporites* Sah & Kar 1970

Type Species *Striacolporites striatus* Sah & Kar 1970

Generic Description (Sah and Kar 1970, p. 135): Pollen grains tricolporate, oval in equatorial and subcircular in polar view, exine striato-reticulate.

Botanical Affinity *Bauhinia* (Caesalpiniaceae).

Species Recorded from India

Striacolporites cephalus Sah & Kar 1970 (Plate 18, Fig. 15)

Description (Sah and Kar 1970, p. 136): Pollen grains subcircular in polar view, 70–75 × 58–65 μm. Tricolporate, colpi membrane faintly granular. Sexine thicker than nexine, striato-reticulate.

Occurrence Early Eocene, Laki Series, Kutch, Gujarat (Sah and Kar 1970); Early Eocene, Naredi Formation, Kutch, Gujarat (Kar 1978); Palaeocene, Matanomadh Formation, Matanomadh, Kutch, Gujarat (Saxena 1979).

Striacolporites ovatus Sah & Kar 1970 (Plate 19, Fig. 1)

Description (Sah and Kar 1970, p. 136): Pollen grains oval in equatorial view, 70–80 × 40–52 μm. Tricolporate, apertures ill-developed, sexine thicker than nexine, striato-reticulate.

Occurrence Early Eocene, Laki Series, Kutch, Gujarat (Sah and Kar 1970); Early Eocene, Naredi Formation, Kutch, Gujarat (Kar 1978); Palaeocene, Matanomadh Formation, Matanomadh, Kutch, Gujarat (Saxena 1979).

Striacolporites striatus Sah & Kar 1970 (Plate 19, Fig. 2)

Description (Sah and Kar 1970, p. 135): Pollen grains oval in equatorial view, 45–57 × 30–48 µm. Tricolporate, pore lalongate. Exine 2–3 µm thick, sexine thicker than nexine, striato-reticulate.

Occurrence Early Eocene, Laki Series, Kutch, Gujarat (Sah and Kar 1970); Early Eocene, Naredi Formation, Kutch, Gujarat (Kar 1978); Palaeocene, Matanomadh Formation, Matanomadh, Kutch, Gujarat (Saxena 1979) Palaeocene, Barmer Sandstone, Barmer, Rajasthan (Jain et al. 1973).

Genus *Symplocoipollenites* Potonié 1951 ex Potonié 1960

Type Species *Symplocoipollenites vestibulus* Potonié 1951 ex Potonié 1960

Generic Description (Potonié 1960, p. 106): Lenticular (oblate) pollen; outline finely dentate, equator more or less triangular, exine granulate to rugulate, colpi short, vestibulum distinct.

Species Recorded from India

Symplocoipollenites baksii Ramanujam 1987 (Plate 19, Fig. 3)

Description (Ramanujam 1987, p. 33): Pollen grains isopolar, amb triangular, 24–29 µm, angulaperturate, colpi very short, narrow, streak-like, ora lalongate; exine 1.8 µm thick, tectate, surface punctate to microreticulate.

Occurrence Miocene, Warkalli Beds, Kerala (Ramanujam 1987).

Symplocoipollenites bhavnagarensis Samant 2000 (Plate 19, Fig. 4)

Description (Samant 2000, pp. 108, 110): Pollen grains triangular in polar view; 28–31 × 35–38 µm in size, tricolporate; colpi short, open; ora distinct, lalongate; exine thick, punctate to reticulate.

Occurrence Early Eocene, Kharsalia Clay Formation, Bhavnagar, Cambay Basin, Gujarat (Samant 2000).

Symplocoipollenites gracilis Venkatachala & Rawat 1972 (Plate 19, Fig. 5)

Description (Venkatachala and Rawat 1972, p. 315): Pollen grain triangular in polar view, sides straight to slightly concave; angulaperturate; 20–26.4 × 21.2–28.4 µm, tricolporate. Colpi obscure to short, brevicolpate, up to 6 µm deep in polar view, ora lalongate, encircled by nexinous thickening. Exine about 2.6 µm;

tectate, sexine thinner than nexine, surface scabrate, scabrae minute, closely placed, less than 1 μm in diameter.

Occurrence Palaeocene-Eocene, Cauvery Basin, Tamil Nadu (Venkatachala and Rawat 1972).

Symplocoipollenites paravurensis Saxena 1992 (Plate 19, Fig. 6)

= *Symplocoipollenites punctatus* Venkatachala and Rawat (1972)

= *Symplocoipollenites punctatus* Rao and Ramanujam (1982)

Description (Venkatachala and Rawat 1972, p. 315): Pollen grains distinctly triangular in polar view with straight to slightly convex sides and rounded angles, angulaperturate; 23.7–30.4 × 24.5–30.4 μm; tricolporate. Colpi very short, slit-like to slightly opened, extending usually within pore cavity, sometimes slightly deep, up to 4 μm in polar view; ora distinct, lalongate, oval shaped, surrounded by nexinous thickening. Exine about 1.5 μm thick, tectate, sexine thicker than nexine, punctate, puncta minute, less than 1 μm in diameter.

Occurrence Palaeocene-Eocene, Cauvery Basin, Tamil Nadu (Venkatachala and Rawat 1972); Miocene, Quilon Beds, Paravur, Kerala (Rao and Ramanujam 1982).

Remarks *Symplocoipollenites punctatus* Rao & Ramanujam (1982) has following diagnosis "Pollen grains isopolar, amb triangular, sides convex, polar diameter 16–24 μm, zonaperturate, tricolporate, colpae short, narrow and slit-like with blunt ends, ora prominently thickened and lalongate. Exine 1.5 μm thick, sexine as thick as nexine, surface finely punctate". *Symplocoipollenites punctatus* proposed by Venkatachala and Rawat (1972) and the same species proposed by Rao and Ramanujam (1982) are very similar in characters. Both of these species have therefore, been transferred to *Symplocoipollenites paravurensis* by Saxena (1992).

T

Genus *Talisiipites* Wodehouse 1933

Type Species *Talisiipites fischeri* Wodehouse 1933

Generic Description (Wodehouse 1933, p. 513): Pollen oblately flattened, triangular in polar view, tricolporate with furrows long, narrow and shallow, each enclosing a small germ pore; pores aspidate, surrounded by a subexinous thickening, similar to those of the grains of the Betulaceae; exine thin and rather finely granular, more evidently so near the pores.

Species Recorded from India

Talisiipites wodehousei Dutta & Sah 1970 (Plate 19, Fig. 8)

Description (Dutta and Sah 1970, p. 39): Size range 26–37 μm; amb distinctly triangular, sometimes convexly triangular; 3-angulaperturate, distinctly colporate, pores small, furrows deep, sexine as thick as nexine, surface sculpture scabrate.

Affinity Sapindaceae.

Occurrence Palaeocene, Cherra Formation, Shillong Plateau, Meghalaya (Dutta and Sah 1970).

Genus *Thymelaepollis* Sah & Kar 1970

Type Species *Thymelaepollis crotonoides* Sah & Kar 1970

Generic Description (Sah and Kar 1970, p. 140): Pollen grains circular-subcircular. Polyporate, pores uniformly distributed. Exine thick, baculate-spinose, sometimes with excrescences on top.

Species Recorded from India

Thymelaepollis crotonoides Sah & Kar 1970 (Plate 4, Fig. 2)

Description (Sah and Kar 1970, p. 140): Pollen grains circular-subcircular, 42–54 × 40–53 μm. Polyporate. Exine 2–4 μm thick, mostly baculate, sometimes with excrescences on top.

Occurrence Early Eocene, Laki Series, Kutch, Gujarat (Sah and Kar 1970); Early Eocene, Naredi Formation, Kutch, Gujarat (Kar 1978).

Genus *Tinalipollenites* Mandal & Kumar 2000

Type Species *Tinalipollenites duttae* Mandal & Kumar 2000

Generic Description (Mandal and Kumar 2000, p. 205): Pollen grains triangular-subtriangular in polar view, radially symmetrical. Tricolporate, planaperturate, apertures frequently associated with exinal folds. Exine thick, sexine thicker than nexine, tectum semiperforate, surface scabrate.

Species Recorded from India

Tinalipollenites duttae Mandal & Kumar 2000 (Plate 19, Fig. 9)

Generic Description (Mandal and Kumar 2000, p. 205): Polar axis 30–35 μm (5 specimens), triangular to subtriangular, equatorial axis 28–30 μm (2 specimens), subprolate, holotype 30 × 33 μm. Tricolporate, planaperturate; colpi narrow, 16 μm long, frequently bordered by 2–2.5 μm thick folds; pores lalongate, 5.5 μm wide, simple. Exine 3.5–6.5 μm thick; sexine thicker than nexine, 5 μm thick at poles, forming cap-like structure, thin towards colpi (1.5 μm). Columellae not prominent, 1–3 μm apart, tectum perforate, forming loose fibrilous layer, nexine 1.5 μm, thickened near aperture. Surface uneven, faintly scabrate.

Occurrence Miocene-Pliocene, Namsang Formation, subsurface sediments, Upper Assam (Mandal and Kumar 2000).

Genus *Triangulorites* Kar 1985

Now: *Grevilloideaepites* Biswas 1962

Genus *Tricolpites* Cookson 1947 ex Couper 1953

Type Species *Tricolpites reticulatus* Cookson 1947 ex Couper 1953

Generic Description (Cookson 1947, p. 134): Free anisopolar, tricolpate. Exine variable in thickness and sculpture. Size variable.

Generic Description (Potonié 1960, p. 95): Shape spherical to ovoid, tricolpate, mesocolpium bulging, colpi broad, exine on either side stronger, causing the equator to be trilobate, exine finely reticulate.

Description (Jargen and Dettmann 1989, p. 99): Pollen grains free, tricolpate, isopolar, oblate to subprolate; amb circular to trilobate with convex mesocolpia. Colpi meridionally aligned, parallel sided, but often gaping, margin entire; membrane absent or reduced, without freely standing bacula. Exine stratified, sexine baculate, semitectate with reticulate surface. Reticulum regular and composed of smooth to crested muri and elongate to equi-dimensional lumen that are of uniform size and less than 1 μm in diameter over entire surface of grain.

Species Recorded from India

Tricolpites brevis Sah & Kar 1970

Now: *Intrareticulites brevis* (Sah and Kar 1970) Kar 1985

Tricolpites crassireticulatus Dutta & Sah 1970

Now: *Albertipollenites crassireticulatus* (Dutta and Sah 1970) Mandal & Rao 2001

Tricolpites delicatus (Kar 1979) Mandal and Rao 2001 (Plate 11, Fig. 2)

Description (Mandal and Rao 2001, p. 362): Pollen grain prolate, 27–36 μm in equatorial view. Tricolpate, colpi long, 24 μm. Exine 2 μm thick, sexine and nexine not separable, semitectate, columellae distinct, 1.5 μm long, slender, closely placed. Surface microreticulate, lumina irregular in shape, less than 1 μm in size, muri thin and low.

Occurrence Oligocene, Kutch Basin (Kar 1979).

Remarks *Retitricolpites delicatus* Kar (1979) was transferred to *Tricolpites* by Mandal and Rao (2001).

Tricolpites densiornatus Venkatachala & Rawat 1972 (Plate 11, Fig. 3)

Description (Venkatachala and Rawat 1972, pp. 302, 303): Pollen grains roundly triangular, 35.4–39.6 × 35–41.7 μm; tricolpate. Colpi long, wedge-shaped, 15 μm deep in polar view, not reaching the poles. Exine about 3.3 μm thick, slightly thickening towards apertural region, sexine thicker than nexine, closely pilate forming a pseudoreticulum.

Occurrence Subsurface Palaeocene-Eocene sediments, Cauvery Basin (Venkatachala and Rawat 1972).

Tricolpites foxii (Biswas 1962) Ramanujam 1966 (Plate 11, Fig. 4)

Description (Ramanujam 1966, p. 162): Pollen isopolar, prolate spheroidal, 34 × 25 μm; 3-zonicolpate. Amb prominently 3-lobed, lobes widely spaced. Colpi deep, long, ends pointed. Exine 1–8 μm thick, surface minutely reticulate.

Occurrence Lower-Middle Eocene, Um Sohryngkew River Section, Meghalaya (Biswas 1962); Miocene, Neyveli lignite, Tamil Nadu (Ramanujam 1966).

Tricolpites incognatus Kar & Jain 1981 (Plate 11, Fig. 5)

Description (Kar and Jain 1981, p. 120): Pollen grains subtriangular-subcircular in polar view, 18–22 μm. Tricolpate, colpi long, funnel-shaped in equatorial view. Sexine as thick as nexine, scrobiculate.

Description (Mandal and Rao 2001, p. 362): Pollen subtriangular in polar view, 22 × 24 μm. Tricolpate, colpi 7.5 μm deep, gapping at equator, colpi margin irregular, triangular-shaped apocolpial area slightly thickened (not parasyncolpate). Exine 1.5 μm, sexine thicker than nexine, uniformly thick, tectate, columellae 0.5 μm long, slender, uniformly spaced. Surface uneven, uniformly microreticulate, lumina circular, nearly 0.5 μm, muri low.

Occurrence Miocene, Warkalli Formation, Kerala (Kar and Jain 1981).

Tricolpites levis Sah & Dutta 1966

Now: *Ladakhipollenites levis* (Sah and Dutta 1966) Mathur & Jain 1980

Tricolpites margocolpites Venkatachala & Rawat 1972 (Plate 11, Fig. 6)

Description (Venkatachala and Rawat 1972, p. 303): Pollen grains triangular in polar view, sides straight to convex; 35.7–37 × 37 μm; tricolpate, angulaperturate. Margocolpate, colpi long, 1.5–2 μm deep in polar view, reaching almost to the poles, wide in the equatorial region, margo about 2.7 μm thick. Exine 1.5 μm thick, thickening towards the pore region; tectate, sexine as thick as nexine or slightly thicker, intrapunctate, surface finely reticulate.

Occurrence Subsurface Palaeocene-Miocene sediments, Cauvery Basin (Venkatachala and Rawat 1972).

Tricolpites microreticuloides Ramanujam 1966 (Plate 11, Fig. 7)

Description (Ramanujam 1966, p. 162): Pollen grains isopolar, prolate spheroidal, 26 × 24 μm. Trizonicolpate. Poles rounded. Colpi long, almost reaching the poles, of uniform breadth for greater part. Exine 2.5 μm thick, columellae distinct, surface microreticulate.

Occurrence Miocene, Neyveli lignite, Tamil Nadu (Ramanujam 1966).

Tricolpites paucireticulatus Sah & Kar 1974 (Plate 11, Fig. 8)

Description (Sah and Kar 1974, p. 169): Pollen grain subcircular-circular, 38–46 μm; tricolpate, colpi broad, colpi margin laevigate. Exine reticulate only in middle part of mesocolpial region.

Description (Mandal and Rao 2001, p. 363): Pollen grain circular-subcircular in polar view, 50 × 54 μm. Tricolpate, colpi 16.5 μm deep, end pointed, wide at equator. Exine 2.5 μm thick at mesocolpia and gradually thins out, colpi margin (0.5 μm), nexine as thick as sexine, semitectate, columellae distinct at mesocolpia, 1 μm in length and width, indistinct near colpi margins. Surface reticulate, lumina nearly circular, larger (1 μm) at mesocolpia and apocolpia, almost absent at colpi margins, muri simplicolumellate, thin narrow, smooth.

Occurrence Early Eocene Palana lignite, Rajasthan (Sah and Kar 1974)

Tricolpites retibaculatus Saxena 1979

Now: *Albertipollenites retibaculatus* (Saxena 1979) Mandal & Rao 2001

Genus *Tricolporocolumellites* Kar 1985

Type Species *Tricolporocolumellites pilatus* Kar 1985

Generic Description (Kar 1985, p. 122): Pollen grains subcircular-oval. Tricolporate, brevicolpate, pore margin thickened. Exine columellate, pila forming negative reticulum in surface view.

Botanical Affinity Bombacaceae.

Species Recorded from India

Tricolporocolumellites eocenicus Samant & Phadtare 1997 (Plate 19, Fig. 10)

Description (Samant and Phadtare 1997, p. 63): Pollen grains oblate-spheroidal to suboblate, 66–70 × 52–61 μm in size, tricolporate, brevicolpate, colpi 8–10 μm long and 4–6 μm wide, tenuimarginate; ora endexinous, almost lalongate, 9–11 μm in diameter, costate; exine about 2.5 μm thick, semitectate, columellate, sexine and nexine of same thickness except around ora where nexine is distinctly thicker; foveolate, columellae on surface focus give striate appearance; striations generally oblique to the poles.

Occurrence Early Eocene, Tarkeshwar Formation, Rajpardi, Cambay Basin, Gujarat (Samant and Phadtare 1997).

Tricolporocolumellites pilatus Kar 1985 (Plate 19, Fig. 11)

Description (Kar 1985, p. 122, 123): Pollen grains circular-oval, 78–90 × 76–85 μm. Tricolporate, brevicolpate, colpi oval-elliptical in shape, length almost equal to colpi, colpi and pore placed crosswise, pore margin generally thickened, Exine 2–3 μm thick, sexine much thicker than nexine, columellate, pila 4–6 μm long, about 1 μm broad, pila provides negative reticulum in surface view.

Occurrence Eocene, subsurface sediments, Rataria, Kutch, Gujarat (Kar 1985); Early Eocene, Tarkeshwar Formation, Rajpardi, Cambay Basin, Gujarat (Samant and Phadtare 1997).

Genus *Tricolporopilites* (Kar 1985)

Type Species *Tricolporopilites robustus* (Kar and Saxena 1981) Kar 1985

Generic Description (Kar 1985, p. 121): Pollen grains triangular to subcircular in polar and subcircular – oval in equatorial views. Tricolporate, colpi short to long, pore well-developed, margin generally thickened, exine heavily pilate, pila robustly built, inter-pilar space granulate, sculptural elements furnish negative reticulum in surface view.

Emended generic description (Singh and Misra 1991a, p. 63, 65): Pollen isopolar, circular to more or less triangular in polar view; tri- to rarely tetrabrevicolporate; colpi short, narrow, slightly gaping in polar view, indistinct due to heavy ornamentation, colpus margin uneven; pore more or less circular, costate; exine incompletely tectate to occasionally completely tectate, tectal elements forming pila, clava, gemmae, verrucae and rugulae which are supported by groups of columellae; sexine consisting of a thin columellate layer and tectal elements, nexine smooth to undulating, tectal elements in surface view appear as irregular patches with a dupli- or multicolumellate structure separated by narrow grooves; ornamentation (tectal patches) uniform at both the poles or coarser/finer at one pole.

Botanical Affinity Alangiaceae.

Remarks Kar (1985) transferred *Retitrescolpites robustus* Kar and Saxena (1981) to *Tricolporopilites*.

Species Recorded from India

Tricolporopilites differentialis Singh & Misra 1991a (Plate 19, Fig. 12)

Description (Singh and Misra 1991a, p. 67): Pollen isopolar, 50–85 μm, circular in polar view; tri- to rarely tetrabrevicolporate; colpi fairly narrow, slightly gaping in polar view, with uneven margin, indistinct due to heavy ornamentation; pore mere or less circular, costate, 5–8 × 9–10 μm in diameter; exine 5–10 μm thick, thickness of exine at one pole (10 μm) is more or less double the thickness of exine at other pole (5 μm), tectum incomplete, sexine 2–3 times thicker than nexine, the latter smooth to undulating, tectal elements forming clava, gemmae, verrucae and rugulae are supported by the groups of columellae; surface ornamentation (tectal patches) distinctly closer at a pole with greater exine thickness, dupli- to multicolumellate patches are separated by narrow grooves.

Occurrence Miocene, Cuddalore Formation, subsurface sediments, Neyveli Lignitefield, Tamil Nadu (Singh and Misra 1991a; Singh et al. 1992).

Tricolporopilites robustus (Kar and Saxena 1981) Kar 1985 (Plate 19, Fig. 13)

Description (Kar 1985, p. 122): Pollen grains subtriangular in polar and subcircular in equatorial views, 53–70 × 49–68 μm. Tricolporate, brevicolpate, pore generally distinct, margin thickened. Exine up to 3.5 μm thick, sexine much thicker than nexine, pilate, pila closely placed, broad at top, gradually tapered at base, inter-pilar space granulose, sculptural elements appear as mud-cracks in surface view.

Occurrence Middle-Late Eocene, subsurface sediments, Rataria, Kutch, Gujarat (Kar and Saxena 1981; Kar 1985); Early Eocene, Tarkeshwar Formation, Rajpardi, Cambay Basin, Gujarat (Samant and Phadtare 1997).

Tricolporopilites tectatus Singh & Misra 1991a (Plate 19, Fig. 14)

Description (Singh and Misra 1991a, p. 69, 70): Pollen isopolar, more or less circular in polar view, 45–55 μm, tribrevicolporate; colpi narrow, margin more or less uneven; pore circular, costate, 6–9 μm across; exine 4–5 μm thick, consisting of a complete tectum, thin columellate layer and separated verrucate elements; surface ornamentation indistinct or faint, formed of dupli- to multicolumellate tectal patches separated by narrow grooves that do not penetrate the tectum.

Occurrence Miocene, Cuddalore Formation, subsurface sediments, Neyveli Lignitefield, Tamil Nadu (Singh and Misra 1991a; Singh et al. 1992).

Tricolporopilites uniformis Singh & Misra 1991a (Plate 19, Fig. 15)

Description (Singh and Misra 1991a, p. 65, 67): Pollen isopolar, circular in polar view, 45–95 μm, tri- to tetrabrevicolporate; colpi narrow, slightly gaping in polar view, indistinct due to heavy ornamentation, colpus margin uneven; pore circular, 4–9 × 7–12 μm in diameter, costate; exine 4–9 μm thick, consisting of a nexine and structured sexine (tectal elements and a columellate layer), tectum incomplete, sexine equal to or 2–3 times thicker than nexine, latter smooth; tectal elements forming gemmae, verrucae and rugulae which are supported by the groups of columellae, in surface view these elements form patches of uniform shape and size at both the poles, dupli- to multicolumellate patches are separated by narrow grooves.

Occurrence Miocene, Cuddalore Formation, subsurface sediments, Neyveli Lignitefield, Tamil Nadu (Singh and Misra 1991a; Singh et al. 1992); Early Eocene, Tarkeshwar Formation, Rajpardi, Cambay Basin, Gujarat (Samant and Phadtare 1997).

Genus *Tricolporopollis* Dutta & Sah 1970

Type Species *Tricolporopollis decoris* Dutta & Sah 1970

Synonym 1985—*Retitribrevicolporites* Kar, *Palaeobotanist* 34: 59, 80.

Generic Description (Dutta and Sah 1970, p. 41): Pollen grains spheroidal-subspheroidal in polar view and more or less subspheroidal to sometimes ellipsoidal in equatorial view; 3-aperturate, apertures subequatorial, distinctly colporate, brevicolpate, ora more or less circular to slightly lalongate; sexine as thick as nexine or thinner or sometimes thicker; surface sculpture reticulate, pitted-reticulate, or sub-reticulate, or foveolate.

Remarks Kar (1985, p. 43) proposed *Retitribrevicolporites* with following diagnosis: "Pollen subcircular, tribrevicolporate, colpi and pore more or less of the same size, pore margin thickened. Exine reticulate". He proposed *Retitribrevicolporites matanomadhensis* as type species of this genus, which exhibits foveolate exine. Considering the diagnosis the genus *Retitribrevicolporites* Kar (1985) becomes junior synonym of *Tricolporopollis* Dutta & Sah (1970).

Species Recorded from India

Tricolporopollis alleppeyensis Rao & Rajendran 1996 (Plate 20, Fig. 1)

Description (Rao and Rajendran 1996, p. 73): Pollen grains rounded triangular, associated with folds, some specimens showing triradiate scar. Size range 57–65 × 54–60 μm. Tricolporate, brevicolpate, pore 8–10 μm in diameter, apertural end thickened. Exine up to 3.5 μm thick, sexine and nexine sometimes not differentiated, sexine up to 2.5 μm thick, perforate, nexine 1 μm thick. Surface showing finely pitted reticulation.

Occurrence Miocene, Quilon Formation, Kannur, Kerala (Rao and Rajendran 1996).

Tricolporopollis cannanorensis Rao & Rajendran 1996 (Plate 20, Fig. 2)

Description (Rao and Rajendran 1996, p. 72, 73): Pollen grains oval-subcircular, associated with folds. Size range 45–65 × 40–60 μm. Tricolporate, brevicolpate, colpi 3–4.5 μm wide, apertural ends thickened. Exine 2–3.5 μm thick, sexine and nexine not differentiated. Surface showing pitted-reticulate ornamentation.

Occurrence Miocene, Quilon Formation, Kannur, Kerala (Rao and Rajendran 1996).

Tricolporopollis globosa Dutta & Sah 1970 (Plate 20, Fig. 4)

Description (Dutta and Sah 1970, p. 42): Size range 27–28 µm; holotype 36 µm, amb more or less spheroidal, prolate-spheroidal in polar view; nearly globular in equatorial view, 3-aperturate, apertures subequatorial, colporate; brevicolpate, pores lalongate, surrounded by conspicuously thickened rings; sexine thinner than nexine; the latter considerably more thick at the apertures; surface sculpture coarsely reticulate.

Occurrence Palaeocene, Cherra Formation, Shillong Plateau, Meghalaya (Dutta and Sah 1970).

Tricolporopollis matanomadhensis (Venkatachala and Kar 1969) Tripathi & Singh 1985 (Plate 20, Fig. 5)

Description (Venkatachala and Kar 1969, p. 169): Pollen grains circular, 28–40 × 25–38 µm. Tribrevicolporate; colpi short, inconspicuous. Pores well-developed, more or less oval elliptical, surrounded with thickened rim, subequatorial in position, exine 2.5–5 µm thick, scrobiculate.

Occurrence Early Eocene, Naredi Formation, Matanomadh, Kutch, Gujarat (Venkatachala and Kar 1969; Kar 1978); Early Eocene, Palana lignite, Palana, Bikaner, Rajasthan (Sah and Kar 1974); Palaeocene, Matanomadh Formation, Kutch, Gujarat (Saxena 1979); Palaeocene-Eocene, subsurface sediments, Kapurdi, Barmer, Rajasthan (Tripathi 1995).

Remarks *Lakiapollis matanomadhensis* Venkatachala and Kar (1969) was transferred to *Tricolporopollis* by Tripathi and Singh (1985).

Tricolporopollis ruber (Dutta and Sah 1970) Tripathi & Singh 1985 (Plate 20, Fig. 6)

Description (Dutta and Sah 1970, p. 42): Size range 55–68 µm; amb more or less spheroidal to prolate-spheroidal, 3-aperturate, apertures subequatorial, colporate; brevicolpate; ora circular to slightly lalongate; sexine as thick as nexine, surface ornamentation distinctly foveolate, muri thick, lumina small, deep and circular in shape.

Occurrence Palaeocene, Cherra Formation, Shillong Plateau, Meghalaya (Dutta and Sah 1970); Palaeocene-Eocene, Jaintia Group, Jaintia Hills, Meghalaya (Tripathi and Singh 1985).

Genus *Trilatiporites* Ramanujam 1966

Now: *Dorreenipites* Biswas 1962

Genus *Triorites* Erdtman 1947 ex Cookson 1950

Type Species *Triorites magnificus* Cookson 1950

Generic Description (Cookson 1950, p. 175): Large, 55 × 71 μm, oblate, the equatorial diameter ranging from 53 to 88 μm, triangular, with straight to slightly concave sides.

Exine thick, 9–10.5 μm, of clearly defined layers of these two inner are nexinous (ecto- and endonexine) with an approximate width of 2.5 μm and two outer are sculptinous (sexine & ?perine). The inner layer of sculptine is broad, about 4.5 μm and has a pitted or somewhat spongy consistency, this outer layer of sculptine, which is about 2.5 μm thick is highly referective and does not take the stain like rest of the exine. It is baculate and in surface view appears reticuloid. In some specimens it is partially or completely destroyed, thus exposing the spongy layer.

Another unusual feature of this sporomorph is position of the true apertures. These appear to be situated at the base of the cavities formed by forward extension of the exine. Moreover, the sculptinous layers turn into line these extra-apertural cavities, thereby completely enveloping the distal extensions of the ectoexine.

Species Recorded from India

Triorites bellus Sah & Kar 1970

Now: *Grevilloideaepites eocenicus* (Biswas 1962) Singh & Misra 1991b

Triorites communis Sah & Dutta 1966 (Plate 26, Fig. 2)

Description (Sah and Dutta 1966, p. 82): Size range 34–53 μm; amb triangular, sides between ora convex in polar view; triorate, ora distinctly pouted, circular, diameter about 8–12 μm, usually surrounded by a collar-like thickening; exine up to 6 μm thick, sexine thicker than nexine, finely granular.

Occurrence Early Eocene, Cherra Formation, South Shillong Plateau, Meghalaya (Sah and Dutta 1966).

Triorites hirsutus Sah & Kar 1974 (Plate 26, Fig. 3)

Description (Sah and Kar 1974, p. 174): Pollen grains triangular, 38–50 μm, 3-orate, ora protruding. Exine pilate-baculate, forming negative reticulum on surface view.

Occurrence Early Eocene, Palana lignite, Palana, Bikaner, Rajasthan (Sah and Kar 1974).

Triorites inferius Dutta & Sah 1970

Now: *Grevilloideaepites inferius* (Dutta and Sah 1970) Singh & Misra 1991b

Triorites kutchensis (Sah and Kar 1970) Saxena 1992 (Plate 26, Fig. 8)

Description (Sah and Kar 1970, p. 139): Pollen grains mostly triangular, sometimes subtriangular, 3-orate, ora distinct, without thickened margin, ora very big. Exine pilate-baculate, pila/bacula closely placed.

Remarks *Triorites triangulus* Sah and Kar (1970) is a junior homonym of *Triorites triangulus* Sah (1967), hence Saxena (1992) replaced it with a new name, viz. *Triorites kutchensis* Saxena (1992).

Occurrence Early Eocene, Laki Series, Kutch, Gujarat (Sah and Kar 1970).

Triorites minutus Sah & Kar 1970 (Plate 26, Fig. 4)

Description (Sah and Kar 1970, p. 139): Pollen grains generally triangular with more or less straight inter-oral margin. Ora distinct without thickened margin. Exine 1.5–3 μm thick, sexine thicker than nexine, scrobiculate.

Occurrence Early Eocene, Laki Series, Kutch, Gujarat (Sah and Kar 1970).

Triorites protrudus Samant 2000 (Plate 26, Fig. 5)

Description (Samant 2000, p. 110): Pollen grains rounded triangular, angulaperturate, 26–27 × 24–26 μm in size; triporate, pores distinctly protruding, pore margins thin, irregular; exine about 2 μm thick, sexine and nexine of about same thickness except at aperture region where sexine is distinctly thick, psilate.

Occurrence Early Eocene, Kharsalia Clay Formation, Bhavnagar, Cambay Basin (Samant 2000).

Triorites pseudoreticulatus Venkatachala & Rawat 1972 (Plate 26, Fig. 6)

Description (Venkatachala and Rawat 1972, p. 319): Pollen grain subspheroidal in polar view; 17.16 × 17.16 μm; triporate, angulaperturate. Pores circular, 2.5–

3.3 μm in diameter, margin thickened. Exine about 2 μm thick, tectate; sexine thicker than nexine, formed of pila with slightly rounded heads; surface reticulate; lumina as wide as muri, about 1 μm in diameter; reticulations more prominent in the polar region.

Occurrence Palaeocene-Eocene, Cauvery Basin, Tamil Nadu (Venkatachala and Rawat 1972).

Triorites sahii Samant 2000 (Plate 26, Fig. 7)

Description (Samant 2000, p. 114): Pollen grains rounded triangular; sides convex; 60–65 × 65–67 μm in size, triporate; pores slightly protruding, about 10 μm wide, tenuimarginate, margin smooth to irregular, exine about 3 μm thick, nexine thicker than sexine, however, at pore margins nexine is distinctly thicker than sexine, psilate.

Occurrence Early Eocene, Kharsalia Clay Formation, Bhavnagar, Cambay Basin (Samant 2000).

Triorites tubiferus Venkatachala & Rawat 1972 (Plate 26, Fig. 9)

Description (Venkatachala and Rawat 1972, p. 318): Pollen grains triangular in polar view; sides between apertures convex to straight, angulaperturate; 37 × 38.3 μm; triporate. Pores distinctly pouting, ora circular, about 2.6–3 μm thick; sexine thinner than nexine, granulate, grana closely placed, less than 1 μm in diameter.

Occurrence Palaeocene-Eocene, Cauvery Basin, Tamil Nadu (Venkatachala and Rawat 1972).

Genus *Triporopollenites* Pflug & Thomson in Thomson and Pflug (1953)

Type Species *Triporopollenites coryloides* Pflug & Thomson in Thomson and Pflug (1953)

Generic Description (Pflug & Thomson in Thomson and Pflug 1953, p. 82): Triporate pollen, amb more or less triangular, never circular; pores equatorial, at the corners, never with atrium, vestibulum or post-vestibulum; endexine and ectexine always tightly appressed; endoporus less than three times the exoporus; Isopolar; pore canal less than 0.3 μm, cylindrical, never beak-shaped, no inter-loculum, solutions-wedge, oculus, paravestibulum, endannulu; occasionally with annulus or labrum.

Species Recorded from India

Triporopollenites meenkunnuensis Rao & Rajendran 1996 (Plate 26, Fig. 10)

Description (Rao and Rajendran 1996, p. 73): Pollen grains subtriangular in polar view. Size range 45–52 × 40–45 µm. Triporate, angulaperturate, pores up to 8 µm wide, surrounded by thick annulus. Exine 4 µm thick, sexine thicker than nexine, sexine 3 µm thick, intectate, perforated, nexine 1 µm thick, smooth. Distal surface showing distinct reticulate ornamentation.

Occurrence Miocene, Quilon Formation, Kannur, Kerala (Rao and Rajendran 1996).

Triporopollenites robustus Kar & Jain 1981 (Plate 26, Fig. 11)

Description (Kar and Jain 1981, p. 123): Pollen grains triporate, pore margin thickened, 45–65 µm, exine up to 3 µm thick, laevigate.

Occurrence Neogene, around Kollam and Varkala, Kerala (Kar and Jain 1981).

Triporopollenites vimalii Sah & Dutta 1966 (Plate 26, Fig. 12)

Description (Sah and Dutta 1966, p. 82): Size range 26–30 µm; amb triangular, sides straight to convex; triorate, ora circular, slightly aspidate, 3–4 µm in diameter; exine up to 1.5 µm, undifferentiated, nexine thickened at the pore region; ornamentation obscure to faintly granulate.

Occurrence Early Eocene, Cherra Formation, South Shillong Plateau, Meghalaya (Sah and Dutta 1966); Palaeocene, Cherra Formation, South Shillong Plateau, Meghalaya (Dutta and Sah 1970).

Genus *Trisyncolpites* Kar 1979

Type Species *Trisyncolpites ramanujamii* Kar 1979

Generic Description (Kar 1979, p. 28): Pollen grains trisynmargocolporate, margocolpi broad, thickened, providing the appearance of triradiate ridge in equatorial view. Exine psilate-baculate, sometimes retipilate-retibaculate, nexine very much thicker than sexine, infrapunctate.

Species Recorded from India

Trisyncolpites ramanujamii Kar 1979 (Plate 12, Fig. 6)

Description (Kar 1979, p. 29): Pollen grains subcircular in polar and elliptical in equatorial view, 51–87 μm. Trisynmargocolporate, margocolpi broad, thickened, united to provide a pseudotriradiate ridge-like structure in equatorial view. Exine pilate-baculate, in some specimen retipilate-retibaculate, nexine almost double than sexine, infrapunctate, pores lalongate.

Occurrence Maniyara Fort Formation, Sarangwara, Kutch, Gujarat (Kar 1979).

Genus *Tuberculozonisulcites* Kar & Sharma 2001

Type Species *Tuberculozonisulcites retibaculatus* Kar & Sharma 2001

Generic Description (Kar and Sharma 2001, p. 132): Pollen grains subcircular in proximo-distally flattened specimens, subtriangular-triangular in folded ones, zonisulcate, tuberculate, sulcus at equatorial margin, exine 2–4 μm thick, retibaculate.

Species Recorded from India

Tuberculozonisulcites retibaculatus Kar & Sharma 2001 (Plate 7, Fig. 15)

Description (Kar and Sharma 2001, p. 132): Pollen grains generally found in form of folded specimens, size range of folded specimens 54–70 × 45–54 μm, size of proximo-distally flattened specimens 64–50 μm, zonisulcate, sulcus more or less parallel to the margin, dividing the grain into two unequal halves, tubercles confined to equatorial margin in proximo-distally flattened specimens. In case of folded specimens three tubercles present at apices, fourth one situated in the middle, tubercles rounded at apices, tubercles 6–8 μm long, 5–11 μm broad, exine 2–4 μm thick, retibaculate, bacula closely placed, forming false reticulation in surface view.

Occurrence Late Palaeocene-Early Eocene, Palana Formation, Bikaner, Rajasthan (Kar and Sharma 2001).

U

Genus *Umbelliferoipollenites* Venkatachala & Kar 1969

Type Species *Umbelliferoipollenites ovatus* Venkatachala & Kar 1969

Generic Description (Venkatachala and Kar 1969, p. 164): Pollen grains spindle-oval in shape, generally constricted in the middle. Dicolporate. Exine 2–3 μm thick, laevigate and finely intramicroreticulate.

Emended generic description (Samant and Phadtare 1997, p. 64): Pollen grains generally subprolate to perprolate, broad along the poles, often constricted in the middle, usually bi- to tricolporate; colpi distinct to indistinct; ora endexinous, often indistinct, circular to lalongate with uneven nexinal boundary; exine about 2–3 μm thick, generally psilate to variously sculptured. SEM observations show colpi distinctly to indistinctly fossaperturate; exine generally psilate to punctate in polar areas, and distinctly sculptured on rest of the pollen surface.

Botanical Affinity Apiaceae.

Species Recorded from India

Umbelliferoipollenites constrictus Venkatachala & Kar 1969 (Plate 8, Fig. 6)

Description (Venkatachala and Kar 1969, p. 165): Pollen grains with equally rounded lateral ends and marked constriction in middle, 25–30 × 8–14 μm. Two colporate, colpi extending more than three-fourth the radius along longitudinal axis. Pores indistinct. Exine laevigate and intrastructured.

Occurrence Eocene, Laki Series, Matanomadh, Kutch, Gujarat (Venkatachala and Kar 1969; Sah and Kar 1970).

Umbelliferoipollenites ovatus Venkatachala & Kar 1969 (Plate 8, Fig. 7)

Description (Venkatachala and Kar 1969, p. 165): Pollen grains oval with equally broad lateral ends, 25–35 × 10–18 μm. Two colporate, colpi extending more than three fourth of the pollen grain longitudinally. Pore distinct, lolongate. Exine 2–3 μm thick, laevigate and very finely intra-microreticulate.

Occurrence Eocene, Laki Series, Matanomadh, Kutch, Gujarat (Venkatachala and Kar 1969; Sah and Kar 1970).

Umbelliferoipollenites typicus Samant & Phadtare 1997 (Plate 8, Fig. 8)

Description (Samant and Phadtare 1997, p. 65): Pollen grains oval in shape, 25–36 × 10–18 μm in size, tricolporate; colpi medium to brevicolpate; ora endexinous, circular to lalongate; exine about 2 μm thick, tectate, columellate, sexine thicker than nexine, polar area almost psilate, rest of the pollen finely reticulate-punctate to irregularly verrucate.

Occurrence Early Eocene, Tarkeshwar Formation, Rajpardi, Cambay Basin, Gujarat (Samant and Phadtare 1997); Early Eocene, Kharsalia Clay Formation, Bhavnagar, Cambay Basin, Gujarat (Samant 2000).

V

Genus *Varispinitriporites* Kar 1985

Type Species *Varispinitriporites ratariaensis* (Kar and Saxena 1981) Kar 1985

Generic Description (Kar 1985, p. 125): Pollen grains subcircular-circular. Triporate, pore margin thickened, exine ornamented with two kinds of spines, bigger ones sparsely placed, smaller ones closely placed.

Species Recorded from India

Varispinitriporites ratariaensis (Kar and Saxena 1981) Kar 1985 (Plate 26, Fig. 13)

Description (Kar and Saxena 1981, p. 114): Pollen grains subcircular, 55–62 × 50–60 μm, triporate, pore margin thickened. Exine up to 2 μm thick, baculate, interbacular space granulose.

Restated Description (Kar 1985, p. 126): Pollen grains subcircular, 55–62 × 50–60 μm. Triporate, pore distinct, 8–15 μm, equally spaced, margin thick, spinose, spines of two types, bigger ones 3–6 μm long, 2–3 μm broad, tips somewhat pointed, spines placed 8–12 μm apart, smaller spines 2–3 μm long, about 1 μm broad, very closely placed.

Occurrence Eocene, subsurface sediments, Rataria, Kutch, Gujarat (Kar and Saxana 1981; Kar 1985).

Remarks *Semitectotriporites ratariaensis* Kar and Saxena (1981) was transferred to *Varispinitriporites* by Kar (1985).

Genus *Verrucolporites* Sah & Kar 1970

Type Species *Verrucolporites verrucus* Sah & Kar 1970

Generic Description (Sah and Kar 1970, p. 132): Pollen grain oval-elliptical in equatorial and subcircular in polar view. 3-colporate, colpi long, pores generally indistinct. Exine thick, verrucose.

Species Recorded from India

Verrucolporites verrucus Sah & Kar 1970 (Plate 20, Fig. 7)

Description (Sah and Kar 1970, p. 132): Pollen grains oval-elliptical in equatorial and subcircular in polar view. 30–50 × 25–44 μm, 3-colporate. Colpi long, pore indistinct, Exine strongly verrucose.

Occurrence Early Eocene, Laki Series, Kutch, Gujarat (Sah and Kar 1970); Early Eocene, Palana lignite, Palana, Bikaner, Rajasthan (Sah and Kar 1974); Palaeocene Tura Formation, Nongwal Bibra, Garo Hills, Meghalaya (Singh 1977); Early-Middle Eocene, Naredi and Harudi formations, Kutch, Gujarat (Kar 1978); Palaeocene, Matanomadh Formation, Matanomadh, Kutch, Gujarat (Saxena 1979).

W

Genus *Warkallipollenites* Ramanujam & Rao in Thanikaimoni et al. 1984

Type Species *Warkallipollenites erdtmanii* Ramanujam & Rao in Thanikaimoni et al. 1984

Generic Description (Ramanujam and Rao in Thanikaimoni et al. 1984, p. 86): Pollen grains Isopolar, circular to trilobate in polar view, tricolpate, colpus long. Sexine densely clavate, clavae surmounted by 3–4 spinules; nexine thinner than sexine.

Botanical Affinity and Ecology *Aegialitis rotundifilia* (Plumbaginaceae); mangrove shrub.

Species Recorded from India

Warkallipollenites erdtmanii Ramanujam & Rao in Thanikaimoni et al. (1984) (Plate 11, Fig. 9)

Description (Ramanujam and Rao in Thanikaimoni et al. 1984, p. 86): Pollen grains Isopolar, circular to trilobed in polar view, equatorial dimension 40–52 μm; tricolpate, colpus long, widened at equator, margin thin, border of colpus appearing fringed with closely arranged sculptural elements. Exine intectate, sexine densely clavate, clavae 3 × 1 μm, surmounted by 3–4 spinules.

Occurrence Miocene, Warkalli Beds, Kerala (Ramanujam and Rao in Thanikaimoni et al. 1984; Ramanujam 1987).

Warkallipollenites thanikaimonii Ramanujam 1987 (Plate 11, Fig. 10)

Description (Ramanujam 1987, p. 30): Pollen grains isopolar, amb more or less subcircular to three-lobed, sides convex, 45–55 μm, tricolpate, longicolpate, colpi gaping at equatorial margin, colpus margin ragged, fringed with sculptural elements; exine up to 5 μm thick (including sculpture), crassisexinous, intectate, prominently clavate-baculate, processes up to 4 μm long, densely placed in a distinct radiating pattern from polar region as seen in polar region, radial rows of bacules simple or forked, heads of processes generally pentagonal in surface view.

Occurrence Miocene, Warkalli Beds, Kerala (Ramanujam 1987).

References

Ambwani K, Bande MB, Prakash U (1981) Pollen grains of *Ctenolophonidites* from the Neyveli lignites of South India. Palaeobotanist 27(1):100–106

Ambwani K, Singh RS (1996) *Clavadiporopollenites raneriensis* gen. et sp. nov. from the Tertiary sediments of Bikaner District, Rajasthan, India. Palaeobotanist 43(3):139–142

Baksi SK (1962) Palynological investigation of Simsang River Tertiaries, South Shillong Front, Assam. Bull Geol Mining Metal Soc India 26:1–22

Baksi SK, Deb U (1980) Palynostratigraphic zonation of the Late Cretaceous-Palaeogene sequence of Bengal Basin. Geophytology 10(2):199–224

Baksi SK, Deb U, Siddhanta BK (1979) On *Crotonipollis*—A new genus from the Palaeocene-Eocene of India. Ind J Earth Sci 6(2):232–236

Baksi SK, Venkatachala BS (1970) *Meyeripollis*, a new genus from the Tertiary sediments of Assam. J Geol Soc India 11(1):81–83

Biswas B (1962) Stratigraphy of the Mahadeo, Langpar, Cherra and Tura formations, Assam, India. Bull Geol Mining Metal Soc India 25:1–48

Clarke RT, Frederiksen NO (1968) Some new sporomorphs from the Upper Tertiary of Nigeria. Grana Palynologica 8:210–224

Cookson IC (1947) Plant microfossils from the lignites of Kerguelen Archipelago. B.A.N.Z. Antarctic research expedition, 1929–31. Rep Ser A2(8):127–142

Cookson IC (1950) Fossil pollen grains of Proteaceous type from Tertiary deposits in Australia. Aust J Sci Res Ser B Biol Sci 3(2):166–177

Cookson IC, Pike KM (1954) Some dicotyledonous pollen types from Cainozoic deposits in the Australian region. Aust J Bot 2(2):197–219

Couper RA (1953) Upper Mesozoic and Cainozoic spores and pollen grains from New Zealand. New Zealand Geol Surv Palaeontol Bull 22:1–77

Couper RA (1960) New Zealand Mesozoic and Cainozoic plant microfossils. New Zealand Geol Surv Palaeontol Bull 32:1–87

Dutta SK, Sah SCD (1970) Palynostratigraphy of the Tertiary sedimentary formations of Assam-5. Stratigraphy and palynology of South Shillong Plateau. *Palaeontographica.* Abteilung B 131 (1–4) 1–72

Erdtman G (1947) Suggestions for the classification of Fossil and recent pollen grains and spores. Svensk Botanisk Tidskrift Utgifven af Svenska Botaniska Foreningen. Stockholm 41(1) 104–114

Gonzalez-Guzman AE (1967) A palynological study on the upper Los Cuervos and Mirador formations (Lower and Middle Eocene, Tibu area, Columbia. Thesis, vol 68, E.J. Brill, Leiden, pp 1-68.

Jain KP, Kar RK, Sah SCD (1973) A palynological assemblage from Barmer, Rajasthan. Geophytology 3(2):150–165
Jansonius J, Hills LV (1976) Genera file of fossil spores. Special Publication, Department of Geology, University of Calgary, Canada, pp 1–3287
Jansonius J, Hills LV (1981) Genera file of fossil spores-suppliment. Special Publication, Department of Geology, University of Calgary, Canada, pp 3801–3932
Jansonius J, Hills LV (1987) Genera file of fossil spores-suppliment. Special Publication, Department of Geology, University of Calgary, Canada, pp 4361–4575
Jargen DM, Dettmann ME (1989) Taxonomic revision of *Tricolpites reticulates* Cookson ex Couper 1953 with notes on biogeography of *Gunnera* L. Pollen Spores 31:97–112
Kar RK (1978) Palynostratigraphy of the Naredi (Lower Eocene) and the Harudi (Middle Eocene) formations in the district of Kutch, India. Palaeobotanist 25:161–178
Kar RK (1979) Palynological fossils from the Oligocene sediments and their biostratigraphy in the district of Kutch, western India. Palaeobotanist 26(1):16–49
Kar RK (1985) The fossil floras of Kachchh-IV. Tertiary palynostratigraphy. Palaeobotanist 34:1–280
Kar RK (1992) Occurrence of *Dipterocarpus* type of pollen from the Miocene sediments of Kerala, South India. J Palynol 28:79–85
Kar RK (1995) Diporocolpis: A new type of aperture from the Early Eocene sediments of Rajasthan, India. The Palaeobotanist 42(3):380–386
Kar RK (1996) On the Indian origin of *Ocimum* (Lamiaceae): a palynological approach. Palaeobotanist 43(3):43–50
Kar RK, Bhattacharya M (1992) Palynology of Rajpardi lignite, Cambay Basin and Gujra Dam and Akri lignite, Kutch Basin. The Palaeobotanist 39(2):250–263
Kar RK, Jain KP (1981) Palynology of Neogene sediments around Quilon and Varkala, Kerala coast, South India-2. Spores and pollen grains. The Palaeobotanist 27(2):113–131
Kar RK, Kumar M (1986) Palaeocene palynostratigraphy of Meghalaya. Pollen Spores 28:177–217
Kar RK, Kumar M (1987) *Neocouperipollis*—A new name for *Couperipollis* Venkatachala and Kar. Palaeobotanist 35(2):171–174
Kar RK, Mandal J, Sarkar S, Kumar M (1994) *Pilatrisyncolpites triangulatus* gen. et sp. nov. from the Oligocene of Upper Assam. Geophytology 23(2):287–289
Kar RK, Saxena RK (1981) Palynological investigation of a bore core near Rataria, southern Kutch, Gujarat. Geophytology 11(2):103–124
Kar RK, Sharma P (2001) Palynostratigraphy of Late Palaeocene and Early Eocene sediments of Rajasthan, India. *Palaeontographica*. Abteilung B 256(4–6):123–157
Krutzsch W (1962) Stratigraphisch bzw. botanisch wichtige neue Sporen und Pollenformen aus dem deutschen Tertiäir. Geologie 11:265–238
Krutzsch W (1970) Atlas der mittel-und jungtertiaren dispersen Sporen - und Polle –sowie der Microplankton – formen des nördlichen MitteleuropasVII. – Gustav Fischer Verlag, Jena, 1–175
Kulkarni AR, Phadtare NR, Dalvi N (1985) Monocotyledonous pollen grains from Ratnagiri lignite. In: Verghese TM (ed) Recent advances in pollen research. Allied Publishers Private Limited, pp 295–313
Kumar M (1996) Palynostratigraphy and palaeoecology of Early Eocene palynoflora of Rajpardi lignite, Bharuch District, Gujarat. The Palaeobotanist 43(3):110–121
Kumar A, Takahashi K (1991) Palynology of the Tertiary sediments of southern Assam, India. Bull Faculty Lib Arts Nagasaki Univ (Nat Sci) 31(2):515–659
Mandal J (1987) Palynological study of Sutunga coal seam, Jaintia Hills, Meghalaya. Palaeobotanist 35(2):196–199
Mandal J (1997) Palynofossils from the Tertiary (Barail Group) of Nagaland: palaeoecological interpretation and age. Palaeobotanist 45:98–108
Mandal J (2000) Occurrence of *Pilatrisyncolpites* from the Early Eocene of Kutch Basin and its implication. Geosci J 21(1):69–72

Mandal J, Chandra A, Kar RK (1994) Palynofossils from the Kadamtala coal, Middle Andaman, India. Geophytology 23(2):209–214

Mandal J, Guleria JS (2006) Palynology of Vastan lignite (Surat District), Gujarat: its age, palaeoecology and depositional environment. Palaeobotanist 55(1–3):51–66

Mandal J, Kumar M (2000) Stratigraphical significance of some angiosperm pollen from Tinali Oilfield, Upper Assam, India. Palaeobotanist 49:197–207

Mandal J, Rao MR (2001) Taxonomic revision of tricolpate pollen from Indian Tertiary. Palaeobotanist 50(2–3):341–368

Mandaokar BD (2007) Depositional environment and age of the Bhuban Formation of Chawngte area, Mizoram, India. Gondwana Geol Mag 22(1):37–46

Mathur YK, Chopra AS (1987) Palynofossils from the Cenozoic subsurface sediments of the Bengal Basin, India. Geosci J 1(8):109–152

Mathur YK, Jain AK (1980) Palynology and age of the Dras Volcanics near Shergol, Ladakh, Jammu and Kashmir, India. Geosc J 1(1):55–74

Mathur YK, Mathur K (1969) Studies in the fossil flora of Kutch (India)-3. On the palaeopalynoflora in the Pliocene sediments of Naera-Baraia area, Kutch. Bull Geol Mining Metal Soc India 42:1–12

Mathur YK, Mathur K (1980) Barail (Laisong) palynofossils and Late Oligocene nannofossils from the Andaman Island, India. Geosc J 1(2):51–65

Mehrotra NC (1983) Palynology of Mikir Formation in the type area. Geosci J 4(1):1–34

Misra BK, Singh A, Ramanujam CGK (1996) Trilatiporate pollen from Indian Palaeogene and Neogene sequences: evolution, migration and continental drift. Rev Palaeobot Palynol 91(1–4):331–352

Morley RJ (1982) Fossil pollen attributable to *Alangium* Lamarck (Alangiaceae) from the Tertiary of Malesia. Rev Palaeobot Palaeopalynol 36:65–94

Muller J (1968) Palynology of the Pedawan and Plateau sandstone formations (Cretaceous-Eocene) in Sarawak, Malaysia. Micropaleontology 14:1–37

Nagy E (1962) New pollen species from the Lower Miocene of the Bakony Mountain (Varpalota) of Hungary. Acta Bot 8(1–2):153–163

Nandi B, Sharma R (1984) Palynology and biostratigraphy of the Boldamgiri Formation, Garo Hills, Meghalaya. In: Sharma AK et al (eds) Proceedings of the symposium on evolutionary botony and biostratigraphy Calcutta 1979, A.K. Ghosh Commemoration volume, current trends in life sciences vol 10, pp 565–580

Naskar P, Baksi SK (1978) Palynological investigation of Akli lignite, Rajasthan, India. Palaeobotanist 25:314–329

Naumova SN (1939) The spores and pollen of the coals of U.S.S.R. In: Proceedings of the 17th international geological congress, Gonti, Moscow, 1937, vol 1, pp 353–364

Navale GKB, Misra BK (1979) Some new pollen grains from Neyveli lignite, Tamil Nadu, India. Geophytology 8(2):226–239

Nichols DJ, Tate Ames H, Traverse A (1973) On *Arecipites* Wodehouse, *Monocolpopollenites* Thomson & Pflug, and the species "*Monocolpopollenites tranquilus*". Taxon 22:241–256

Pflanzel G (1956) Das Alter Braunkohlen des Meissners, der Floze 2 and 3 des Hirschberges und eines benachbarten Kohlenlagers bei Laudenbach. Notizbl. hess Landesamt. Bodenforsch. Wiesbaden 84:232–244

Phadtare NR, Kulkarni AR (1984) Affinity of the genus *Quilonipollenites* with the Malaysian palm Eugeissona Griffith. Pollen et Spores 26:217–226

Phadtare NR, Thakur B (1990) Fossil pollen of *Alangium* from the Eocene lignite of Gujarat, India, with comments on its stratigraphic antiquity. Rev Palaeobot Palynol 63(3–4):281–297

Pierce RL (1961) Lower Upper Cretaceous plant microfossils from Minnesota. Bull Geol Surv Minnessota Univ 42:1–86

Potonié R (1931) Zur Microskopie der Braunkohlem. Tertiären Spore und Blütenstaubformrn IV. Braunkohle 30(27):554–556

Potonié R (1934) Zur Mikrobotanik des eocanen Humodile des Geiseltals (in Zur Mikrobotanik del' Kohlen and ihrer Verwandten). Arb Inst Palaeobot 4:25–125

Potonié R (1951) Revision stratigraphisch wichtiger Sporomorphen des mitteleuropaischen Tertiars. Palaeontographica B 91:131–151
Potonié R (1958) Synopsis der Gattungen der Sporae dispersae II. Teil. Sporites (Nachtrage), Saccites, Aletes, Praecolapates, Polyplicates, Monocolpates. Beihefte zum Geologischen Jahrbuch 31:1–114
Potonié R (1966) Synopsis der Gattungen der sporae dispersae IV Teil: Nachtrage zu allen Gruppen (Turmae). Beihefte zum Geologischen Jahrbuch 72:1–244
Potonié R (1970) Synopsis der Gattungen der sporae dispersae. V Teil. Nachtrage zu allen gruppen (Turmae). Beihefte zum Geologischen Jahrbuch 87:1–172
Potonié R, Sah SCD (1960) Sporae dispersae of the lignites from Cannanore beach of the Malabar Coast of India. Palaeobotanist 7(2):121–135
Ramanujam CGK (1966) Palynology of the Miocene lignite from South Arcot District, Madras, India. Pollen et Spores 8(1):149–203
Ramanujam CGK (1982) Tertiary palynology and palynostratigraphy of southern India. In Cenozoic Stratigraphy and Palynology in India, vol 1. Palaeontological Society of India, Special Publication, pp 57–64
Ramanujam CGK (1987) Palynology of the Neogene Warkalli Beds of Kerala State in South India. J Palaeontol Soc India 32:26–46
Ramanujam CGK, Rao KP (1973) A study of the pollen grains of *Ctenolophonidites* from the Warkalli deposits of South India with a note on the geological history of *Ctenolophon*. Palaeobotanist 20(2):210–215
Ramanujam CGK, Rao KP (1977) A palynological approach to the study of Warkalli deposits of Kerala in South India. Geophytology 7(2):160–164
Ramanujam CGK, Reddy PR, Sarma PS (1985) Additions to the palynoflora of Neyveli lignite, Tamil Nadu. J Palaeontol Soc India 30:49–53
Ramanujam CGK, Reddy PR, Sarma PS (1988) *Marginipollis* from the clay and lignite of South Arcot District, Tamil Nadu. Spec Publ Geol Surv India 11(2):271–276
Rao GM, Reddy PR, Ramanujam CGK (1993) Miocene spore and pollen complex from a bore well at Thakkazhi in Alleppey District, Kerala. Gondwana Geol Mag 4–5:80–86
Rao KP, Ramanujam CGK (1975) A palynological approach to the study of Quilon Beds of Kerala State in South India. Curr Sci 44:730–732
Rao KP, Ramanujam CGK (1978) Palynology of the Neogene Quilon Beds of Kerala State in South India-1. Spores of pteridophytes and pollen of monocotyledons. Palaeobotanist 25:397–427
Rao KP, Ramanujam CGK (1982). Palynology of the Quilon Beds of Kerala State in South India-II. Pollen of dicotyledons and discussion. Palaeobotanist 30(1):68–100
Rao MR (1990) Palynological investigation of Arthungal Borehole, Alleppey District, Kerala. Palaeobotanist 38:243–255
Rao MR (1995) Palynostratigraphic zonation and correlation of the Eocene-Early Miocene sequence in Alleppey District, Kerala, India. Rev Palaeobot Palynol 86(3–4):325–348
Rao MR, Rajendran CP (1996) Palynological investigation of Tertiary lignite and associated sediments from Cannanore, Kerala Basin, India. Palaeobotanist 43(2):63–82
Rao MR, Saxena RK, Singh HP (1985) Palynology of the Barail (Oligocene) and Surma (Lower Miocene) sediments exposed along Sonapur-Badarpur Road Section, Jaintia Hills (Meghalaya) and Cachar (Assam)—Part V. Angiospermous pollen grains. Geophytology 15(1):7–23
Rawat MS, Mukherjee JS, Venkatachala BS (1977) Palynology of the Kadi Formation, Cambay Basin, India. In: Venkatachala BS, Sastri VV (eds) Proceedings of the 4th Colloquium on Indian Micropalaeontology and Stratigraphy, Dehradun, 1974–75. Institute of Petroleum Exploration, Oil and Natural Gas Commission, Dehradun, pp 179–192
Sah SCD (1967) Palynology of an Upper Neogene profile from Rusizi Valley (Burundi). Musée Royal de l' Afrique Centrale Tervuren, Belgique Annales- Séries in 8°. Sci Géol 57:1–173
Sah SCD, Dutta SK (1966) Palynostratigraphy of the sedimentary formations of Assam-1. Stratigraphical position of the Cherra Formation. Palaeobotanist 15(1–2), pp 72–86

Sah SCD, Dutta SK (1968) Palynostratigraphy of the Tertiary sedimentary formations of Assam-2. Stratigraphic significance of spores and pollen in the Tertiary succession of Assam. Palaeobotanist 16:177–195

Sah SCD, Kar RK (1970) Palynology of the Laki sediments in Kutch-3. Pollen from the boreholes around Jhulrai, Baranda and Panandhro. Palaeobotanist 18(2):127–142

Sah SCD, Kar RK (1974) Palynology of the Tertiary sediments of Palana, Rajasthan. Palaeobotanist 21(2):163–188

Sah SCD, Singh RY (1974) Palynological biostratigraphy of the Tura Formation in the type area. In: Sah SCD, Cross AT (eds) Symposium on Stratigraphical Palynology Lucknow 1971. Special Publication 3, Birbal Sahni Institute of Palaeobotany, Lucknow, pp 76–98

Salujha SK, Kindra GS, Rehman K (1972) Palynology of the South Shillong front-part I: the Palaeogene of Garo Hills. In: Ghosh AK. et al (eds) Proceedings of the symposium on Paleopalynology and Indian Stratigraphy Calcutta 1971. Botany Department, Calcutta University, Kolkata, pp 265–291

Samant B (2000) Palynostratigraphy and age of the Bhavnagar lignite, Gujarat, India. Palaeobotanist 49(1):101–118

Samant B, Phadtare NR (1997) Stratigraphic palynoflora of the Early Eocene Rajpardi lignite, Gujarat and the lower age limit of the Tarkeshwar Formation of South Cambay Basin, India. *Palaeontographica*. Abteilung B 245(1–6):1–108

Samant B, Tapaswi PM (2001) Palynology of the Early Eocene Surat lignite deposits of Gujarat, India. J Palaeontol Soc India 46:121–132

Sarkar S, Prasad V (2002) *Ocimum* pollen grains from the Subathu Formation (Late Ypresian) of Shimla Hills, Himachal Pradesh, India. Palaeobotanist 51(1–3):165–167

Sarkar S, Singh HP (1988) Palynological investigation of the Subathu Formation (Eocene) in the Banethi-Bagthan area of Himachal Pradesh. *Palaeontographica*. Abteilung B 209:29–109

Saxena RK (1979) Palynology of the Matanomadh Formation in type area, north-western Kutch, India (Part 2). Systematic description of gymnospermous and angiospermous pollen grains. Palaeobotanist 26(2):130–143

Saxena RK (1992) Replacement names for later homonyms of ten Indian Tertiary palynofossils. Taxon 41(3):532–533

Saxena RK (2010) *Monocolpopollenites* Pflug and Thomson (monosulcate arecaceous pollen) from India. Geophytology 38(1–2):105–120

Saxena RK (2011) Validation of the names of two genera and thirtynine species of fossil spores and pollen recorded from the Tertiary sediments of India. Taxon 60(3):860–865

Saxena RK, Khare S (1996) *Gemmatriporopollis*, a new pollen genus from Neyveli lignite mines and Jayamkondacholapuram Well-12 in Tamil Nadu, India. Geophytology 26(1):129–131

Saxena RK, Khare S (2004) Palynological investigation of the Jayamkondacholapuram Well 12, Tiruchirapalli District, Tamil Nadu, India. Geophytology 34(1–2):73–93

Saxena RK, Khare S, Misra NK (1991) *Echimonoporopollis*, a new pollen genus from the Neyveli Formation of Jayamkondacholapuram Well-12, Tiruchirapalli District, Tamil Nadu. Palaeobotanist 39(1):46–49

Saxena RK, Misra BK (1987) *Spinulotricolporites*, a new pollen genus from the Matanomadh Formation (Palaeocene), Kutch, India. ISG Bull 3:19–22

Saxena RK, Misra NK (1990) Palynological investigation of the Ratnagiri Beds of Sindhu Durg District, Maharashtra. Palaeobotanist 38:263–276

Saxena RK, Singh HP (1981) *Pinjoriapollis*, a new fossil pollen from the Pinjor Formation (Upper Siwalik) exposed near Chandigarh. Curr Sci 50(9):418–419

Saxena RK, Tripathi SKM, Prasad V (1996) Palynofloral investigation of the Tura Formation (Palaeocene) in Nongwal Bibra area, East Garo Hills, Meghalaya. Geophytology 26(1):19–31

Saxena RK, Trivedi GK (2009) Palynological investigation of the Kopili Formation (Late Eocene) in North Cachar Hills, Assam, India. Acta Palaeobot 49(2):253–277

Siddhanta BK (1986) The age of Neyveli lignite with reference to stratigraphy and palynology. Ind Miner 40:61–82

Singh A, Misra BK (1991a) New colporate pollen taxa from Neyveli lignite, South India. Rev Palaeobot Palynol 67(1–2):59–74
Singh A, Misra BK (1991b) Revision of some Tertiary pollen genera and species. Rev Palaeobot Palynol 67(3–4):205–215
Singh A, Misra BK (1991c) A new spinose monosulcate genus *Spinomonosulcites* and an emendation of spinose porate *Acanthotricolpites*. Rev Palaeobot Palynol 67(3–4):217–227
Singh A, Misra BK, Singh BD, Navale GKB (1992) The Neyveli lignite deposits (Cauvery Basin), India: organic composition, age and depositional pattern. Int J Coal Geol 21:45–97
Singh HP, Saxena RK (1981) Palynology of the Upper Siwalik sediments in Una District, Himachal Pradesh. Geophytology 11(2):173–181
Singh HP, Singh RY, Sah SCD (1976) Palynological correlation of the Palaeocene subsurface assemblages from Garo Hills, Meghalaya. J Palynol 11:43–64
Singh HP, Tripathi SKM (1986) Observations on some Tertiary zonisulcate pollen grains. Palaeobotanist 35(1):48–52
Singh RS (1990) Palynology of Langrin Coalfield, South Shillong Plateau, Meghalaya. Palaeobotanist 38:217–228
Singh RY (1975) Morphological study of the Retialetes complex from Indian Tertiaries. Geophytology 5(1):98–104
Singh RY (1977) Stratigraphy and palynology of the Tura Formation in the type area-Part II (Descriptive palynology). Palaeobotanist 23(3):189–205
Singh RY, Dogra NN, Vimal KP (1985) Palynology of the Barail sediments in the states of Assam and Meghalaya, India. J Palynol 21:28–55
Singh RY, Dogra NN, Modak SK (1986) Palynostratigraphy of the Oligocene sediments of Namdang River, Assam. J Palynol 22(2):105–124
Srivastava SK (1966) Upper Cretaceous microflora (Maestrichtian), Alberta, Canada. Pollen Spores 8:497–552
Srivastava SK (1969) Some angiosperm pollen from the Edmonton Formation (Maestrichtian), Alberta, Canada. In: Santapu H et al (eds) J. Botanical Society of Bengal, Calcutta, Sen Memorial Volume, pp 47–67
Thanikaimoni G, Caratini C, Venkatachala BS, Ramanujam CGK Kar RK (1984) Selected Tertiary angiospermous pollen from India and their relationship with African Tertiary pollen. Institut Francais de Pondichéry, vol 19. Travaux de la Section Scientfique et Technique, 92p
Thiergart F (1937) Die pollen flora der Niedorlausitzer Braunkohle, besondern in profil der Gurube Marga bei Senfftenberg. Jb. preuss. Geol. Landest. 58
Thomson PW, Pflug H (1953) Pollen und sporen des mitteleuropäiaschen Rertiärs. *Palaeontographica* Abt. 94(B):1–138
Traverse A (1955) Polen analysis of Brandon lignite of Vermont. US Breau Mines Rep Invest 51:1–107
Tripathi SKM (1994) New angiosperm pollen from subsurface Early Palaeogene sediments of Barmer District, Rajasthan, India. Palaeobotanist 42(1):61–65
Tripathi SKM (1995) Palynology of subsurface Palaeocene-Eocene sediments near Kapurdi, Barmer District, Rajasthan, India. The Palaeobotanist 43(1):45–53
Tripathi SKM, Saxena RK, Prasad V (2000) Palynological investigation of the Tura Formation (Early Eocene) exposed along Tura-Dalu Road, West Garo Hills, Meghalaya, India. Palaeobotanist 49(2):239–251
Tripathi SKM, Singh HP (1984) Two new pollen genera from the Lower Tertiary sediments of Meghalaya. Palaeobotanist 32(2):153–157
Tripathi SKM, Singh HP (1985) Palynology of the Jaintia Group (Palaeocene-Eocene) exposed along Jowai-Sonapur Road, Meghalaya, India–Part I. Systematic palynology. Geophytology 15 (2):164–187
Tripathi SKM, Singh UK, Sisodia MS (2003) Palynological investigation and environmental interpretation on Akli Formation (Late Palaeocene) from Barmer Basin, Rajasthan, India. Palaeobotanist 52(1–3):87–95

Tripathi SKM, Srivastava D (2010) Palynological investigations, facies analysis and palaeoenvironmental interpretations from Late Palaeocene to Early Eocene lignites and associated sediments of Barmer, western India. Palaeobotanist 59:1–32

Trivedi GK, Saxena RK (2000) Palynofloral investigation of the Kopili Formation (Late Eocene) exposed near Umrongso in North Cachar Hills District, Assam, India. The Palaeobotanist 49 (2):269–280

van der Hammen T (1956) A palynological systematic nomenclature. Boletin Geologico Bogota 4 (2–3):63–101

Varma CP, Rawat MS (1963) A note on some diporate grains recovered from Tertiary horizons of India and their potential marker value. Grana Palynologica 4(1):130–139

Venkatachala BS (1973) Palynological evidence on the age of Cuddalore Sandstone. Geophytology 3(2):145–149

Venkatachala BS, Caratini C, Tissot C, Kar RK (1989) Palaeocene-Eocene marker pollen from India and tropical Africa. Palaeobotanist 37(1):1–25

Venkatachala BS, Kar RK (1968) Fossil pollen comparable to pollen of *Barringtonia* from the Laki sediments of Kutch. Pollen Spores 10(2):335–339

Venkatachala BS, Kar RK (1969) Palynology of the Tertiary sediments of Kutch-1. Spores and pollen from Borehole No. 14. Palaeobotanist 17(2):157–178

Venkatachala BS, Rawat MS (1972). Palynology of the Tertiary sediments in the Cauvery Basin-1. Palaeocene-Eocene palynoflora from the subsurface. In: Ghosh AK et al (eds)—Proceedings of the seminar on Paleopalynology and Indian Stratigraphy, Calcutta, 1971. Botany Department, Calcutta University, Kolkata, pp 292–335

Venkatachala BS, Rawat MS (1973) Palynology of the Tertiary sediments in the Cauvery Basin-2. Oligocene-Miocene palynoflora from the subsurface. Palaeobotanist 20(2):238–263

Venkatachala BS, Saxena RK, Singh HP, Kar RK, Tripathi SKM, Kumar M, Sarkar S, Mandal J, Rao MR, Singh RS, Mandaokar BD, Ambwani K (1996) Indian Tertiary angiosperm pollen: a critical assessment. Palaeobotanist 43(2):106–138

Von Hoeken-Klinkenberg PMJ (1964) A palynological investigation of some Upper Cretaceous sediments in Nigeria. Pollen Spores 6:209–231

Von Hoeken-Klinkenberg PMJ (1966) Maestrichtian Palaeocene and Eocene pollen and spores from Nigeria. Leidse Geol. Mededel 38:37–48

Wodehouse RP (1933) Tertiary pollen. II. The oil shale of the Eocene Green River Formation. Bull Torrey Bot Club 60:479–524

Chapter 3
Discussion

3.1 Palynofloras and Palynological Biostratigraphy from Indian Palaeogene-Neogene Sediments

Palynological information from Palaeogene-Neogene sediments of India is available from five distinct areas. These are, (1) Northern India (Jammu and Kashmir, Himachal Pradesh, Punjab, Haryana and Uttar Pradesh); (2) Eastern and north-eastern India (Arunachal Pradesh, Assam, Bihar, Jharkhand, Meghalaya, Mizoram, Nagaland, Tripura and West Bengal); (3) Southern India (Andhra Pradesh, Karnataka, Kerala and Tamil Nadu); (4) Western India (Goa, Gujarat, Maharashtra, and Rajasthan); and (5) Central India (Madhya Pradesh). Additionally, some information is also available from offshore basins, e.g. Bombay Offshore, Krishna-Godavari Basin, Cauvery Basin, and Andaman and Nicobar Islands. Palynological information available from these regions along with related aspects is briefly discussed below.

Northern India

Western Himalayas: Palaeogene-Neogene sediments exposed in northern India are found along the southern margin of Himalayas and range in age from Palaeocene to Middle Pleistocene (Fig. 3.1). These sediments are collectively known as Subathu-Murree-Siwalik. Subathu Formation represents the Early Palaeogene rocks in this region and is exposed in a narrow belt running along the foot-hills from Jammu in the northwest to central Nepal in the southeast. Early Palaeogene rocks of Himachal Pradesh are divided into Subathu, Dagshai and Kasauli formations. Good numbers of palynological papers have been published from Subathu Formation (Mathur 1963a, b, 1964; Salujha et al. 1969; Singh et al. 1978; Singh and Sarkar 1987a, b, 1992; Sarkar and Singh 1988; Sarkar 1991, 1997). Dinoflagellate cysts

© The Editor(s) (if applicable) and The Author(s), under exclusive license to Springer Nature Switzerland AG 2020

S. K. M. Tripathi, *Catalogue of Selected Angiosperm Pollen Grains from Palaeogene and Neogene Sediments of India*, Society of Earth Scientists Series,
https://doi.org/10.1007/978-3-030-42435-0_3

Fig. 3.1 Map of northern region of india. After Saxena and Tripathi (2012)

dominate palynological assemblages therefore, it has been inferred that deposition of Subathu Formation took place in shallow marine environment.

Tuffaceous shales of Dras Volcanics exposed near Shergol, southeast Ladakh yielded some angiospermic pollen. Based on these records, Mathur and Jain (1980) assigned Palaeocene age to this sequence. Bhandari et al. (1977) described palynofossils from the Ladakh Molasse Group (Palaeocene to Miocene) exposed in Wakka Chu near Kargil, Jammu and Kashmir. In Jammu region, the Subathu Formation is overlain by Murree sediments, which have been equated with Dharmsala Formation of Kangra Valley and Dagshai and Kasauli formations of Shimla Hills, Himachal Pradesh. Mathur and Venkatachala (1979), Dogra et al. (1985) and Berry et al. (1996a, b) reported occurrence of diversified palynofloras from Dharmsala Formation of Himachal Pradesh. Mathur (1984) reported variety of palynofossils from sediments of the Upper Murree of Jammu.

The Kasauli Formation (Early Miocene) has been palynologically worked out in Solan District (Singh et al. 1973; Singh and Khanna 1980; Kapoor et al. 1988; Singh and Sarkar 1994) and Sirmaur District (Singh and Sarkar 1984b). Diverse spore/pollen assemblages constitute the palynoflora from Kasauli Formation. The assemblage suggests a fluvio-deltaic environment of deposition in a fast sinking basin. The Kasauli assemblage indicates moist subtropical climate.

The Siwalik group (Middle Miocene to Middle Pleistocene) overlies the Dharmsala/Kasauli sediments. Sediments of this group are made up of thick molasse deposits and are exposed in outer Himalayan ranges extending from Potwar Plateau (Pakistan) in the west to Mishmi Hills, Arunachal Pradesh in the east. Considering the lithological characters and fauna, the Siwalik Group is divided into three subgroups, Lower, Middle and Upper, which are further divided into formations. Palynofloras from Siwalik Group have been recorded from Jammu (Mathur 1984), Himachal Pradesh (Nandi and Bandyopadhyay 1970; Nandi 1975, 1980; Mathur 1984; Singh and Saxena 1981; Rangarao et al. 1981; Saxena and Singh 1982a; Saxena and Bhattacharyya 1987; Phadtare et al. 1994; Sarkar and Singh 1994), Haryana (Singh et al. 1973; Saxena and Singh 1981, 1982b; Saxena et al. 1987) and Uttar Pradesh (Nandi 1972, 1980; Mathur 1984).

Palynological assemblages from Siwalik Group recorded by Venkatachala (1972), Ghosh (1977) and Mathur and Venkatachala (1979) deal with descriptions of the palynofloras and its palaeoecological interpretations. Nandi (1975), for the first time, used Siwalik palynofossils in biostratigraphy and divided the sequence of Jwalamukhi area of Himachal Pradesh into four zones. Zone-I comprises lower and partly middle part of Lower Siwalik, Zone-II is represented by rest of middle and upper parts of Lower Siwalik, Zone-III mainly covers the Middle Siwalik and Zone-IV is represented by uppermost part of Middle Siwalik and Upper Siwalik. Pteridophytic spores dominate the palynoflora from Jwalamukhi area whereas, gymnospermic pollen are less in number. In general, pteridophytic spores and angiospermic pollen constitute the Lower Siwalik palynoflora and that of Middle Siwalik is represented by the gymnospermic pollen, pteridophytic spores and angiospermic pollen. The lower part of the Upper Siwalik has yielded few palynofossils. Nandi (1980) traced lateral continuity of these four zones in six Siwalik sections: two in Punjab, three in Himachal Pradesh and one in Uttar Pradesh. Nandi (1975) did not recover palynofossils from Upper Siwalik however, palynological assemblages from these sediments were later recorded by Saxena and Singh (1980, 1981, 1982a, b), Singh and Saxena (1980, 1981), Saxena et al. (1984) and Saxena and Bhattacharyya (1987). Saxena and Singh (1982b) identified two distinct assemblages in the Upper Siwalik of Hoshiarpur-Una section of Punjab and Himachal Pradesh.

Based on relative frequency of ecologically significant palynotaxa, Phadtare et al. (1994) recognized four major stages of vegetation succession in the Upper Siwalik sequence of Haripur Khol area in Sirmaur District, Himachal Pradesh. These are: (A) dry grassland during pre-Tatrot time, (B) marshy grassland during

Lower Tatrot time, (C) predominantly ponding conditions during Upper Tatrot time, and (D) wet open and mixed flora during post-Tatrot time (Pinjor Formation).

Palynological studies suggest that during deposition of Kasauli Formation (Early Miocene) of northwest India enjoyed tropical to subtropical moist climate and environment of deposition ranged from brackish water to fluvio-deltaic. It is also inferred that deposition of lower part of Upper Dharmsala took place under shore-line whereas; during the later part, a coastal-fluvial condition prevailed. The climate remained tropical to subtropical with temperate elements coming from nearby upland areas indicating that the basin was protected by land (Saxena 2000a). Jammu area was occupied by mangrove vegetation during the deposition of Upper Murree sediments (Early Miocene) and the area experienced predominant fresh-water influx. Studies suggest that inland areas had streams, lakes, swamps and marshes having trees along with water and shade loving herbs and climbing ferns. During later period, with more influx of fresh water, mangrove swamps vanished and Savanna type of vegetation was established. The floral assemblage indicates a tropical climate.

Studies indicate that during the deposition of Lower Siwalik (Middle Miocene) flora remained tropical to subtropical. Grasses became prominent and widespread. During Middle Siwalik (Late Miocene-Early Pliocene) the flora was subtropical to temperate. Shrubs and trees were developed, grasses became widespread and upland areas supported temperate vegetation. The climate during Upper Siwalik (Plio-Pleistocene) became distinctly temperate and Coniferous plants and temperate angiosperms became more developed.

Eastern India

Eastern Himalayas (Arunachal Pradesh): Neogene sediments are well developed in Arunachal Pradesh (Fig. 3.2). However, some reports confirming Early Palaeogene deposits have also been published. Different workers under different names describe Neogene rocks of Arunachal Pradesh. Karunakaran and Rangarao (1979) divided these sediments into Dafla (Miocene), Subansiri (Late Miocene-Pliocene) and Kimin (Pleistocene) formations. Dutta and Singh (1980) described these sediments as Siwaliks and correlated them with sequences of Brahmaputra Valley of northeast India.

Palynological studies on Palaeogene-Neogene sequences of Arunachal Pradesh have been carried out by Jain and Dutta (1978), Dutta (1980), Dutta and Singh (1980), Prasad and Day (1986), Singh and Tripathi (1990) Tripathi and Singh (1992) and Mandaokar (2000b). Jain and Dutta (1978), for the first time, described Eocene palynofossils containing dinoflagellate cysts, spores and pollen grains from West Siang District, Arunachal Pradesh. Subsequently, Early Palaeogene palynofossils from the same district were reported by Prasad and Day (1986) and Tripathi and Singh (1992). Palynofossils of Oligocene age were described from the coal bearing sediments of Tikak Parbat Formation of Changlang District (Mandaokar

Fig. 3.2 Map of eastern region of india. After Saxena and Tripathi (2012)

2000b). Dutta and Singh (1980) published a palynofloral assemblage from Neogene sediments of Kameng District and divided them into three parts: Lower, Middle and Upper. Lower part contains reworked Permian and some Eocene palynofossils; Middle part is dominated by Late Palaeogene palynotaxa; and Upper part mainly contains Miocene taxa and few Permian and Eocene forms. It appears that either the material for these sediments was provided by both Permian and Eocene rocks or Permian rocks were re-deposited in Eocene times, which provided material for sediments containing both reworked Eocene and Permian elements.

Bengal Basin: Palynological data from Bengal Basin have been published by Deb (1970), Baksi (1972, 1973), Baksi and Deb (1980) Mathur and Chopra (1987) and Bera and Banerjee (1995). Based on palynological studies, Deb (1970) divided the sequence of Bodra Well-1 in southern part of Bengal Basin into three distinct palynozones ranging in age from Miocene to Quaternary. These authors compared the assemblages with those from Bengal, Assam and Meghalaya. Baksi (1972) studied eleven wells in the Bengal Basin and identified seven palynological zones representing the Upper Cretaceous to Pleistocene sediments. Palaeogene sediments of Bengal Basin are represented by Jalangi, Sylhet Limestone, Kopili, Burdwan and Memari (basal part) formations. These sequences are divided into four palynozones that are rich in pteridophytic spores and angiospermic pollen. Excepting the basal part, majority of sediments of Zone-5 represent the Memari Formation. This zone is rich in Palm pollen however, some dinoflagellate cysts also occur. The overlying Zone-6 represents Pandua/Malta formations (Late Miocene-Early Pliocene) and is dominated by conifer pollen and fern spores. The occurrence of conifer pollen and a few dinoflagellate cysts characterize zone-7, representing Debagram/Ranaghat formations. Palynological assemblages from the studied sequence indicate prevalence of warm and humid subtropical climate during Cretaceous time (Zone-1) and swampy conditions during deposition of Palaeocene-Early Eocene sediments representing Zone-2. Palynoflora of Zone-3 (Middle-Late Eocene) indicates humid rain forest vegetation with occasional brackish water conditions. Zone-4 (Oligocene) sediments are indicative of a regressive phase. Conifer pollen appeared in Zone-5 (Miocene) and became predominant in Zone-6 indicating cooling of climate and an elevated topography.

Baksi and Deb (1980) established eight palynological zones in subsurface sediments of Bengal Basin. This study is, in fact, elaboration of the work carried out earlier by Baksi (1973) but the biostratigraphic divisions proposed in this publication are limited to Cretaceous-Palaeogene only. This investigation proposes two zones for Palaeocene sediments, three zones in Eocene sediments and one zone in Oligocene sediments.

Northeastern India (Assam, Meghalaya, Mizoram, Nagaland and Tripura)

Detailed palynological studies have been carried out from northeastern part of India, particularly, Meghalaya and Assam (Fig. 3.2). Palaeogene-Neogene sediments of Meghalaya and Assam are developed in shelf and geosynclinal facies and are named differently in both of these two facies. In shelf facies, the oldest stratigraphic unit is named as Jaintia Group (Palaeocene-Eocene). It is best developed on the southern slopes of the Shillong Plateau and is divided into three formations: Therria Formation, Sylhet Limestone, and Kopili Formation. In geosynclinal facies lowermost Tertiary rocks are undivided and are named as Disang Group.

The Jaintia Group is conformably overlain by the Barail Group (Oligocene). This group is not divided in Shelf facies but in the geosynclinal facies it is divided into three formations and is differently named in Surma Valley, Upper Assam and Naga Hills. In the Surma Valley they are divided into Laisong, Jenam and Renji formations. In Upper Assam and Naga Hills, the geosynclinal sediments are divided into Naogaon, Baragolai and Tikak Parbat formations. Surma Group unconformably overlies the Barail Group and is divided into Bhuban and Bokabil formations in geosynclinal facies of Surma Valley but is undivided in shelf facies as well as in the geosynclinal facies of Upper Assam and Naga Hills. Tipam Group conformably overlies the Surma Group and is divided into Tipam Sandstone and Girujan Clay formations in both the facies. The Surma and Tipam groups are of Miocene age. Unconformably overlying the Tipam Group is the Dupitila Group which is named as Namsang Beds in the shelf facies and geosynclinal sediments of Upper Assam and Naga Hills. The Dupitila Group is of Mio-Pliocene age. This group is unconformably overlain by Dihing Group (Pliocene). In the shelf facies the group is named as Dhekiajuli Beds.

First palynological report from Meghalaya was published by Ghosh (1941). He recorded single bisaccate conifer pollen. Prof. Birbal Sahni initiated palynostratigraphic work in the area under a project sponsored by Burmah Oil Company in connection with an oil exploration programme. Sahni et al. (1947) published the results of this project. Baksi (1962) and Biswas (1962) conducted detailed palynostratigraphic studies on the Palaeogene sediments of Meghalaya. These were followed by more significant studies from Palaeogene-Neogene sequences of northeast India (Sah and Dutta 1966, 1968, 1974; Dutta and Sah 1970, 1974, Banerjee 1973; Baksi 1974a, b; Sah and Singh 1974; Singh et al. 1976; Singh 1977a, b; Sah et al. 1980; Mehrotra 1981, 1983; Mehrotra and Sah 1982; Tripathi and Singh 1984b; Salujha and Kindra 1986; Kar and Kumar 1986; Saxena et al. 1987; Kumar and Takahashi 1991; Kumar 1993, 1994; Misra et al. 1996a, b; Saxena et al. 1996; Rao 2000; Tripathi et al. 2000).

Biswas (1962) published first detailed account of palynology on Palaeogene sediments of northeast India. He discussed the stratigraphical status of some of the Early Tertiary sediments of Meghalaya (Langpar, Cherra and Tura formations) and described palynofossil assemblages. In the same year, Baksi (1962) published a palynological report from measured sections of sediments of Simsang River Section, Meghalaya and recognized four palynological zones in the section ranging in age from Middle Eocene to Miocene. He compared the palynofissils of these zones with contemporaneous assemblages from other areas. Sah and Dutta (1966) recorded palynofossils from Cherra Formation, Meghalaya (Palaeogene) and discussed the stratigraphy of Palaeogene-Neogene sediments of Assam (Sah and Dutta 1966). These authors concluded that palynological studies can be potentially used in correlation and dating of strata. At a later stage, significant studies on Early Palaeogene sediments were carried out from Cherra (Sah and Dutta 1974), Tura (Sah and Singh 1974) and Therria formations (Tripathi and Singh 1984a, b, 1985) of Meghalaya and Mikir Formation of Assam (Mehrotra 1981, 1983).

Dutta and Sah (1970) established three palynozones in Cherra Formation and one palynozone in the overlying Lakadong Limestone Formation. Sah and Singh (1974) established four palynozones in the Tura Formation and found palynofloral similarity of lower three palynozones with those of Cherra Formation. Mehrotra (1981) established five palynozones in the Mikir Formation, Assam. Tripathi and Singh (1984b) divided Therria formation of Meghalaya into three palynozones and correlated them with those from Cherra, Tura and Mikir formations. Therria Formation is overlain by Sylhet Limestone Formation (Early-Middle Eocene). Tripathi and Singh (1984b) reported poor palynoflora from this formation.

Late Eocene sediments in Jaintia Hills, Meghalaya are represented by the Kopili Formation which overlies the Sylhet Limestone Formation. Palynofloras from this formation are recorded by Sein and Sah (1974), Tripathi and Singh (1984a, b, 1985) and Trivedi (1985). Tripathi and Singh (1984b) divided this formation into two palynozones. Trivedi and Saxena (2000) and Saxena and Trivedi (2009) described palynological assemblages from this formation exposed in Assam. Palynological studies from Barail (Oligocene) and Surma (Early Miocene) groups exposed in Jaintia Hills, Meghalaya were made by Rao (1983, 1986), Rao et al. (1985), Singh et al. (1987) and Kumar et al. (2012). Based on these studies, Saxena et al. (1987) divided each of these groups into three cenozones. Sediments of Barail Group exposed around Sonapur, Meghalaya have yielded rich palynofloras and are divided into three zones, each corresponding with Laisong, Jenam and Renji formations (Singh et al. 1987).

Surma Group (lower part of Miocene) represents Zone IV of Simsang River Section, Meghalaya (Baksi 1962). The Zone is dominated by conifer pollen but also shows frequent occurrence of spores having affinity with pteridophytic families Schizaeaceae and Ceratopteridaceae. Banerjee (1964) also studied the palynoflora of this section and identified two broad divisions. He noticed restricted occurrence of coniferous pollen in the upper part. Salujha et al. (1973) palynologically demarcated Bhuban and Bokabil formations of the Surma Group exposed in South Shillong Plateau. Saxena et al. (1987) divided the Bhuban Formation exposed along Sonapur-Badarpur Road, Meghalaya and Assam into three zones. The Bhuban palynoflora is dominated by pteridophytic spores and gymnospermic pollen followed by dinoflagellate cysts and angiospermic pollen. Palynoflora from the overlying Bokabil Formation is marked by absence of phytoplanktons, presence of fresh-water elements, pteridophytic spores and gymnospermic pollen suggesting regression of sea and existence of highland adjacent to the site of deposition (Salujha et al. 1973; Saxena et al. 1987).

The Tipam Group conformably overlies the Surma Group. Palynology of the Tipam Group has been worked out by Sah and Dutta (1968), Banerjee et al. (1973), Shrivastava et al. (1974), Singh and Tewari (1979), Sah et al. (1980) and Singh and Saxena (1984). Palynological assemblages from this group are dominated by pteridophytic spores. Palynofloras from Girujan Clay, Namsang and Dhekiajuli formations of Moran and Nahorkatiya Wells in Upper Assam was studied by Nandi (1981). She recognized two zones in this sequence, the lower zone corresponding to the lower part of Girujan Clay and upper zone corresponding to the upper part of

Girujan Clay and Namsang and Dhekiajuli formations. Palynofloras indicate tropical to subtropical climate with warm, humid, swampy conditions during Girujan Clay sedimentation and warm, dry, inland conditions during the sedimentation of Namsang and Dhekiajuli formations. Sedimentation during end of Girujan Clay witnessed a period of uplift and erosion followed by deposition of continental sediments of Dupitila/Namsang Group. These sediments are extensively spread over in Surma Valley and also in Upper Assam. Sah and Singh (1977) reported poor palynoflora from Namsang sediments (Pliocene) along with recycled Permian and Eocene palynofossils. Singh and Saxena (1984) conducted palynological studies on the Namsang Beds of Jorajan Well-3, Upper Assam and noticed the absence of recycled palynomorphs. Palynological information from post-Surma sediments is scanty and hence is of limited stratigraphical value.

Palynological studies on Palaeogene-Neogene sediments of Assam and Meghalaya resulted into records of palynofloras as well as their application in biostratigraphic zonation, correlation and dating of sediments. Palynozones established in these areas proved useful in correlation of sediments in Garo, Khasi, Jaintia and Mikir Hills as well in inter-basinal correlation. These studies also helped in inferring the palaeoclimate and environments of deposition. Better Palynological information is available from Palaeogene sediments than to those from Neogene.

Palynological reports on Palaeogene-Neogene sediments of Nagaland are few and mostly deal with description of palynofossils (Mandal 1997; Mandaokar 2002b). These sediments in Mizoram have been worked out by Aswal (1993), Banerjee and Nandi (1994) Hait and Banerjee (1994) and Mandaokar (2000a, 2002a, c, 2003, 2004, 2007, 2008). Aswal (1993) established four palynozones in Late Oligocene to Middle Miocene sediments exposed in Kharkhublen Anticline, Mizoram. Hait and Banerjee (1994) recorded rich palynofloras from Miocene lignites of Mizoram. Miocene sediments of Mizoram, represented by Dulte, Keifang and Bhuban formations, were palynologically studied by Mandaokar (2000a, 2002a, c, 2003, 2004, 2007, 2008). He recorded rich palynofloras indicating tropical to subtropical humid climate having fresh-water, mangrove and coastal elements. Palynological studies from Tripura have been carried out by Salujha and Kindra (1984) and Salujha et al. (1977, 1979, 1980). Salujha et al. (1977) studied the sediments of Surma (Bokabil) and Tipam groups from Gojalia Anticline structure and identified two palynozones. Salujha et al. (1980) studied palynoflora from the Bokabil and Tipam sediments of Tulamara Anticline in South Tripura. Palynological assemblages from both of these areas are dominated by pteridophytic spores followed by angiospermic and gymnospermic pollen.

Southern India

Andhra Pradesh (Krishna-Godavari Basin): Pericratonic Krishna-Godavari Basin is developed along the eastern coast of India and is extended in the Bay of Bengal (Fig. 3.3). Well developed in the eastern part of the basin,

Palaeogene-Neogene sequences are divided into several formations. Venkatachala and Sharma (1984, 1988), Ramanujam et al. (1989a, b, 1997), and Rawat et al. (1991) have carried out palynological studies from sediments of this basin.

Venkatachala and Sharma (1984) studied the palynofossils from subsurface sediments of Narsapur Well-1 and identified six palynozones ranging in age from Palaeocene to Miocene. Based on spore-pollen assemblages, Rawat et al. (1991) proposed ten formal biozones in Late Cretaceous to Pleistocene sediments. Subsurface Miocene sediments from this basin were palynologically studied by Ramanujam et al. (1989a, b). These workers suggested the presence of mangrove, back mangrove, fresh-water swamp and water-edge, and sand dune and salt marsh ecological groups. They inferred that a tropical humid climate with heavy precipitation prevailed along the east coast of southern India during the Miocene,

Tamil Nadu

Cauvery Basin: Palaeocene to Miocene Bore-hole samples from Cauvery Basin covering Karaikal, Mannargudi and Madanam have provided significant palynological information (Venkatachala and Rawat 1972, 1973). The sequence is divided into seven palynozones. Of these, one zone represents Palaeocene, three zones Eocene, one zone Oligocene and two zones Miocene. Tertiary palynoflora of Cauvery Basin is dominated by angiospermic pollen. Warm and humid climate with heavy precipitation and deposition under mangrove swamp conditions with local pockets of marine influence is interpreted. These studies proved very useful in correlation of Palaeogene-Neogene sediments of South India.

Neyveli Lignite: Lignites associated with sandstone and clays of Cuddalore Sandstone Formation have been worked out palynologically very well and significant reports have been published (Navale 1962; Ramanujam 1966, 1967a; Deb 1972; Deb et al. 1973; Venkatachala 1973; Navale and Misra 1979; Ramanujam et al. 1984, 1985, 1988; Ramanujam and Reddy 1984). Popularly known as Neyveli lignites, the lignite deposits were considered to be a part of the Cuddalore Formation. Siddhanta (1986) divided this formation into two parts, the lower Neyveli Formation and upper Cuddalore Formation. Saxena (1992a) recorded rich palynological assemblages from the Neyveli Formation of Neyveli mines I and II in Cuddalore Distict and Jayamkondacholapuram Well—12 in Tiruchirapalli District. Based on these studies, Neyveli Formation was divided into three biozones. In Jayamkondacholapuram Well-12, all three palynozones were encountered whereas in Neyveli mines I and II only upper zone was traced. Saxena (1992a) correlated the biozones of Neyveli Formation with those of Late Palaeocene-Middle Eocene sediments of Kutch, Rajasthan, Bengal Basin, Cauvery Basin and Garo, Khasi and Jaintia Hills of Meghalaya hence considered the Neyveli Formation as synchronous to these sediments. The Neyveli palynoassemblages are diverse and include the elements of fresh-water swamps, water-edge elements, back mangrove pollen and beach elements. The palynoflora suggests a warm and humid tropical climate with

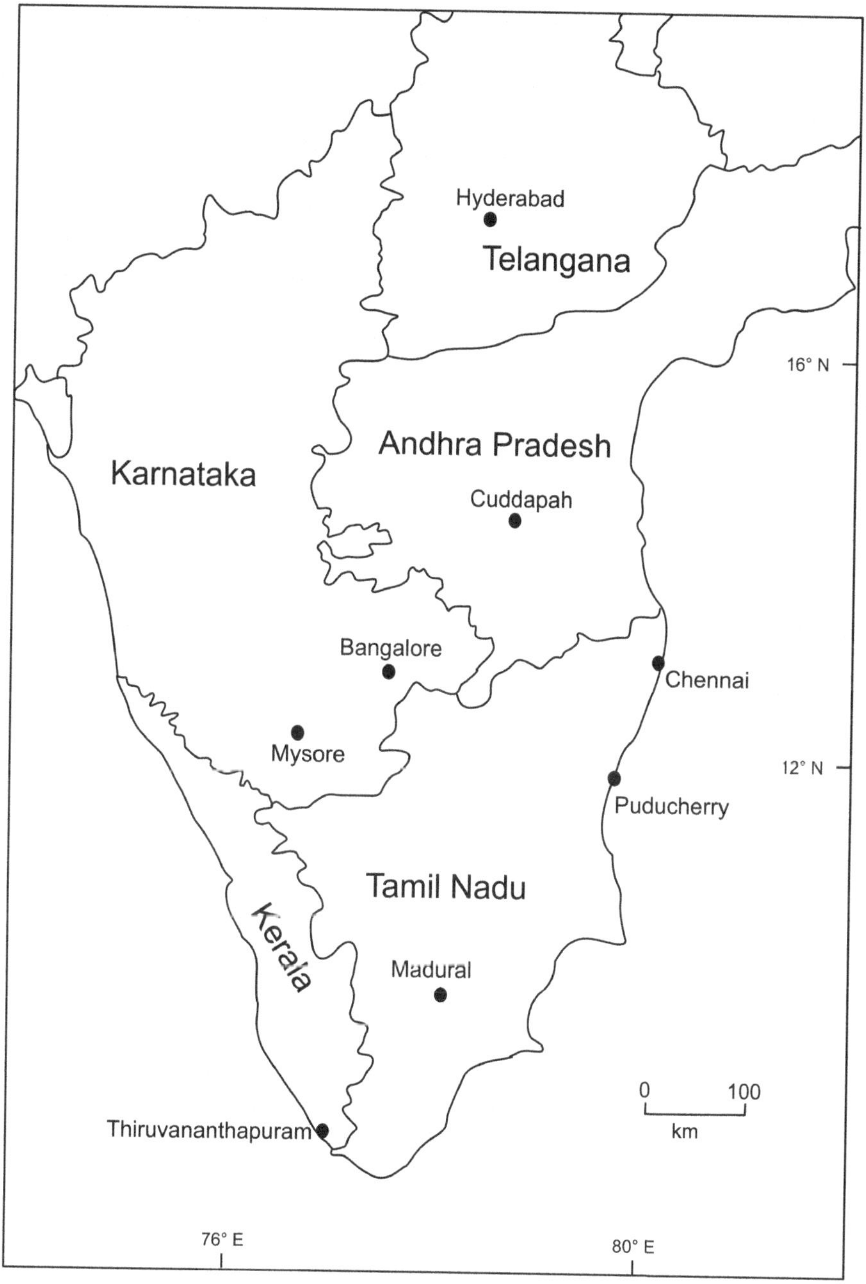

Fig. 3.3 Map of southern region of india. After Saxena and Tripathi (2012)

plenty of rainfall during the sedimentation of Neyveli Formation. The environment of deposition was inferred to be coastal ranging from back-mangrove to mangrove with short transgressive phase just before deposition of lignite.

The Neyveli Formation, a subsurface stratigraphic unit, contains thick workable lignite deposits. The age of lignite is controversial. Based on palynological studies, Miocene age has been given to Neyveli lignite (Ramanujam 1966; Navale and Misra 1979; Misra et al. 1996a, b). Considering the palynoflora and geological evidences, age of the lignite is inferred to be Palaeocene-Eocene (Venkatachala 1973; Deb et al. 1973; Siddhanta 1986; Saxena 1992a).

Kerala: Neogene sediments of Kerala are represented by Quilon and Warkalli beds. Palynofloras from these beds have been published by Potonié and Sah (1960), Ramanujam (1960, 1967b, 1972, 1982, 1987), Ramanujam and Rao (1973, 1977), Rao and Ramanujam (1975, 1978, 1982) and Kar and Jain (1981). Subsurface Neogene sediments from Alappuzha (Alleppey) District have been studied by Rao (1990, 1995, 1996, 2001) and Rao and Nair (1998). Quilon Beds have yielded diversified palynofloras represented by angiospermic pollen, pteridophytic spores, dinoflagellate cysts, acritarchs and variety of fungal remains. The assemblage indicates that these are shallow marine deposits. Warkalli Beds overlie Quilon Beds. Warkalli palynoflora is continental and is more or less similar to that of Quilon beds (Ramanujam 1982). It is inferred that both Quilon and Warkalli Beds were deposited during Early to Middle Miocene under different environmental conditions. Palynoflora from Warkalli Formation indicates a complex coastal vegetation thriving under tropical humid climate with heavy precipitation.

Western India

Kutch Basin (Gujarat): Complete sequence of Palaeocene to Pliocene sediments is developed in western and southern parts of the Kutch Basin (Fig. 3.4). Palaeocene sediments of Kutch Basin are represented by Matanomadh Formation. Mathur (1966) was first to report palynofossils from these sediments. Detailed palynology and biostratigraphy on Matanomadh Formation and that from Eocene lignites along with associated sediments from Matanomadh mines of Kutch basin have been published by Saxena (1977, 1978, 1979, 1980, 1981), Kar (1985) and Runcie et al. (2013). The Matanomadh Formation is divisible into two members, of which the lower one is devoid of palynofossils. The overlying member has been divided into three zones (Saxena 1981). However, Kar (1985) divided this member into four zones. Palynological assemblage from this formation is dominated by angiospermic pollen followed by pteridophytic spores and gymnospermic pollen. Saxena (1980) observed that the Matanomadh Formation is contemporaneous to Cherra and Tura formations of Meghalaya, as the palynofloras from all these formations are very similar. Depositional environment for the Matanomadh Formation is interpreted to be fluviatile to estuarine. Matanomadh Formation is unconformably succeeded by Naredi Formation (Early Eocene) which is rich in spore/pollen assemblages

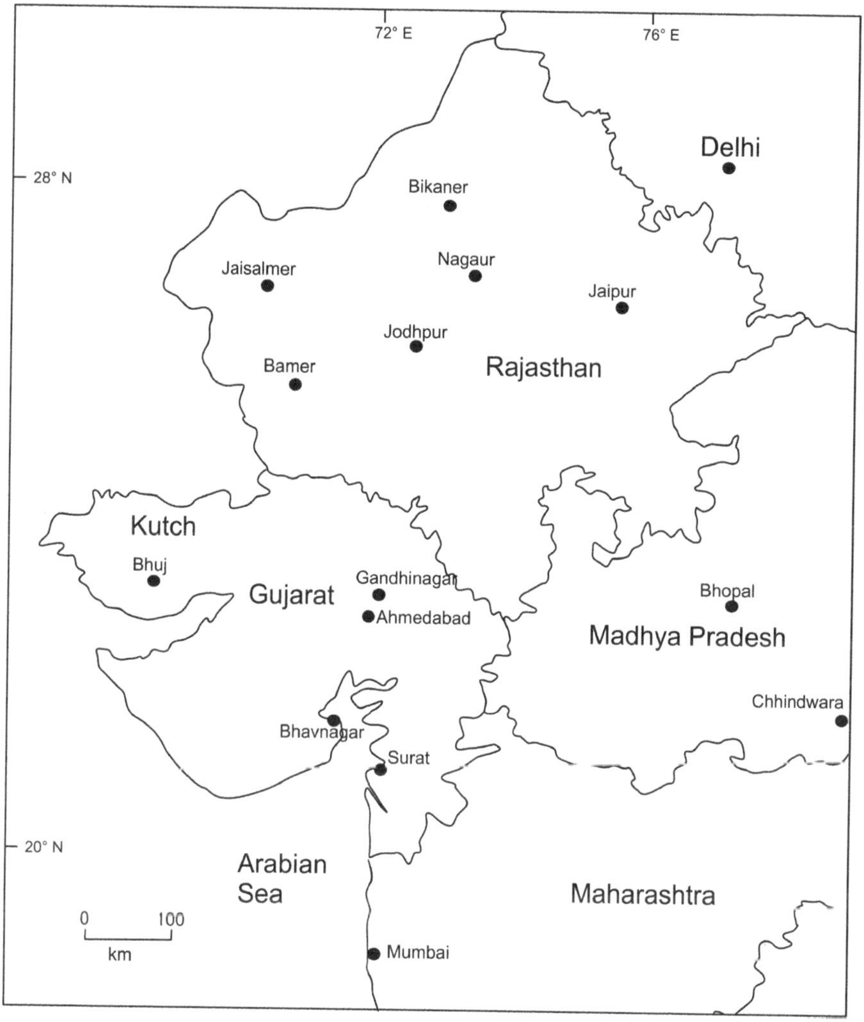

Fig. 3.4 Map of western region of india. After Saxena and Tripathi (2012)

indicating estuarine environment of deposition (Mathur 1963a, b; Venkatachala and Kar 1969; Sah and Kar 1970; Kar 1978, 1985). Middle Eocene sediments are represented by Harudi and Fulra Limestone formations. The Harudi assemblage is poor and the Fulra limestone is devoid of palynofossils. Fulra Limestone is conformably overlain by Maniyara Fort Formation (Oligocene). This unit has been palynologically worked out by Kar (1977, 1979, 1985). The Maniyara Fort Formation is overlain by Khari Nadi Formation which is succeeded by Chhasra Formation. These two litho-units are of Miocene age and contain poor palynofossil assemblages. Palynological information from the Pliocene sediments of Naera-Baraia area has been published by Mathur and Mathur (1969).

Cambay Basin (Gujarat): Resting unconformably over the Deccan Trap, Tertiary sequences in Cambay Basin are completely covered by Recent alluvium. These sequences range in age from Palaeocene to Plio-Pleistocene and have been divided into several formations. The oldest Palaeogene rocks in the basin are represented by Olpad Formation which is unconformably overlain by Cambay Formation (Palaeocene-Early Eocene). A sequence, named as Kadi Formation, constituted by coaly sand has been identified within the Cambay Shale in the Ahmedabad-Mehsana block. This formation is overlain by, Tharad, Kalol, Ankleshwar, Bhavnagar, Tarapur and other formations. Palynological studies in Cambay Basin have been carried out by Rawat et al. (1977), Venkatachala and Chowdhary (1977), Mathur and Chowdhary (1977), Koshal and Uniyal (1984, 1986), Koshal (1988), Shanmukhappa (1991), Kar and Bhattacharyya (1992), Shanmukhappa and Koshal (1993), Samant (1994, 2000), Gupta et al. (1996), Kumar (1996), Samant and Phadtare (1997), Shanmukhappa et al. (2000), Samant and Tapaswi (2001), Mandal and Guleria (2006), Shanmukhappa and Uniyal (2008), Tripathi and Srivastava (2012) and Monga et al. (2015).

Sediments of Kadi Formation yielded rich palynological assemblages dominated by angiospermic pollen particularly the mangroves (Rawat et al. 1977; Venkatachala and Chowdhary 1977). Kadi Formation has been dated as Early Eocene. This dating is based on palynofloral resemblance with Cauvery Basin, Tamil Nadu. Mathur and Chowdhary (1977) inferred shallow marine depositional regime for Kalol Formation (Middle-Late Eocene) and identified many transgressive-regressive phases during the deposition.

Palynological studies have been carried out on Early Palaeogene sediments of Gujarat exposed in Rajpardi (Kar and Bhattacharya 1992; Kumar 1996; Samant and Phadtare 1997) and Bhavnagar (Samant 1994, 2000). Samant and Phadtare (1997) described rich palynological assemblage from the Early Eocene sediments of Rajpardi lignite of Gujarat and identified four palynozones in the sequence. Palynological studies from Cambay Shale of Vastan, Surat are scarce (Mandal and Guleria 2006; Tripathi and Srivastava 2012).

Rajasthan Basin: Palaeogene sediments in northwestern Rajasthan are mostly covered with sand. These sediments form a part of Western Rajasthan Shelf and are divided into several sub-basins located in Jaisalmer, Barmer, Bikaner-Nagaur and Sanchor (Fig. 3.4). In Jaisalmer Basin, these sediments are represented by Sanu, Khuiala and Bandah formations. In Barmer Basin, the oldest Tertiary sediments are named as Akli Formation which is succeeded by Mata-ji-ka Dungar and Kapurdi formations. These sediments range in age from Late Palaeocene to Eocene. Rich palynological assemblages are recorded from the Palana lignite of Bikaner and the Akli Formation of Barmer.

Palynological studies in Rajasthan Basin have been taken up in Barmer (Bose 1952; Jain et al. 1973; Naskar and Baksi 1978; Tripathi 1994, 1995, 1997; Tripathi et al. 2003, 2009; Tripathi and Srivastava 2010), Jaisalmer (Lukose 1974), Bikaner-Nagaur (Singh and Dogra 1988; Kar 1995a, 1996; Ambwani and Singh 1996; Kar and Sharma 2001; Tripathi et al. 2008; Kumar et al. 2016) and Palana beds (Rao and Vimal 1950, 1952; Sah and Kar 1974). Jain et al. (1973) recorded

Palaeocene palynoflora from Barmer Sandstone. Lukose (1974) recorded a palynological assemblage from Palaeocene-Early Eocene succession of Jaisalmer. This assemblage is dominated by angiospermic pollen. Sah and Kar (1974) reported the occurrence of diversified Eocene assemblage from Palana lignite of Bikaner. Singh and Dogra (1988) identified five zones in the Palaeocene to Early Eocene deposits of the Bikaner Basin. The lower SP-1 and SP-2 zones represent the Early Palaeocene, SP-3 and SP-4 zones correspond to the Late Palaeocene and SP-5 zone represents Early Eocene. Rich palynofloral assemblages have been recorded from Early Palaeogene subsurface sediments of Barmer. Rock samples from bore-holes drilled near Kapurdi and Jalipa near Barmer yielded Late Palaeocene-Early Eocene palynological assemblages which were dominated by angiospermic pollen (Tripathi 1994, 1995). The recorded palynofloras are overwhelmingly represented by morphologically diversified forms, which, in all possibility, appear to be related with the family Arecaceae. Tripathi (1997) established two informal palynozones in two of these bore-hole sequences and demarcated Palaeocene and Eocene sediments.

The Akli Formation (Late Palaeocene-Early Eocene), exposed in open-cast lignite mine located near Giral in Barmer yielded a rich palynoflora comprised of dinoflagellate cysts, pteridophytic spores and angiospermic pollen (Tripathi et al., 2003). Continuing these studies further, Tripathi et al. (2009) divided Akli Formation into two informal palynozones. The lower zone is characterized by high frequency of dinoflagellate cysts and pteridophytic spores whereas the upper zone is distinctly dominated by angiosperm pollen grains having affinity with coastal elements, particularly the palms. Further studies enabled Tripathi and Srivastava (2010) to identify three formal assemblage zones in the Akli Formation.

Maharashtra: Neogene sediments in Maharashtra are represented by lignite beds and associated clays described as Sindhudurg Formation (Ratnagiri Beds). Except for a few exposures, these beds are generally found in wells. Phadtare and Kulkarni (1980, 1984) worked out the palynofossils from beds of a well near Golap in Ratnagiri District. These beds have been assigned Neogene age on the basis of close similarity of palynoflora with those from Neyveli and Warkalli lignites. Sindhudurg Formation, Sindhudurg was palynologically investigated by Saxena and Misra (1990), Saxena et al. (1992), Saxena (2000b), Kumaran et al. (2004) and Rao (2004). Rich and diversified assemblages indicating Miocene age were recovered.

Bombay Offshore Basin: Located at the continental Shelf of western India, the Bombay Offshore Basin is one of the major oil producing basins of India. Sediments of this basin unconformably overlie the Deccan Trap and range in age from Palaeocene to Recent. Considering the structural configuration and type of sediments, the Bombay Offshore Basin has been divided into different blocks. The lithostratigraphic units in these blocks are named differently. The oldest Tertiary litho-unit in the basin is Panna Formation (Palaeocene-Early Eocene) which is succeeded by Bassein Formation (Eocene to Early Oligocene). Sediments ranging in age from Late Oligocene to Recent are named as Alibagh, Mahim, Tapti and Chinchini formations. Palynological information from Panna Formation has been published by Rawat, Misra et al. (1996), Rawat et al. (1996a), Mehrotra et al.

(2001) and (2002). Rawat et al. (1996b) established two palynozones in the Mahim area of the basin. The assemblage is dominated by coastal angiospermic pollen. The environment of deposition for this formation is suggested to vary from marginal marine to fluvial swampy conditions.

Central India

Madhya Pradesh: Some palynofossils are recorded from Late Cretaceous to Palaeocene sediments of the Intertrappean beds exposed near Mohgaon Kalan in Chhindwara District (Chitaley 1951, 1957; Chitaley and Patel 1972). The palynoflora is comprised of pteridophytic spores and a few gymnosperm and angiosperm pollen. These reports are preliminary in nature.

Andaman and Nicobar Islands

Most of the sediments in Andaman are Palaeogene to Neogene in age (Fig. 3.5). Palynological reports from Andaman—Nicobar Basin are few and are confined to Middle Andaman and Baratang islands. No palynological work has been done from Nicobar Islands. The first palynological paper published on palynofossils is from Port Blair Formation of Baratang Island. Jafar and Tripathi (2001) however, pointed out that this section belongs to Baratang Formation. Angiosperm pollen reported by Banerjee (1966) indicated Palaeogene age on the basis of comparison of the palynofossils with that from Assam. Mathur and Mathur (1980) reported a palynoflora from upper part of Baratang Formation, which closely resembles with Palaeogene palynofossils from Assam and Bengal Basin. Mandal et al. (1994) described Early Eocene palynofossils from Kadamtala Coal of Middle Andaman. Mandal et al. (2003) published spores, pollen grains and dinoflagellate cysts of Eocene age from Baratang Formation along with reworked Permian to Cretaceous palynofossils. These authors observed that Chindwin Basin of Burma mainly supplied the sediments containing reworked palynofossils.

3.2 Indian Palaeogene-Neogene Floras and Climate

Indian subcontinent exhibits equatorial biotic records of extreme climatic events of early Palaeogene times. The vegetation underwent significant changes during northward movement of Indian subcontinent through various palaeoclimatic belts after its break-up from main Gondwanaland during Aptian time. By end of Cretaceous Period India witnessed decline of Gymnosperms having Gondwanic affinity and was occupied by diversified and rich megathermal angiosperms.

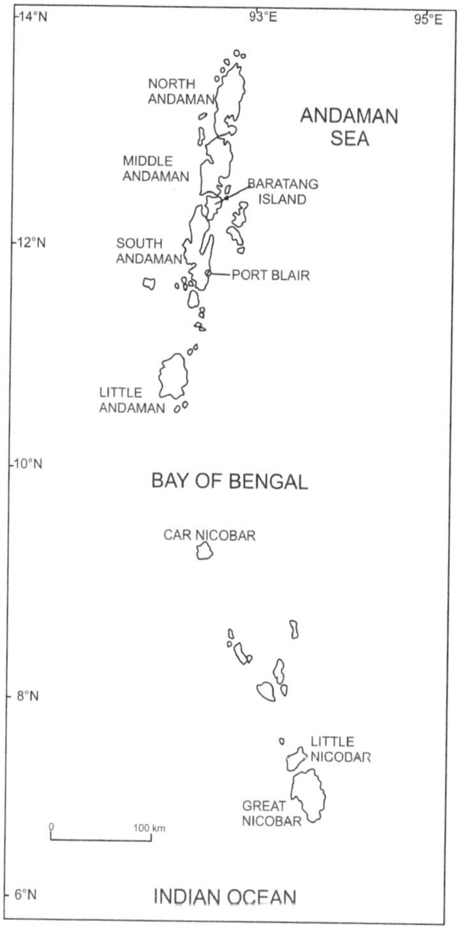

Fig. 3.5 Map of Andaman and Nicobar Islands

The Early Palaeogene flora contains some pollen having affinity with Gondwanaland as well as the endemic elements evolved during isolated movement of the subcontinent across the equatorial climatic zone through Palaeocene (Morley 2000). Early Eocene was the time when Indian Plate was situated at the equator resulting into a very hot and wet climate. During this period, dense tropical rain forests with diverse plants occupied the Indian subcontinent and extensive peat swamps were formed. During Late Eocene and Oligocene floristic diversity decreased. Onset of monsoon and seasonality started appearing in northern part of India during Early Miocene. In Middle Miocene a wet climate prevailed and the area was occupied by evergreen rain forests. Plants of this flora had affinity with those of SE Asia (Mehrotra 2003). Late Miocene and Pliocene had more seasonal climate while northwestern part of Indian subcontinent experienced drier climate. This climatic change caused disappearance of rain forests from most of the areas and the plants were restricted to few local pockets only (Morley 2000; Prasad et al. 2009). Climatic change caused extensive damage to Indian Neogene flora resulting into extinction of many genera and species.

During movement of Indian subcontinent to the higher latitudes, the palynofossils show great similarity with those from Africa, Australia and Madagascar at different times. Pollen assemblages from India show that its northward journey provided an opportunity for exchange of floral elements between adjoining landmasses culminating into rapid evolution of new plant taxa. Consequent to the Palaeocene Eocene Thermal Maxima (PETM) and Early Eocene Climatic Optima (EECO), the equatorial rain forests became widespread in Indian subcontinent and diversification of tropical angiosperms took place. Palynological assemblages from Palaeogene and Neogene sediments of India are dominated by tropical angiosperm pollen suggesting prevalence of warm humid tropical climate. The India–Asia collision played a significant role in the dispersal and distribution of flora and fauna in Asia during Palaeogene and Neogene (Bossuyt and Milinokovitch 2001; Cooper et al. 2001; Clementz et al. 2010). Palynological studies throw light on the biogeographical patterns in flora. Studies carried out by Morley (2000) and Conti et al. (2002) suggest that many angiosperms did not originate in the SE Asian region, but dispersed into the area from western Gondwanaland.

Late Cretaceous and early Palaeogene are very significant from the viewpoint of floral change as, during this period the latitudinal position of India favored a colossal evolutionary pattern in tropical angiospermic flora. Huge depositions of coal and lignites during this period in coastal lowland areas of Indian western and northeastern margins provide evidence of warm and humid climate. These successions have yielded rich and diversified palynofloras clearly indicating extensive spread of tropical rain forests. Prasad et al. (2009) observed that evergreen vegetation thriving in high precipitation zone of southwestern parts of Western Ghats show close similarity with fossil pollen of Late Palaeocene and Early Eocene times. These pollen have mostly been recorded from western and northeastern India. During this time, India experienced warm and humid climate with high precipitation.

Palynological studies show that Late Palaeocene of Indian subcontinent is marked with abundance of morphologically diverse Palm pollen. Apart from these, dicotyledonous families are also very well represented. Spread of the pollen having affinity with Bombacaceae, Caesalpiniaceae, Ctenolophonaceae, Alangiaceae, Annonaceae and Proteaceae is noticed. High frequency of stephanocolpate pollen characterizes the Indian Early Palaeogene. These pollen have been described under different names. Saxena (1982) has tried to streamline the taxonomy of pollen of this group. Ambwani et al. (1981) commented upon diversification of *Ctenolophon* (Ctenolophonaceae), a polycolpate pollen. Warm and humid climate supporting the diverse rain forest vegetation continued until Late Eocene and Oligocene although, the palynomorph diversity declined (Kar 1985; Morley 2000).

Frequently occurring in Indian Tertiarysediments, pollen grains showing affinity with the family Arecaceae are assigned to different species of *Acanthotricolpites, Arengapollenites, Clavapalmaedites, Dicolpopollis, Echimonoporopollis, Kapurdipollenites, Longapertites, Quilonipollenites, Neocouperipollis, Monocolpopollenites (Palmaepollenites), Palmidites, Proxapertites, Spinizonocolpites* and *Spinomonosulcites*. Pollen grains of the family Bombacaceae are also abundant in Indian Tertiary sediments. Most of these pollen belong to *Bombax-* and *Durio-*types. Fossil pollen having affinity with *Bombax*-type are represented by *Bombacacidites*. *Bombax-* type pollen did not evolve rapidly in India. These are less common in Eocene and Oligocene but frequently occur in Miocene sediments (Mandal 2005). The *Durio-* type of pollen are represented by *Lakiapollis ovatus, Dermatobrevicolporites dermatus, Tricolporocolummelites* spp., *Tricolporopollis matanomadhensis* and *Tricolporopollis ruber*. In Palaeocene and Eocene sediments pollen grains of Durio- type are abundant and are widely represented.

Pollen grains of the family Alangiaceae are also quite common in early Palaeogene sediments of western and northeastern India and in Miocene of south India. These are described as *Alangiopollis, Lanagiopollis, Densiverrupollenites* and *Pellicieroipollis*. Phadtare and Thakur (1990) noticed that the section *Marlea* of this family possibly existed in India during Late Palaeocene and was accompanied by members of the section *Conostigma*. These authors opined that *Alangium* appeared during Miocene.

3.3 Frequently Found Pollen/Pollen Groups in the Palynofloras

Remarks on Pollen Having Affinity with the Family Arecaceae

Different species of *Spinizonocolpites* are related to modern brackish water Palm *Nypa* (Muller 1968). Prior to the recognition of extant *Nypa* from Asia, Parsons (1757) described its fossil fruits from the Eocene of London Clay. This genus is placed under the family Arecaceae. The genus *Nypa*, represented by one living species (*N. fruticans*), is restricted to a narrow tropical area of southeast Asia, including India (Sundarbans and Andaman Islands) and Bangladesh in mangrove ecosystem. These plants favor quiet estuarine or shallow lagoons with fresh-water influxes. It does not grow on shores with much wave action and in hyper-saline conditions. Different species of *Spinizonocolpites* are widely recorded from Late Cretaceous to Eocene sediments covering wide geographical areas (Singh 1999). From Indian subcontinent, these are reported from Late Cretaceous to Pliocene sediments (Baksi and Deb 1980; Kar 1985; Kar and Jain 1981; Prasad and Dey 1986; Ramanujam 1987; Ramanujam and Rao 1977; Rao and Ramanujam 1975, 1978; Dutta and Sah 1970; Sarma et al. 1984; Singh 1990; Srisailam and Ramanujam 1982; Venkatachala 1973; Venkatachala and Rawat 1972; Venkatachala and Sharma 1984).

Proxapertites, based on its morphological features, has been interpreted to represent an extinct group of palms possibly related to *Nypa* (Muller 1968). The genus *Acanthotricolpites* Kar (1985), in all likelihood related to *Nypa*, is also profusely recorded from Early Palaeogene of western and eastern India. Great diversity of *Nypa*-like Indian fossil pollen has been noticed by various workers (Kar 1985; Kar and Kumar 1986; Frederiksen 1994; Tripathi 1994; Tripathi et al. 2003). The aperture character of *Kapurdipollenites* is similar to that of *Spinizonocolpites* suggesting that the former may also be affiliated to *Nypa*. These palynological studies suggest that during Late Cretaceous and Palaeocene more diversified mangrove habitats were occupied by the plants that produced pollen akin to *Spinizonocolpites*. Diversity history of the *Spinizonocolpites* complex appears to have been governed by the exploitation of various brackish water habitats and also by the presence of more effective competitors (Frederiksen 1994).

Clavadiporopollenites Ambwani and Singh (1996) is a unique pollen grain described from subsurface Early Eocene sediments of Bikaner, western India. In this pollen the two sunken and annulated pores are located at a juxtaposed position. The pores are surrounded by a group of clavae of varying size. Ambwani and Singh (1996) suggested that this genus possesses exine processes which are very similar to that found in *Retiverrumonosulcites* Tripathi (1994). These authors surmised that both of these pollen genera appear to be related to the family Arecaceae.

Kar (1995) described three peculiar pollen types, viz. *Psiladiporocolpites pachyexinus, Piladiporocolpites caratinii* and *Retidiporocolpites excellensus* from subsurface sediments of Bikaner, western India. These pollen have variable exine ornamentation (laevigate, gemmate and reticulate) and are diporate. The two pores are connected by a colpus like aperture. In all probability, these pollen also bear affinity with family Arecaceae.

Pollen grains assignable to *Proxapertites* van der Hammen (1956) are very common in Indian Tertiary sediments. Possessing an extended sulcus (zonisulcus), the grains are provided with micro-reticulate to coarsely reticulate exine. The two halves in this pollen are weakly joined resulting into frequent breakage of the grains into two saucer-shaped split parts. Having similar nature of aperture, forms described under *Kapurdipollenites* [*Kapurdipollenites baculatus* Tripathi (1994), *Kapurdipollenites gemmatus* Tripathi (1994), *Kapurdipollenites ovatus* Tripathi and Srivastava (2010) and *Kapurdipollenites clavatus* Tripathi and Srivastava (2010)] are related to the family Arecaceae. Morphological features of *Retiverrumonosulcites* Tripathi (1994) strongly suggest its affinity to the family Arecaceae. It is a Monocolpate pollen having various kinds of processes.

Remarks on Polycolpate Pollen

Polycolpate pollen grains, very common in Indian Palaeogene-Neogene sediments, are described under various genera. Diagnoses of these genera are very confusing and overlapping. Saxena (1982) evaluated the taxonomic and nomenclatural aspects of polycolpate pollen from India. He suggested that exine ornamentation and exine thickness should be considered for generic differentiation while, shape of pollen, aperture characters and exine stratification should be accounted for species differentiation.

Pollen grains of the family Ctenolophonaceae are polycolpate with various kinds of nexinal thickenings in apocolpial region. Pollen of this family are described under *Ctenolophonidites* and *Prectenolophonidites* and *padappakkarapollis*. Phadtare and Samant (1996) observed a decreasing trend in nexinal thickening in pollen occurring in younger Tertiary sediments. These authors noticed that by the end of Miocene these pollen became extinct and the family disappeared from Indian subcontinent. Ctenolophonaceae is a monogeneric family and is represented by two recent species viz., *Ctenolophon englerianus* and *Ctenolophon parvifolius*. The family has a disjunctive distribution, *C. englerianus* occurring in tropical West Africa and *C. parvifolius* in Malaysia. Pollen of *Ctenolophon* is very distinct. Germeraad et al. (1968), Ramanujam and Rao (1973), Muller (1981), Thanikaimoni et al. (1984) and Srivastava (1988) studied fossil forms and evolutionary history of this genus. Germeraad et al. (1968) described fossil pollen of *Ctenolophon englerianus* from Maastrichtian onwards from Africa, South America, and India and those of *Ctenolophon parvifolius* from the Palaeocene onwards in Africa, India and Malaysia. He postulated the probable origin of *Ctenolophon* in the Upper Cretaceous, possibly in Africa, followed by origin of the *englerianus* type and the *parvifolius* type. The *englerianus* type is found throughout the Tertiary and Quaternary of Africa. One species related to the *englerianus* type migrated to South America in the Palaeocene and soon became extinct (Van der Hammmen 1989) while *parvifolius* type migrated to Malaysia in Eocene and became extinct in Africa. Ramanujam and Rao (1973) and Rajendran et al. (1989) discussed in detail the fossil pollen records from Indian sediments. These studies indicate that probably *englerianus* type migrated from Africa to India in the Palaeocene (Van der Hammen 1989). Srivastava (1988) does not support the Miocene records of *Ctenolophonidites* from India, but different species of this genus have been well documented from Miocene deposits of Kerala and Tamil Nadu (Ramanujam and Rao 1973; Navale and Misra 1979; Ambwani et al. 1981; Kar and Jain 1981; Rajendran et al. 1989).

Remarks on Pollen of Dipterocarpaceae

Most of the modern dipterocarps are confined to wet climates. Fossil records of Dipterocarpaceae from India are frequent from Miocene and younger sediments (Prasad 1993; Guleria 1996; Khan and Bera 2010). Based on rich representation of dipterocarpaceous fossils, some workers opine that members of this family migrated into India from SE Asia once land connection between Indian and Asian plate was established (Lakhanpal 1970; Sasaki 2006). Contrary to it, other workers think that plants of Dipterocarpaceae have a Gondwanic origin and reached Asia through the moving Indian plate (Ashton 1982, 1988; Ashton and Gunatilleke 1987; Dayanandan et al. 1999; Ducousso et al. 2004). The earliest confirmed fossil dipterocarp from SE Asia is reported from Oligocene (34–23 Ma) sediments of Borneo which is known for the diversity of extant dipterocarps (Muller 1981). The oldest record of *Dipterocarpus* pollen from Indian subcontinent is from the sub-surface Early Eocene sediments of Mannargudi area, Cauvery Basin, Tamil Nadu (Acharya 2000). Other record of pollen of this family is from Oligocene of Assam. Mention of this record was made by Morley (2000, p. 167) referring it to a personal communication with Muller and Baksi. During Miocene the pollen having affinity with the family Dipterocarpaceae became frequent (Kar 1992; Rao and Rajendran 1996; Rao 1996, 2004). Prasad et al. (2009) assigned the affinity of some Palaeogene palynofossils with pollen of the family Dipterocarpaceae. These are—*Albertipollenites kutchensis*, *Dipterocarpuspollenites retipilatus*, *Foveotricolpites alveolatus*, *Intrareticulites brevis* and *Tricolpites reticulatus*. Studies on Palynological and fossil resin chemistry suggest that plants of this family appeared during Early Eocene on the Indian continent (Dutta et al. 2011). Ashton (1988) suggested that dipterocarps dispersed from India into Asia during the middle Eocene (49–41 Ma) when collision between the Indian and Asian plate was completed.

References

Acharya M (2000) Early Eocene palynofossils from subsurface of Mannargudi area, Tamil Nadu, India. Geophytology 28(1–2):19–30

Ambwani K, Singh RS (1996) *Clavadiporopollenites raneriensis* gen. et sp. nov. from the Tertiary sediments of Bikaner District, Rajasthan, India. Palaeobotanist 43(3):139–142

Ambwani K, Bande MB, Prakash U (1981) Pollen grains of *Ctenolophonidites* from the Neyveli lignites of South India. Palaeobotanist 27(1):100–106

Ashton PS (1982) Dipterocarpaceae. In: van Steenis CGGJ (ed) Flora Malesiana, series 1: *Spermatophyta*, vol 9, pp 237–552

Ashton PS (1988) Dipterocarp biology as a window to the understanding of tropical forest structure. Annu Rev Ecol Syst 19:347–370

Ashton PS, Gunatilleke CVS (1987) New light on the plant geography of Ceylon. Historical plant geography. J Biogeogragr 14:249–285

Aswal HS (1993) Palynostratigraphy of Kharkhublen Anticline (central part), central and southern districts, Mizoram, India. In: Biswas SK et al (eds) Proceedings of the second seminar on Petroliferous basins of India, vol 1. Indian Petroleum Publishers, Dehradun, pp 477–491

Baksi SK (1962) Palynological investigation of Simsang River Tertiaries, South Shillong Front, Assam. Bull Geol Mining Metall Soc India 26:1–22

Baksi SK (1972) On the palynological biostratigraphy of Bengal Basin. In: Ghosh AK et al (eds) Proceedings of the seminar on Paleopalynology and Indian Stratigraphy Calcutta 1971. Botany Department, Calcutta University, Kolkata, pp 188–206

Baksi SK (1973) On the palynological biostratigraphy of Cenophytic sediments of the Bengal Basin. In: Palynology of the Cenophytic, proceedings of the 3rd international palynological conference, Novosibirsk, 1971. Academy of Science, U.S.S.R. Moscow, pp 78–88

Baksi SK (1974a) Significant pollen taxa in the stratigraphical analysis of the Tertiary sediments of Assam. In: Surange KR et al (eds) Aspects and Appraisal of Indian Palaeobotany. Birbal Sahni Institute of Palaeobotany, Lucknow, pp 502–515

Baksi SK (1974b) Stratigraphical position of Cherra Formation of South Shillong Plateau. In: Surange KR et al (eds) Aspects and Appraisal of Indian Palaeobotany. Birbal Sahni Institute of Palaeobotany, Lucknow, pp 534–549

Baksi SK, Deb U (1980) Palynostratigraphic zonation of the Late Cretaceous-Palaeogene sequence of Bengal Basin. Geophytology 10(2):199–224

Banerjee D (1964) A note on the microflora from Surma (Miocene) of Garo Hills, Assam. Bull Geol Mining Metal Soc India 29:1–8

Banerjee D (1966) A note on Tertiary microflora from Andaman Islands, India. Pollen et Spores 8 (1):205–212

Banerjee D (1973) Palynological evidences and the Tertiary sediments of India. J Palaeontol Soc India 17:1–6

Banerjee S, Nandi B (1994) Palynology of the Middle Bhuban Formation near Kolasib, northern Mizoram. Geophytology 23(2):215–220

Banerjee D, Misra CM, Koshal VN (1973) Palynology of the Tertiary subcorps of Upper Assam. The Palaeobotanist 20(1):1–6

Bera S, Banerjee M (1995) Eocene palynoassemblage from lignite deposits of West Bengal, India with remarks on environment of deposition. Indian Journal of Earth Science 22(4):149–152

Berry CM, Mehrotra NC, Nautiyal DD (1996a) Hydrocarbon potential in the Gulf of Cambay—A palynological perspective. In: Pandey J et al (eds) Contributions to the 15th Indian Colloquium on Micropalaeontology and Stratigraphy, Dehradun, 1996. Oil and Natural Gas Corporation, Dehradun, pp 707–710

Berry CM, Misra CM, Chopra AS (1996b) Palynostratigraphy, paleoenvironment of Changartalai well in Punjab Basin. In: Pandey J et al (eds) Contributions to the 15th Indian Colloquium on Micropalaeontology and Stratigraphy, Dehradun, 1996. Oil and Natural Gas Corporation, Dehradun, pp 643–645

Bhandari LL, Venkatachala BS, Singh P (1977) Stratigraphy, palynology and palaeontology of Ladakh Molasse Group in the Kargil area. In: Venkatachala BS, Sastri VV (eds)—Proceedings of the 4th Colloquium on Indian Micropalaeontology and Stratigraphy, Dehradun, 1974–75. Institute of Petroleum Exploration, Oil and Natural Gas Commission, Dehradun, pp 127–133

Biswas B (1962) Stratigraphy of the Mahadeo, Langpar, Cherra and Tura formations, Assam, India. Bull Geol Mining Metal Soc India 25:1–48

Bose MN (1952) Plant remains from Barmer District, Rajasthan. J Sci Ind Res 11B(5):185–190

Bossuyt F, Milinokovitch MC (2001) Amphibians as indicators of Early Tertiary "out of India" dispersal of vertebrates. Science 292:93–95

Chitaley SD (1951) Fossil microflora from the Mohgaon Kalan beds of the Madhya Pradesh, India. Proc Natl Inst India 17(5):373–383

Chitaley SD (1957) Further report of the fossil microflora from the Mohgaon Kalan beds of the Madhya Pradesh, India. Proceedings of the National Institute of India 23B(2–4):69–79

Chitaley SD, Patel MZ (1972) Microflora of Intertrappean and Lameta formations of India. In: Ghosh AK et al (eds) Proceedings of the seminar on Paleopalynology and Indian Stratigraphy Calcutta 1971. Botany Department, Calcutta University, Kolkata, pp 212–219

Clementz M, Bajpai S, Ravikant V, Thewissen JGM, Saravanan N, Singh IB, Prasad V (2010) Early Eocene warming events and the timing of terrestrial faunal exchange between India and Asia. Geology. https://doi.org/10.1130/G31585.1

Conti E, Eriksson T, Schönenberger J, Systma K, Baum DA (2002) Early Tertiary out of-India dispersal of Crypteroniaceae: evidence from phylogeny and molecular dating. Evolution 56:1931–1942

Cooper A, Lalueza-Fox C, Anderson S, Rambaut A, Austin J, Ward R (2001) Complete mitochondrial genome sequences of two extinct moas clarify ratite evolution. Nature 409:704–707

Dayanandan S, Aston PS, Williams SM, Primack, RB (1999) Phylogeny of the tropical tree family Dipterocarpaceae based on Nucleotide sequence of the chloroplast rbcL gene. Am J Bot 86:1182–1190

Deb U (1970) Palynological investigation of Tertiary sediments of Bengal Basin, south of Calcutta. Q J Geol Mining Metal Soc India 42(3):127–140

Deb U (1972) Some pollen grains from the Neyveli lignite. In: Ghosh AK et al (eds) Proceedings of the seminar on Paleopalynology and Indian Stratigraphy Calcutta 1971. Botany Department, Calcutta University, Kolkata, pp 220–228

Deb U, Baksi SK, Ghosh AK (1973) On the age of Neyveli lignite-a palynological approach. Q J Geol Mining Metal Soc India 45:23–38

Dogra NN, Singh RY, Misra PS (1985) Palynology of Dharamsala Beds, Himachal Pradesh. J Palaeontol Soc India 30:63–77

Ducousso M, Bena G, Bourgeois C, Buyck B, Eyssartier G, Vincelette M, Rabevohitra R, Randrihasipara L, Dreyfus B, Prin Y (2004) The last common ancestor of Sarcolaenaceae and Asian dipterocarp trees was ectomycorrhizal before the India-Madagascar separation, about 88 million years ago. Mol Ecol 13:231–236

Dutta SK (1980) Palynostratigraphy of the sedimentary formations of the Arunachal Pradesh-2. Palynology of the Siwalik equivalent rocks of Kameng District. Geophytology 10(1):5–13

Dutta SK, Sah SCD (1970) Palynostratigraphy of the Tertiary sedimentary formations of Assam-5. Stratigraphy and palynology of South Shillong Plateau. *Palaeontographica.* Abteilung B 131 (1–4):1–72

Dutta SK, Sah SCD (1974) Palynostratigraphy of the sedimentary formations of Assam, India-4. Age of the Laitryngew-Mawkma coal-bearing sandstones and their relationship with the Cherra Formation. Palaeobotanist 21(1):48–51

Dutta SK, Singh HP (1980) Palynostratigraphy of sedimentary formations in Arunachal Pradesh-1. Palynology of Siwalik rocks of Lesser Himalayas, Kameng District. In: Bharadwaj DC et al (eds.) Proceedings of the 4th international Palynological conference, Lucknow, 1976–77, vol 2. Birbal Sahni Institute of Palaeobotany, Lucknow, pp 617-626

Dutta S, Tripathi SKM, Mallick M, Mathews Runcie P, Paul F Greenwood, Rao MR, Summons Roger E (2011a) Eocene out-of-India dispersal of Asian Dipterocarps. Rev Palaeobot Palynol 166:63–68

Frederiksen NO (1994) Middle and late Palaeocene angiosperm pollen from Pakistan. Palynology 18:91–137

Germeraad JG, Hopping CA, Muller J (1968) Palynology of Tertiary sediments of tropical areas. Rev Palaeobot Palynol 6:189–348

Ghosh AK (1941) Fossil pollen in the Tertiary rocks of Assam. Sci Cult 6:674

Ghosh AK (1977) Palynology of the Siwaliks. Adv Pollen-Spore Res 2:14–19

Guleria JS (1996) Occurrence of Dipterocarpus in the Mar Formation of Bikaner, Rajasthan, western India. Palaeobotanist 43:49–53

Gupta A, Kotnala SK, Poovendan A (1996) Palynostratigraphy and paleoenvironmental analysis of subsurface Paleogene sequence of Jambusar Dabka area, South Cambay Basin. In: Pandey J et al (eds) Contributions to the 15th Indian Colloquium on Micropalaeontology and Stratigraphy, Dehradun, 1996. Oil and Natural Gas Corporation, Dehradun, pp 615–633

Hait AK, Banerjee M (1994) Palynology of lignite sediments from Mizoram, eastern India with remarks on age and environment of deposition. J Palynol 30(1–2):113–135

Jafar SA, Tripathi SKM (2001) Late Triassic Palynomorphs from Andaman—Nicobar Basin, Andaman Sea, India. Mod Geol 24:205–219

Jain KP, Dutta SK (1978) Lower Tertiary dinoflagellates, spores and pollen grains from Siang District, Arunachal Pradesh. J Palaeontol Soc India 21–22:106–111

Jain KP, Kar RK, Sah SCD (1973) A palynological assemblage from Barmer, Rajasthan. Geophytology 3(2):150–165

Kapoor R, Dogra NN, Singh RY (1988) Palynology of the Kasauli Formation in the type area in Solan District, Himachal Pradesh. J Palaeontol Soc India 33:105–115

Kar RK (1977) Palynostratigraphy of Maniyara Fort Formation (Oligocene) in the district of Kutch, western India. Geophytology 7(1):121–122

Kar RK (1978) Palynostratigraphy of the Naredi (Lower Eocene) and the Harudi (Middle Eocene) formations in the district of Kutch, India. Palaeobotanist 25:161–178

Kar RK (1979) Palynological fossils from the Oligocene sediments and their biostratigraphy in the district of Kutch, western India. The Palaeobotanist 26(1):16–49

Kar RK (1985) The fossil floras of Kachchh-IV. Tertiary palynostratigraphy. The Palaeobotanist 34:1–280

Kar RK (1992) Occurrence of *Dipterocarpus* type of pollen from the Miocene sediments of Kerala, South India. J Palynol 28:79–85

Kar RK (1995) Some new spore-pollen genera from Early Eocene sediments of Rajasthan. J Palynol 31:161–170

Kar RK (1996) On the Indian origin of *Ocimum* (Lamiaceae): a palynological approach. Palaeobotanist 43(3):43–50

Kar RK, Bhattacharya M (1992) Palynology of Rajpardi lignite, Cambay Basin and Gujra Dam and Akri lignite. Kutch Basin. Palaeobotanist 39(2):250–263

Kar RK, Jain KP (1981) Palynology of Neogene sediments around Quilon and Varkala, Kerala coast, South India-2. Spores and pollen grains. Palaeobotanist 27(2):113–131

Kar RK, Kumar M (1986) Palaeocene palynostratigraphy of Meghalaya. Pollen Spores 28:177–217

Kar RK, Sharma P (2001) Palynostratigraphy of Late Palaeocene and Early Eocene sediments of Rajasthan, India. *Palaeontographica*. Abteilung B 256(4–6):123–157

Karunakaran C, Ranga Rao A (1979) Status of exploration of hydrocarbons in the Himalayan region—contributions to stratigraphy and structure. Geol Surv India Misc Publ 41(5):1–66

Khan ML, Bera S (2010) Record of fossil fruit wing of Shorea Roxb. from the Neogene of Arunachal Pradesh. Curr Sci 98:1573–1575

Koshal VN (1988) Palynofossils and palaeoenvironment of the Tertiary subsurface sediments in Ahmedabad-Mehsana Block, Cambay Basin, western India. In: Whyte P et al (eds) The Palaeoenvironment of East Asia from the Mid-Tertiary, Palaeoclimatology and Palaeobotany, vol 1. Centre of Asian Studies, Univesity of Hong Kong, Hong Kong, pp 621–672

Koshal VN, Uniyal SN (1984) Palaeocene—Early Eocene palynofossils in the subsurface of North Cambay Basin, Gujarat (western India). In: Badve RM et al (eds) Proceedings of the 10th Indian Colloquium on the Indian Micropalaeontology and Stratigraphy, Pune, 1982. Maharashtra Association for the Cultivation of Science, Pune, pp 233–243

Koshal VN, Uniyal SN (1986) Palynostratigraphy of the Cenozoic succession of Cambay Basin, Gujarat. Bull Geol Mining Metal Soc India 54:208–226

Kumar A (1993) Biozonation and depositional environment of the Tertiary outcrop section along the Silchar–Haflong Road, Cachar District, southern Assam. In: Biswas SK et al (eds) Proceedings of the second seminar on Petroliferous basins of India, vol 1. Indian Petroleum Publishers, Dehradun, pp 459–476

Kumar A (1994a) Palynology of the Tertiary sediments exposed along the Silchar-Haflong Road Section, southern Assam. Palaeontographica Indica 2:1–241

Kumar M (1996) Palynostratigraphy and palaeoecology of Early Eocene palynoflora of Rajpardi lignite, Bharuch District, Gujarat. Palaeobotanist 43(3):110–121

Kumar A, Takahashi K (1991) Palynology of the Tertiary sediments of southern Assam, India. Bull Faculty of Liberal Arts Nagasaki Univ. (Nat Sci) 31(2): 515–659

Kumar M, Spicer RA, Spicer TE, Mehrotra RC, Mehrotra NC (2012) Sedimentology, palynostratigraphy and palynofacies of the late Oligocene Makum Coalfield, Assam, India: a window on lowland tropical vegetation during the most recent episode of significant global warmth. Palaeogeogr Palaeoclimatol Palaeoecol 342–343:143–162

Kumar M, Spicer RA, Spicer TE, Shukla A, Mehrotra RC, Monga P (2016) Palynostratigraphy and palynofacies of the Early Eocene Gurha lignite mine, Rajasthan, India. Palaeogeogr Palaeoclimatol Palaeoecol 461:98–108

Kumaran KPN, Shindikar M, Limaye RB (2004) Mangrove associated lignite beds of Malvan, Konkan: evidence for higher sea-level during the Late Tertiary (Neogene) along the west coast of India. Curr Sci 86:335–340

Lakhanpal RN (1970) Tertiary floras of India and their bearing on the historical geology of the region. Taxon 19:675–694

Lukose NG (1974) Palynology of the subsurface sediments of Manhera-Tibba Structure, Jaisalmer, western Rajasthan, India. Palaeobotanist 21(3):285–297

Mandal J (1997) Palynofossils from the Tertiary (Barail Group) of Nagaland: palaeoecological interpretation and age. Palaeobotanist 45:98–108

Mandal J (2005) Bombacaceae pollen from the Indian Tertiary sediments and its bearing on evolution and migration. Rev Palaeobot Palynol 133:277–293

Mandal J, Guleria JS (2006) Palynology of Vastan lignite (Surat District), Gujarat: its age, palaeoecology and depositional environment. Palaeobotanist 55(1–3):51–66

Mandal J, Chandra A, Kar RK (1994) Palynofossils from the Kadamtala coal, Middle Andaman, India. Geophytology 23(2):209–214

Mandal J, Chandra A, Bhattacharyya AP (2003) Palynology of the Baratang Formation, Andaman-Nicobar Islands and the significance of reworked palynomorphs. Palaeobotanist 52 (1–3):97–112

Mandaokar BD (2000a) Palynology and palaeoenvironment of the Bhuban Formation (Early Miocene) of Ramrikawn near Aizawl, Mizoram, India. The Palaeobotanist 49(2):317–324

Mandaokar BD (2000b) Palynology of coal bearing sediments of the Tikak Parbat Formation (Oligocene) from Namchik River Section, Changlang District, Arunachal Pradesh, India. Tertiary Res 20(1–4):37–46

Mandaokar BD (2002a) An interpretation of the palynology and palaeoecology of the Early Miocene Dulte Formation, Mizoram, India. The Palaeobotanist 51(1–3):113–121

Mandaokar BD (2002b) Palynological investigation of the Tikak Parbat Formation (Late Oligocene) of Borjan area, Nagaland, India. Minetech 23(1–2):19–33

Mandaokar BD (2002c) Palynoflora from the Keifang Formation (Early Miocene), Mizoram, India and its environmental significance. J Palaeontol Soc India 47:77–83

Mandaokar BD (2003) Palynology and palaeoecological consideration of Middle Bhuban Formation (Miocene), Lawngtlai, Mizoram, India. Gondwana Geol Mag Spec 6:185–193

Mandaokar BD (2004) Age and depositional environment of the Upper Bhuban Formation of Champhai area (Eastern Mizo Hills), India—A palynological approach. Palaeobotanist 53(1–3):143–153

Mandaokar BD (2007) Depositional environment and age of the Bhuban Formation of Chawngte area, Mizoram, India. Gondwana Geol Mag 22(1):37–46

Mandaokar BD (2008) Palynostratigraphy and Palaeoenvironment of the Upper Bhuban Formation (Early Miocene) of Diltlang, Chhimtuipui District, Mizoram. Gondwana Geol Mag 23(2):135–144

Mathews Runcie P, Tripathi SKM, Banerjee S, Dutta S (2013) Palynology, palaeoecology and palaeodepositional environment of Eocene lignites and associated sediments from Matanomadh mines, Kutch basin, Western India. J Geol Soc India 82:248

Mathur K (1963a) Occurrence of *Pediastrum* in Subathu Formation (Eocene) of Himachal Pradesh. India. Sci Cult 29:250

Mathur YK (1963b) Studies in the fossil microflora of Kutch, India-1 On the microflora and the hystrichosphaerids in the Gypseous shales (Eocene) of western Kutch. India. Proc Natl Inst India 29B(3):357–371

Mathur K (1964) On the occurrence of *Botryococcus* in Subathu beds of Himachal Pradesh, India. Sci Cult 30:607–608

Mathur YK (1966) On the microflora in the Supra-Trappeans of western Kutch, India. Q J Geol Mining Metal Soc India 38(1):33–51

Mathur YK (1984) Cenozoic palynofossils, vegetation, ecology and climate of the North and Northwestern Sub-Himalayan region. In: Whyte RO (ed) The Evolution of the East Asian

Environment, vol 2. Centre of Asian Studies, University of Hong Kong, Hong Kong, pp 504–551

Mathur YK, Chopra AS (1987) Palynofossils from the Cenozoic subsurface sediments of the Bengal Basin, India. Geosci J 1(8):109–152

Mathur YK, Chowdhary LR (1977) Palaeoecology of the Kalol Formation, Cambay Basin, India. In: Venkatachala BS, Sastri VV (eds) Proceedings of the 4th Colloquium on Indian Micropalaeontology and Stratigraphy, Dehradun, 1974–75. Institute of Petroleum Exploration, Oil and Natural Gas Commission, Dehradun, pp 164–178

Mathur YK, Jain AK (1980) Palynology and age of the Dras Volcanics near Shergol, Ladakh, Jammu and Kashmir, India. Geosci J 1(1):55–74

Mathur YK, Mathur K (1969) Studies in the fossil flora of Kutch (India)-3. On the palaeopalynoflora in the Pliocene sediments of Naera-Baraia area, Kutch. Bull Geol Mining Metal Soc India 42:1–12

Mathur YK, Mathur K 1980 Barail (Laisong) palynofossils and Late Oligocene nannofossils from the Andaman Island, India. Geosci J 1(2):51–65

Mathur YK, Venkatachala BS (1979) Palynological studies of the Cenozoic sediments of the Himalayan foothills. Misc Publ Geol Surv India 41(5):103–110

Mehrotra NC (1981) Palynological correlation of Mikir Formation with Lower Palaeogene sediments of Shillong Plateau. Geophytology 11(2):133–142

Mehrotra NC (1983) Palynology of Mikir Formation in the type area. Geosci J 4(1):1–34

Mehrotra RC (2003) Status of plant megafossils during Early Palaeogene in India. In: Wing SL et al (eds) Causes and consequences of Globally Warm Climates in Early Palaeogene: Boulder, Colorado. Geol Soc Am Special Paper 369:413–423

Mehrotra NC, Sah SCD (1982) Palynostratigraphy of Mikir Formation in the type area. Geosc J 3 (2):113–134

Mehrotra NC, Rawat RS, Juyal NP, Swamy SN (2001) Palynostratigraphy and palaeoenvironment of Basal Clastic sediments of Panna Formation of Bombay Offshore. Bull Oil Nat Gas Commission 38(2):7–26

Mehrotra NC, Venkatachala BS, Swamy SN, Kapoor PN (2002) Palynology in hydrocarbon exploration (The Indian Scenario). Part I: Category-I Basins. Memoir Geol Soc India 48:1–161

Misra BK, Singh A, Ramanujam CGK (1996a) Trilatiporate pollen from Indian Palaeogene and Neogene sequences: evolution, migration and continental drift. Rev Palaeobot Palynol 91(1–4):331–352

Misra UK, Shanker K, Patil RS (1996b) Biostratigraphy of the post-Kopili sediments of Garo Hills, Meghalaya. J Geol Soc India 48(1).93–100

Monga P, Kumar M, Prasad V, Joshi Y (2015) Palynostratigraphy, palynofacies and depositional environment of a lignite-bearing succession at Surkha Mine, Cambay Basin, north-western India. Acta Palaeobotanica 55(2):183–207

Morley RJ (2000) Origin and evolution of tropical rain forests. Wiley, Chichester, 362p

Muller J (1968) Palynology of the Pedawan and Plateau sandstone formations (Cretaceous-Eocene) in Sarawak, Malaysia. Micropaleontology 14:1–37

Muller J (1981) Fossil pollen records of extant Angiosperms. Bot Rev 47:1–142

Nandi B (1972) Some observations on the microflora of Middle Siwalik of Mohand (East) Field, Himachal Pradesh. In: Ghosh AK et al (eds) Proceedings of the seminar on Paleopalynology and Indian Stratigraphy Calcutta 1971. Botany Department, Calcutta University, Kolkata, pp 375–383

Nandi B (1975) Palynostratigraphy of the Siwalik Group of Punjab. Himalayan Geol 5:411–423

Nandi B (1980) Further contribution on the palynostratigraphy of the Siwalik Group. In: Bharadwaj DC et al (eds) Proceedings of the 4th international Palynological conference, Lucknow, 1976–77, vol 2. Birbal Sahni Institute of Palaeobotany, Lucknow, pp 727–734

Nandi B (1981) Miofloral investigation on the Neogene formations of Moran and Nahorkatiya wells, Upper Assam, India. Geophytology 11(1):27–40

Nandi B, Bandyopadhyay NN (1970) Preliminary observation on the microfossils and microstructures of Siwalik Lignite from Himachal Pradesh, India. Sci Cult 36:240–242

Naskar P, Baksi SK (1978) Palynological investigation of Akli lignite, Rajasthan, India. Palaeobotanist 25:314–329
Navale GKB (1962) Pollen and spores from Neyveli lignite, South India. Palaeobotanist 10(1–2):87–90
Navale GKB, Misra BK (1979) Some new pollen grains from Neyveli lignite, Tamil Nadu, India. Geophytology 8(2):226–239
Parsons J (1757) An account of some fossil fruits, and other bodies, found in Island of Shepey. Philos Trans 50:396–407
Phadtare NR, Kulkarni AR (1980) Palynological investigation of Ratnagiri lignite. Maharashtra. Geophytology 10(2):158–170
Phadtare NR, Kulkarni AR (1984) Affinity of the genus *Quilonipollenites* with the Malaysian palm *Eugeissona* Griffith. Pollen et Spores 26:217–226
Phadtare NR, Samant B (1996) Morphologic evolution in fossil pollen of *Ctenolophon* and its Indian phytogeographic implications. In: Pandey J et al (eds) Contributions to the 15th Indian Colloquium on Micropalaeontology and Stratigraphy, Dehradun, 1996. Oil and Natural Gas Corporation, Dehradun, pp 671–681
Phadtare NR, Thakur B (1990) Fossil pollen of *Alangium* from the Eocene lignite of Gujarat, India, with comments on its stratigraphic antiquity. Rev Palaeobot Palynol 63(3–4):281–297
Phadtare NR, Kumar R, Ghosh SK (1994) Stratigraphic palynology, floristic succession and the Tatrot/ Pinjor Boundary in Upper Siwalik sediments of Haripur Khol area, District Sirmaur, (H. P.), India. Himalayan Geol 15:68–82
Potonié R, Sah SCD (1960) Sporae dispersae of the lignites from Cannanore beach of the Malabar Coast of India. Palaeobotanist 7(2):121–135
Prasad M (1993) Siwalik (Middle Miocene) woods from the Kalagarh area in the Himalayan foot hills and their bearing on palaeoclimate and phytogeography. Rev Palaeobot Palynol 76:49–72
Prasad B, Dey AK (1986) The occurrence of Eocene sediments in Arunachal Pradesh: palynological evidence. Bull Oil Nat Gas Commission 23(2):67–74
Prasad V, Farooqui A, Tripathi SKM, Garg R, Thakur B (2009) Evidence of Late Palaeocene-Early Eocene equatorial rain forest refugia in southern Western Ghats, India. Journal of Biosciences 34(5):777–797
Rajendran CP, Raha PK, Kar RK (1989) Palynological assemblage from Neogene outcrops of Kerala coast, India. Indian Minerals 43(1):39–46
Ramanujam CGK (1960) Some pteridophytic spores from the Warkalli lignite in South India with special reference to those of Schizaeaceae. J Indian Bot Soc 39(1):46–55
Ramanujam CGK (1966) Palynology of the Miocene lignite from South Arcot District, Madras, India. Pollen et Spores 8(1):149–203
Ramanujam CGK (1967a) Pteridophytic spores from the Miocene Lignite of South Arcot District, Madras. Palynol Bull 2–3:29–40
Ramanujam CGK (1967b) Myrtaceous pollen from the peaty lignite of Alleppey in Kerala State, India. Sci Cult 33:229–230
Ramanujam CGK (1972) Revision of pteridophytic spores from the Warkalli lignite of South India. In: Ghosh AK et al (eds.) Proceedings of the seminar on Paleopalynology and Indian Stratigraphy Calcutta 1971. Botany Department, Calcutta University, Kolkata, pp 248–254
Ramanujam CGK (1982) Tertiary palynology and palynostratigraphy of southern India. Cenozoic Stratigraphy and Palynology in India. Palaeontol Soc India Spec Publ 1:57–64
Ramanujam CGK (1987) Palynology of the Neogene Warkalli Beds of Kerala State in South India. J Palaeontol Soc India 32:26–46
Ramanujam CGK, Rao KP (1973) A study of the pollen grains of *Ctenolophonidites* from the Warkalli deposits of South India with a note on the geological history of *Ctenolophon*. Palaeobotanist 20(2):210–215
Ramanujam CGK, Rao KP (1977) A palynological approach to the study of Warkalli deposits of Kerala in South India. Geophytology 7(2):160–164
Ramanujam CGK, Reddy PR (1984) Palynoflora of the Neyveli lignite-floristic and palaeoenvironmental analysis. J Palynol 20(1):58–74

Ramanujam CGK, Sarma PS, Reddy PR (1984) Quantification of the palynoassemblages of the first and second mine areas of Neyveli lignite. In: Badve RM et al (eds) Proceedings of the 10th Indian Colloquium on the Indian Micropalaeontology and Stratigraphy, Pune, 1982. Maharashtra Association for the Cultivation of Science, Pune, pp 269–275

Ramanujam CGK, Reddy PR, Sarma PS (1985) Additions to the palynoflora of Neyveli lignite, Tamil Nadu. J Palaeontol Soc India 30:49–53

Ramanujam CGK, Reddy PR, Sarma PS (1988) *Marginipollis* from the clay and lignite of South Arcot District, Tamil Nadu. Spec Publ Geol Surv India 11(2):271–276

Ramanujam CGK, Mallesham C, Ramakrishna H (1989a) Some significant palynomorphs from the subsurface Miocene sediments of eastern coast of southern India. In: Patil GV et al (eds) Proceedings of the fifth all india symposium on Palynology, Nagpur, 1979. Department of Botany, Institute of Science, Nagpur, pp 27–30

Ramanujam CGK, Ramakrishna H, Mallesham C (1989b) Palynoassemblage of the subsurface Miocene sediments of the East Coast of southern India: its floristic and environmental significance. In: Biradar NV et al (eds) Proceedings of the special Indian geophytological conference, Pune, 1996. The Palaeobotanical Society, Lucknow, pp 113–117.

Ramanujam CGK, Reddy PR, Ramakrishna H (1997) Dicolpate palm pollen from the Neogene deposits of Godavari-Krishna Basin, AP. J Palynol 33:129–136

Rangarao A, Khan KN, Venkatachala BS, Sastri VV (1981) Neogene/quaternary boundary and the Siwalik. In: Sastry MVA et al (eds) Proceedings of the field conference on neogene-quaternary Boundary in India Chandigarh 1979. Geological Survey of India, Kolkata, pp 131–142

Rao MR (1983) A new record of *Malayaeaspora costata* Trivedi, Ambwani & Kar from Tertiary sediments of Meghalaya and Assam. Geophytology 13(2):241–242

Rao MR (1986) Palynology of the Barail (Oligocene) and Surma (Lower Miocene) sediments exposed along Sonapur-Badarpur Road Section, Jaintia Hills (Meghalaya) and Cachar (Assam)- Part IV. Gymnospermous pollen grains. Geophytology 16(1):89–96

Rao MR (1990) Palynological investigation of Arthungal Borehole, Alleppey District, Kerala. Palaeobotanist 38:243–255

Rao MR (1995) Palynostratigraphic zonation and correlation of the Eocene-Early Miocene sequence in Alleppey District, Kerala, India. Rev Palaeobotany Palynol 86(3–4):325–348

Rao MR (1996) An Early Miocene palynofloral assemblage from Turavur Borehole, Alleppey District, Kerala—its palaeoecological and stratigraphical significance. Geophytology 25(1–2):155–163

Rao MR (2000) Palynological investigation of the Kherapara Formation (Oligocene) exposed along Tura Dalu Road near Kherapara, West Garo Hills District, Meghalaya, India. The Palaeobotanist 49(2):293–309

Rao MR (2001) Palynostratigraphic zonation of the Tertiary sediments of the Kerala Basin, India. In: Goodman DK, Clarke RT (eds) Proceedings of the IX international palynological congress, Houston, 1996. American Association of Stratigraphic Palynologists Foundation, Dallas, USA, pp 277–289

Rao MR (2004) Palynological investigation of the Sindhudurg Formation (Miocene) exposed at Kalviwadi, Sindhudurg District, Maharashtra, India. Palaeobotanist 53(1–3):123–135

Rao MR, Nair KK (1998) Palynological investigation of Miocene sediments exposed in Kannanellur—Kundara area, Quilon District, Kerala. Geophytology 27(1–2):49–59

Rao MR, Rajendran CP (1996) Palynological investigation of Tertiary lignite and associated sediments from Cannanore, Kerala Basin, India. Palaeobotanist 43(2):63–82

Rao KP, Ramanujam CGK (1975) A palynological approach to the study of Quilon Beds of Kerala State in South India. Curr Sci 44:730–732

Rao KP, Ramanujam CGK (1978) Palynology of the Neogene Quilon Beds of Kerala State in South India-1. Spores of pteridophytes and pollen of monocotyledons. Palaeobotanist 25: 397–427

Rao KP, Ramanujam CGK (1982) Palynology of the Quilon Beds of Kerala State in South India-II. Pollen of dicotyledons and discussion. Palaeobotanist 30(1):68–100

Rao AR, Vimal KP (1950) Plant microfossils from Palana Lignite (Eocene), Bikaner. Curr Sci 19:82–84

Rao AR, Vimal KP (1952) Tertiary pollen from lignites from Palana (Eocene) Bikaner. Proceedings of the National Institute of India 18(6):595–601

Rao MR, Saxena RK, Singh HP (1985) Palynology of the Barail (Oligocene) and Surma (Lower Miocene) sediments exposed along Sonapur-Badarpur Road Section, Jaintia Hills (Meghalaya) and Cachar (Assam)-Part V. Angiospermous pollen grains. Geophytology 15(1):7–23

Rawat MS, Mukherjee JS, Venkatachala BS (1977) Palynology of the Kadi Formation, Cambay Basin, India. In: Venkatachala BS, Sastri VV (eds) Proceedings of the 4th Colloquium on Indian Micropalaeontology and Stratigraphy, Dehradun, 1974–75. Institute of Petroleum Exploration, Oil and Natural Gas Commission, Dehradun, pp 179–192

Rawat MS, Sharma KD, Juyal NP (1991) A standard palynozonation for Krishna-Godavari Basin. In: Pandey J, Banerjie V (eds) Proceedings of the conference on integrated exploration—Research achievements and perspectives. Oil and Natural Corporation, Dehradun, pp 281–297

Rawat RS, Swamy SN, Juyal NP, Kapoor PN (1996a) Palynostratigraphic and paleoenvironmental interpretation of Panna Formation in the east of Bombay High. Bull Oil Nat Gas Commission 33(2):123–135

Rawat RS, Misra CM, Sharma J, Roy H (1996b) Palynostratigraphy and Palaeoenvironment of Basal Clastic (Panna Formation) sequence of Mahim area, Bombay Offshore. In: Pandey J et al (eds) Contributions to the 15th Indian Colloquium on Micropalaeontology and Stratigraphy, Dehradun, 1996. Oil and Natural Gas Corporation, Dehradun, pp 657–662

Sah SCD, Dutta SK (1966) Palynostratigraphy of the sedimentary formations of Assam-1. Stratigraphical position of the Cherra Formation. Palaeobotanist 15(1–2):72–86

Sah SCD, Dutta SK (1968) Palynostratigraphy of the Tertiary sedimentary formations of Assam-2. Stratigraphic significance of spores and pollen in the Tertiary succession of Assam. Palaeobotanist 16:177–195

Sah SCD, Dutta SK (1974) Palynostratigraphy of the sedimentary formations of Assam: 3. Biostratigraphic zonation of Cherra Formation of South Shillong Plateau. Palaeobotanist 21 (1):42–47

Sah SCD, Kar RK (1970) Palynology of the Laki sediments in Kutch-3. Pollen from the boreholes around Jhulrai, Baranda and Panandhro. Palaeobotanist 18(2):127–142

Sah SCD, Kar RK (1974) Palynology of the Tertiary sediments of Palana, Rajasthan. Palaeobotanist 21(2):163–188

Sah SCD, Singh RY (1974) Palynological biostratigraphy of the Tura Formation in the type area. In: Sah SCD, Cross AT (eds) Symposium on Stratigraphical Palynology Lucknow 1971. Special Publication 3, Birbal Sahni Institute of Palaeobotany, Lucknow, pp 76–98

Sah SCD, Singh RY (1977) Mesozoic-Cainozoic boundary in Assam. J Geol Soc India 18:445–454

Sah SCD, Singh RY, Singh HP (1980) Palynological zonation of the Tipam Group in Nahorkatiya area of Upper Assam. In: Bharadwaj DC et al (eds) Proceedings of the 4th international Palynological conference, Lucknow, 1976–77, vol 2, Birbal Sahni Institute of Palaeobotany, Lucknow, pp 635–642

Sahni B, Sitholey RV, Puri GS (1947) Palaeobotany in India—6. Correlation of Tertiary succession in Assam by means of microfossils. J Ind Bot Soc 26:261–263

Salujha SK, Kindra GS (1984) Palynostratigraphy of Atharmura Anticline, Tripura. In: Badve RM et al (eds) Proceedings of the 10th Indian Colloquium on the Indian Micropalaeontology and Stratigraphy, Pune, 1982. Maharashtra Association for the Cultivation of Science, Pune, pp 391–409

Salujha SK, Kindra GS (1986) Palynostratigraphy of the Silchar-Haflong Road Traverse, District Cachar. Bull Geol Mining Metal Soc India 54:238–249

Salujha SK, Srivastava NC, Rawat MS (1969) Microfloral assemblage from Subathu sediments of Simla Hills. J Palaeontol Soc India 12:15–40

Salujha SK, Rehman K, Kindra GS (1973) Distinction between the Bhuban and Bokabil sediments on the southern edge of Shillong Plateau based on palynofossil assemblage. Bull Oil Nat Gas Commission 10(1–2):109–117

Salujha SK, Kindra GS, Rehman K (1977) Palynostratigraphy of the Tertiary sediments of the Gojalia Anticline, Tripura—Part I. Discussion. In: Venkatachala BS, Sastri VV (eds) Proceedings of the 4th Colloquium on Indian Micropalaeontology and Stratigraphy, Dehradun, 1974–75. Institute of Petroleum Exploration, Oil and Natural Gas Commission, Dehradun, pp 122–126

Salujha SK, Kindra GS, Rehman K (1979) Palynostratigraphy of Tertiary sediments of the Gojalia Anticline, Tripura-Part 2. Systematic palynology. J Palynol 14(1):71–93

Salujha SK, Kindra GS, Rehman K (1980) Palynostratigraphy of the Tertiary sediments of the Tulamura Anticline, Tripura. In: Bharadwaj DC et al (eds) Proceedings of the 4th international Palynological conference, Lucknow 1976–77, vol 2. Birbal Sahni Institute of Palaeobotany, Lucknow, pp 667–685

Samant B (1994) Age of the Bhaunagar lignite deposit of Gujarat. J Palynol 30(1–2):49

Samant B (2000) Palynostratigraphy and age of the Bhavnagar lignite, Gujarat, India. Palaeobotanist 49(1):101–118

Samant B, Phadtare NR (1997) Stratigraphic palynoflora of the Early Eocene Rajpardi lignite, Gujarat and the lower age limit of the Tarkeshwar Formation of South Cambay Basin, India. *Palaeontographica*. Abteilung B 245(1–6):1–108

Samant B, Tapaswi PM (2001) Palynology of the Early Eocene Surat lignite deposits of Gujarat, India. J Palaeontol Soc India 46:121–132

Sarkar S (1991) Eocene palynofossils from the Kakara Series of the Lesser Himalaya, Himachal Pradesh, India. Rev Palaeobot Palynol 67(1–2):1–11

Sarkar S (1997) Palynostratigraphy and palaeoenvironment of the Subathu Formation (Eocene) of Lesser Himalaya, Himachal Pradesh, India. Ind J Petrol Geol 6(1):99–115

Sarkar S, Singh HP (1988) Palynological investigation of the Subathu Formation (Eocene) in the Banethi-Bagthan area of Himachal Pradesh. *Palaeontographica*. Abteilung B 209:29–109

Sarkar S, Singh HP (1994) Palaeoecology of the Lower Siwalik palynofloras from Kundlu and Nalagarh formations, Himachal Pradesh, India. Himalayan Geol 15:95–106

Sarma PS, Reddy PR, Kalavathi K (1984) Pollen grains referable to monocotyledons from Neyveli lignite, Tamil Nadu. Ind J Bot 7:201–209

Sasaki S (2006) Ecology and physiology of Dipterocarpaceae. In: Suzuki K, Ishii K, Sakurai S, Sasaki S (eds) Plantation technology in tropical forest science. Springer, pp 3–22

Saxena RK (1977) On the stratigraphic status of the Matanomadh Formation, Kutch, India. Palaeobotanist 24(3):211–214

Saxena RK (1978) Palynology of the Matanomadh Formation in type area, north-western Kutch, India (Part 1). Systematic description of pteridophytic spores. Palaeobotanist 25:448–456

Saxena RK (1979) Palynology of the Matanomadh Formation in type area, north-western Kutch, India (Part 2). Systematic description of gymnospermous and angiospermous pollen grains. Palaeobotanist 26(2):130–143

Saxena RK (1980) Palynology of the Matanomadh Formation in type area, north-western Kutch, India (Part 3)-Discussion. The Palaeobotanist 26(3):279–296

Saxena RK (1981) Stratigraphy of the area around Matanomadh in north-western Kutch with special reference to the Matanomadh Formation. Palaeobotanist 27(3):300–313

Saxena RK (1982) Taxonomic study of the polycolpate pollen grains from the Indian Tertiary sediments with special reference to nomenclature. Rev Palaeobot Palynol 37(3–4):283–315

Saxena RK (1992a) Neyveli lignites and associated sediments—Their palynology, palaeoecology, correlation and age. Palaeobotanist 40:345–353

Saxena RK (2000a) Palynology of the Neogene sediments of North-west India. Misc Publ Geol Surv India 64:11–22

Saxena RK (2000b) Palynological investigation of the Sindhudurg Formation in the type area, Sindhudurg District, Maharashtra, India. Bull Oil Nat Gas Corp 37(1):157–166

Saxena RK, Bhattacharyya AP (1987) Palynology of the Siwalik sediments of Kala Amb-Nahan area in Sirmaur District, Himachal Pradesh. Palaeobotanist 35(2):187–195
Saxena RK, Misra NK (1990) Palynological investigation of the Ratnagiri Beds of Sindhu Durg District, Maharashtra. Palaeobotanist 38:263–276
Saxena RK, Singh HP (1980) Occurrence of palynofossils from the Pinjor Formation (Upper Siwalik) exposed near Chandigarh. Curr Sci 49(12):479–480
Saxena RK, Singh HP (1981) *Pinjoriapollis*, a new fossil pollen from the Pinjor Formation (Upper Siwalik) exposed near Chandigarh. Curr Sci 50(9):418–419
Saxena RK, Singh HP (1982a) Palynological investigation of the Upper Siwalik sediments exposed along Hoshiarpur-Una Road Section in Punjab and Himachal Pradesh. Geophytology 12(2):287–306
Saxena RK, Singh HP (1982b) Palynology of the Pinjor Formation (Upper Siwalik) exposed near Chandigarh, India. The Palaeobotanist 30(3):325–339
Saxena RK, Tripathi SKM (2012) Seven decades of Indian spore-pollen floras: a compendium. AASP Contrib Ser 45:1–182 (AASP Foundation, Dallas, Texas, USA)
Saxena RK, Trivedi GK (2009) Palynological investigation of the Kopili Formation (Late Eocene) in North Cachar Hills, Assam, India. Acta Palaeobotanica 49(2):253–277
Saxena RK, Sarkar S, Singh HP (1984) Palynological investigation of Siwalik sediments of Bhakra-Nangal area, Himachal Pradesh. Geophytology 14(2):178–198
Saxena RK, Rao MR, Singh HP (1987) Palynology of the Barail (Oligocene) and Surma (Lower Miocene) sediments exposed along Sonapur-Badarpur Road Section, Jaintia Hills (Meghalaya) and Cachar (Assam)—Part VI, Palynostratigraphic zonation. Palaeobotanist 35(2):150–158
Saxena RK, Misra NK, Khare S (1992) Ratnagiri Beds of Maharashtra—lithostratigraphy, flora, palaeoclimate and environment of deposition. Ind J Earth Sci 19(4):205–213
Saxena RK, Tripathi SKM, Prasad V (1996) Palynofloral investigation of the Tura Formation (Palaeocene) in Nongwal Bibra area, East Garo Hills, Meghalaya. Geophytology 26(1):19–31
Sein MK, Sah SCD (1974) Palynological demarcation of the Eocene-Oligocene sediments in the Jowai-Badarpur Road Section, Assam. In: Sah SCD, Cross AT (eds.) Symposium on Stratigraphical Palynology Lucknow 1971. Special Publication 3. Birbal Sahni Institute of Palaeobotany, Lucknow, pp 99–105
Shanmukhappa M (1991) Palynostratigraphy and palaeoenvironmental analysis of Gandhar area in Cambay Basin. In: Pandey J, Banerjie V (eds) Proceedings of the conference on integrated exploration research, achievements and perspectives. Oil and Natural Corporation, Dehradun, pp 309–320
Shanmukhappa M, Koshal VN (1993) Palynological investigation of Eocene sediments in Gandhar area, Broach Depression, Cambay Basin. In: Biswas SK et al (eds) Proceedings of the second seminar on Petroliferous basins of India, vol 2. Indian Petroleum Publishers, Dehradun, pp 191–202
Shanmukhappa M, Uniyal SN (2008) Biostratigraphy and palaeoenvironmental analysis of subsurface Palaeogene sediments in western part of Ahmedabad Block, North Cambay Basin. Palaeobotanist 57(3):465–477
Shanmukhappa M, Kotnala SK, Poovendan A, Prabhakar PD (2000) Paleogene Palynostratigraphy of Linch-Nandasan Area, Mehsana Block, North Cambay Basin. Bull Oil Nat Gas Corp 37 (2):151–159
Sharma KD (1988) Late Cretaceous—Palaeocene palynological assemblages from the subsurface of Godavary-Krishna Basin. In: Maheshwari HK (ed) Palaeocene of India: proceedings of the symposium on Palaeocene of India: limits and subdivisions Lucknow 1986. Indian Association of Palynostratigraphers, Lucknow, pp 65–67
Shrivastava PK, Ganeshan S, Ray D (1974) Tipam Group in the subsurface of Upper Assam Valley, south of Brahmaputra. J Geol Soc India 15(2):165–181
Siddhanta BK (1986) The age of Neyveli lignite with reference to stratigraphy and palynology. Ind Miner 40:61–82
Singh RY (1977a) Stratigraphy and palynology of the Tura Formation in the type area-Part II (descriptive palynology). The Palaeobotanist 23(3):189–205

Singh RY (1977b) Stratigraphy and palynology of the Tura Formation in type area (Part III): Discussion. Palaeobotanist 24(1):1–12

Singh RS (1990) Palynology of Langrin Coalfield, South Shillong Plateau, Meghalaya. Palaeobotanist 38:217–228

Singh RS (1999) Diversity of *Nypa* in the Indian subcontinent: Late Cretaceous to recent. Palaeobotanist 48(2):147–154

Singh RY, Dogra NN (1988) Palynological zonation of Palaeocene in India with special reference to western Rajasthan. In: Maheshwari HK (ed) Palaeocene of India: proceedings of the symposium on Palaeocene of India: limits and subdivisions Lucknow 1986. Indian Association of Palynostratigraphers, Lucknow, pp 51–64

Singh HP, Khanna AK (1980) Palynology of the Palaeogene marginal sediments of Himachal Pradesh, India. In: Bharadwaj DC et al (eds) Proceedings of the 4th international Palynological conference, Lucknow, 1976–77, vol 2. Birbal Sahni Institute of Palaeobotany, Lucknow, pp 462–471

Singh HP, Sarkar S (1984b) A Kasauli palynoflora from Banethi area of Himachal Pradesh, India. Geophytology 14(1):40–54

Singh HP, Sarkar S (1987a). Palynostratigraphy of Subathu Formation (Eocene) in the Banethi-Bagthan area, Sirmaur District, Himachal Pradesh. Palaeobotanist, 35(2):204–210

Singh HP, Sarkar S (1987b) Significant palynozones of Subathu Formation (Himachal Pradesh) and their bearing on stratigraphy. Palaeobotanist 35(3):249–257

Singh HP, Sarkar S (1992) Palynology and palaeoenvironment of Lower Tertiary sediments around Garkhal, Himachal Pradesh, India. Geophytology 22:181–191

Singh HP, Sarkar S (1994) Palynostratigraphy of the Kasauli Formation (Lower Miocene), Himachal Pradesh, India. Geophytology 24(1):49–54

Singh HP, Saxena RK (1980) Upper Siwalik palynoflora from Gagret-Bharwain Road Section, Himachal Pradesh. Geophytology 10(2):278–279

Singh HP, Saxena RK (1981) Palynology of the Upper Siwalik sediments in Una District, Himachal Pradesh. Geophytology 11(2):173–181

Singh HP, Saxena RK (1984) Palynology of the Neogene sediments of Jorajan Well-3, Upper Assam. In: Sharma AK et al (eds) Proceedings of the symposium on evolutionary botany and biostratigraphy Calcutta 1979, A.K. Ghosh Commemoration volume, current trends in life sciences, vol 10, pp 613–631

Singh RY, Tiwari BS (1979) A note on the palynostratigraphy of the Tertiary sediments of Upper Assam. Bull Ind Geol Assoc 12(2):151–159

Singh T, Tripathi SKM (1990) Siwalik sediments of Arunachal Himalaya: Palynology, palaeoecology and palaeogeography. Palaeobotanist 38:325–332

Singh HP, Khanna AK, Sah SCD (1973) Problems and prospects of Tertiary palynology in northern India. Bull Ind Geol Assoc 6(1):71–77

Singh HP, Singh RY, Sah SCD (1976) Palynological correlation of the Palaeocene subsurface assemblages from Garo Hills, Meghalaya. J Palynol 11:43–64

Singh HP, Khanna AK, Sah SCD (1978) Palynological zonation of the Subathu Formation in the Kalka-Simla area of Himachal Pradesh. Himalayan Geol 8(1):33–46

Singh HP, Rao MR, Saxena RK (1987) Palynology of the Barail (Oligocene) and Surma (Lower Miocene) sediments exposed along Sonapur-Badarpur Road Section, Jaintia Hills (Meghalaya) and Cachar (Assam)-Part VII. Discussion. Palaeobotanist 35(3):331–341

Srisailam K, Ramanujam CGK (1982) Fossil palm pollen grains from the Warkalli sediments of Cannanore area. Rec Geol Surv India 114(5):123–131

Srivastava SK (1988) *Ctenolophon* and *Sclerosperma* Palaeogeography and Semonian Indian Plate position. J Palynol 23(24):239–254

Thanikaimoni G, Caratini C, Venkatachala BS, Ramanujam CGK, Kar RK (1984) Selected Tertiary angiospermous pollen from India and their relationship with African Tertiary pollen, vol 19. Institut Francais de Pondichéry, Travaux de la Section Scientfique et Technique, 92p

Tripathi SKM (1994) New angiosperm pollen from subsurface Early Palaeogene sediments of Barmer District, Rajasthan, India. Palaeobotanist 42(1):61–65

Tripathi SKM (1995) Palynology of subsurface Palaeocene-Eocene sediments near Kapurdi, Barmer District, Rajasthan, India. Palaeobotanist 43(1):45–53

Tripathi SKM (1997) Palynological changes across subsurface Palaeocene-Eocene sediments at Barmer, Rajasthan, India. Palaeobotanist 46(1–2):168–171

Tripathi SKM, Singh HP (1984a) Two new pollen genera from the Lower Tertiary sediments of Meghalaya. Palaeobotanist 32(2):153–157

Tripathi SKM, Singh HP (1984b). Palynostratigraphical zonation and correlation of the Jowai-Sonapur Road Section (Palaeocene-Eocene), Meghalaya, India. In: Tiwari RS et al (eds) Proceedings of the 5th Indian geophytological conference, Lucknow, 1983. Special Publication, Palaeobotanical Society. Lucknow, pp 316–328

Tripathi SKM, Singh HP (1985) Palynology of the Jaintia Group (Palaeocene-Eocene) exposed along Jowai-Sonapur Road, Meghalaya, India-Part I. Systematic palynology. Geophytology 15 (2):164–187

Tripathi SKM, Singh T (1992) Record of Early Tertiary palynotaxa from Siang District, Arunachal Pradesh, India. Palaeobotanist 39(2):149–154

Tripathi SKM, Srivastava D (2010) Palynological investigations, facies analysis and palaeoenvironmental interpretations from Late Palaeocene to Early Eocene lignites and associated sediments of Barmer, western India. Palaeobotanist 59:1–32

Tripathi SKM, Srivastava D (2012) Palynology and Palynofacies of the Early Palaeogene lignite bearing succession of Vastan, Cambay Basin, Western India. Acta Palaeobotanica 52(1):157–175

Tripathi SKM, Saxena RK, Prasad V (2000) Palynological investigation of the Tura Formation (Early Eocene) exposed along Tura-Dalu Road, West Garo Hills, Meghalaya, India. The Palaeobotanist 49(2):239–251

Tripathi SKM, Singh UK, Sisodia MS (2003) Palynological investigation and environmental interpretation on Akli Formation (Late Palaeocene) from Barmer Basin, Rajasthan, India. Palaeobotanist 52(1–3):87–95

Tripathi SKM, Mathur SC, Nama SL, Srivastava D (2008) Palynological studies from Early Eocene sequence exposed near Matasukh, Nagaur District, western Rajasthan, India. In: Trivedi PC (ed) Palaeobotany to Modern Botany. Pointer Publishers, Jaipur, India, pp 49–56

Tripathi SKM, Kumar M, Srivastava D (2009) Palynology of Lower Palaeogene (Thanetian-Ypresian) coastal deposits from the Barmer Basin (Akli Formation, Western Rajasthan, India): Palaeoenvironmental and palaeoclimatic implications. Geologica Acta 7(1–2):147–160

Trivedi GK (1985) Palynology of the Kopili Formation (Late Eocene) exposed along Jowai-Badarpur Road, Meghalaya. J Ind Bot Soc 64:66–72

Trivedi GK, Saxena RK (2000) Palynofloral investigation of the Kopili Formation (Late Eocene) exposed near Umrongso in North Cachar Hills District, Assam, India. Palaeobotanist 49 (2):269–280

Van der Ham RWJM (1989) New observations on the pollen of *Ctenolophon* Oliver (Ctenolophonaceae), with remarks on the history of the genus. Rev Palaeobot Palynol 59 (153):160

van der Hammen T (1956) A palynological systematic nomenclature. Boletin Geologico Bogota 4 (2–3):63–101

Venkatachala BS (1972) Observation on some palynological contributions to Indian stratigraphy. Palaeobotanist 19(3):284–296

Venkatachala BS (1973) Palynological evidence on the age of Cuddalore Sandstone. Geophytology 3(2):145–149

Venkatachala BS, Chowdhary LR (1977). Palaeoecology of the Kadi Formation, Cambay Basin, India. In: Venkatachala BS, Sastri VV (eds) Proceedings of the 4th Colloquium on Indian Micropalaeontology and Stratigraphy, Dehradun, 1974–75. Institute of Petroleum Exploration, Oil and Natural Gas Commission, Dehradun, pp 259–277

Venkatachala BS, Kar RK (1969) Palynology of the Tertiary sediments of Kutch-1. Spores and pollen from Borehole No. 14. Palaeobotanist 17(2):157–178

Venkatachala BS, Rawat MS (1972) Palynology of the Tertiary sediments in the Cauvery Basin-1. Palaeocene-Eocene palynoflora from the subsurface. In: Ghosh AK et al (eds) Proceedings of the seminar on Paleopalynology and Indian Stratigraphy, Calcutta, 1971. Botany Department, Calcutta University, Kolkata, pp 292–335

Venkatachala BS, Rawat MS (1973) Palynology of the Tertiary sediments in the Cauvery Basin-2. Oligocene-Miocene palynoflora from the subsurface. Palaeobotanist 20(2):238–263

Venkatachala BS, Sharma KD (1984) Palynological zonation in subsurface sediments in Narsapur Well No. 1, Godavari-Krishna Basin, India. In: Badve RM et al (eds) Proceedings of the 10th Indian Colloquium on the Indian Micropalaeontology and Stratigraphy, Pune, 1982. Maharashtra Association for the Cultivation of Science, Pune, pp 445–465

Plates

© The Editor(s) (if applicable) and The Author(s), under exclusive license to
Springer Nature Switzerland AG 2020

S. K. M. Tripathi, *Catalogue of Selected Angiosperm Pollen Grains from Palaeogene and Neogene Sediments of India*, Society of Earth Scientists Series,
https://doi.org/10.1007/978-3-030-42435-0

Plate 1

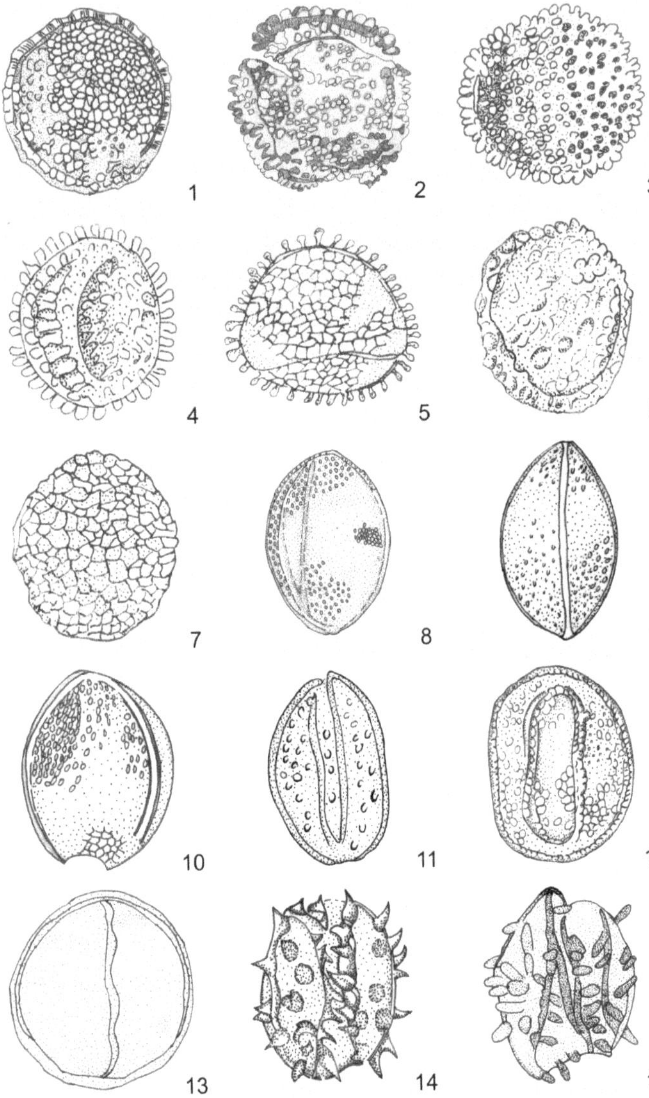

(1) *Assamiapollenites brownii* (Biswas 1962) Singh 1975. (2) *Incrotonipollis burdwanensis* (Baksi et al. 1979) Jansonius and Hills 1981. (3) *Incrotonipollis neyvelii* (Baksi et al. 1979) Jansonius & Hills 1981. (4) *Retipilonapites arcotense* Ramanujam 1966. (5) *Retipilonapites conspicua* Rao & Ramanujam 1978. (6) *Retipilonapites delicatissimus* Ramanujam 1966. (7) *Retipilonapites tertiaris* Rao & Ramanujam 1978. (8) *Arecipites bellus* Sah & Kar 1970. (9) *Arecipites indicus* Venkatachala & Rawat 1972. (10) *Arecipites intrapunctatus* Kar & Saxena 1981. (11) *Arecipites keralensis* Rao & Ramanujam 1978. (12) *Arecipites magnus* Ramanujam 1987. (13) *Arecipites wodehousei* Samant & Phadtare 1997. (14) *Arengapollenites achinatus* Kar 1985. (15) *Arengapollenites clavatus* Samant & Phadtare 1997

Plate 2

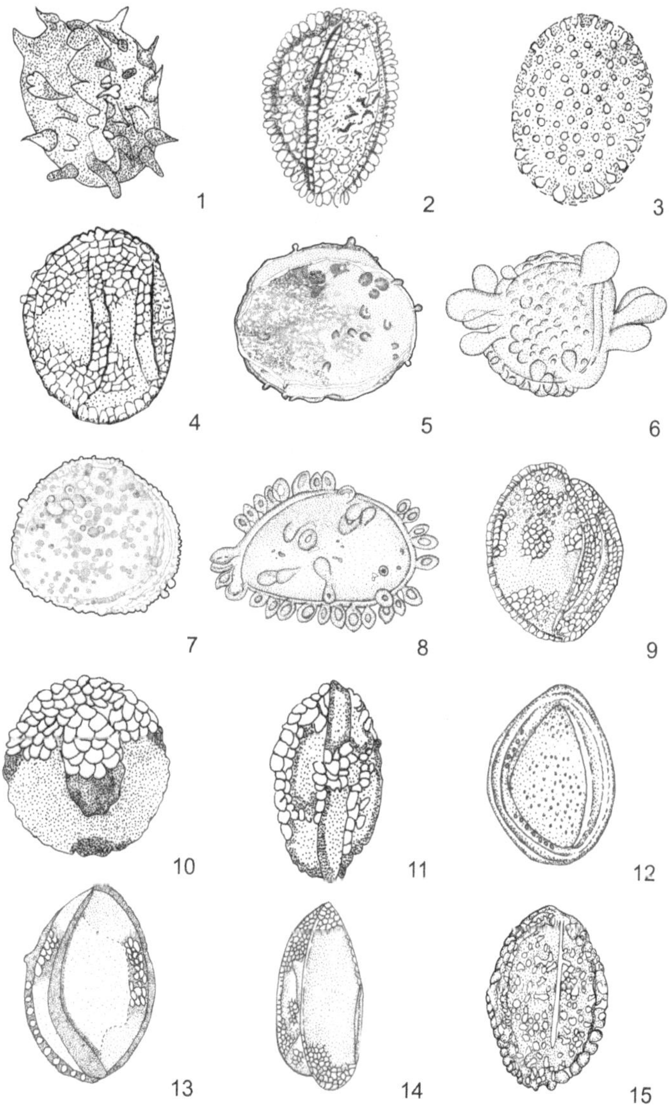

(1) *Arengapollenites ovatus* Kar & Bhattacharya 1992. (2) *Clavapalmaedites hammenii* Rao & Ramanujam 1978. (3) *Clavapalmaedites heteroclavatus* Kulkarni et al. 1985. (4) *Crotonisulcites grandis* Rao & Ramanujam 1978. (5) *Kapurdipollenites baculatus* Tripathi 1994. (6) *Kapurdipollenites clavatus* Tripathi & Srivastava 2010. (7) *Kapurdipollenites gemmatus* Tripathi 1994. (8) *Kapurdipollenites ovatus* Tripathi & Srivastava 2010. (9) *Liliacidites baculatus* Venkatachala & Kar 1969. (10) *Liliacidites circularis* Dutta & Sah 1970. (11) *Liliacidites crassireticulatus* Venkatachala & Rawat 1972. (12) *Liliacidites ellipticus* Venkatachala & Kar 1969. (13) *Liliacidites magnus* Jain et al. 1973. (14) *Liliacidites matanomadhensis* Saxena 1979. (15) *Liliacidites padappakkarensis* Rao & Ramanujam 1978

Plate 3

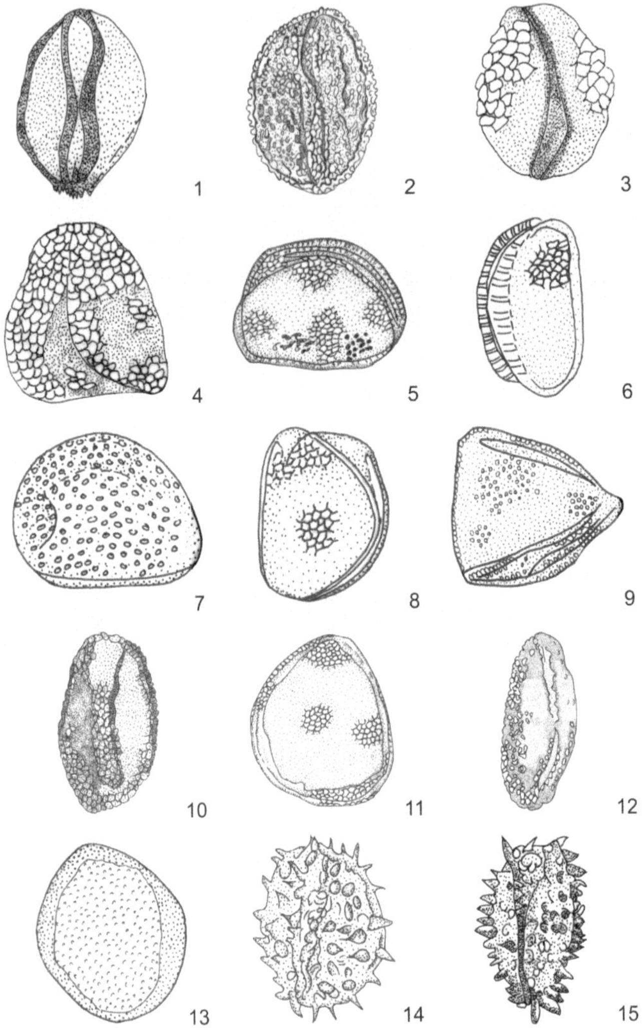

(1) *Liliacidites pilosus* Venkatachala & Rawat 1972. (2) *Liliacidites ramanujamii* Saxena 1992b. (3) *Liliacidites ratnagiriensis* Kulkarni et al. 1985. (4) *Liliacidites reticulatus* Sah & Kar 1974. (5) *Longapertites klinkenbergii* Rao & Ramanujam 1978. (6) *Longapertites ornatus* Kulkarni et al. 1985. (7) *Longapertites psilatus* Mathur & Chopra 1987. (8) *Longapertites retipilatus* Kar 1985. (9) *Longapertites triangulatus* Samant & Phadtare 1997. (10) *Matanomadhiasulcites kutchensis* (Saxena 1979) Kar 1985. (11) *Matanomadhiasulcites major* (Singh in Sah and Singh 1974) Saxena & Khare 2004. (12) *Matanomadhiasulcites maximus* (Saxena 1979) Kar 1985. (13) *Matanomadhiasulcites microreticulatus* (Dutta and Sah 1970) Kar & Kumar 1986. (14) *Spinomonosulcites ankleshwarensis* (Kar and Bhattacharya 1992) Samant & Phadtare 1997. (15) *Neocouperipollis cymbatus* (Venkatachala and Rawat 1972) Saxena & Khare 2004

Plate 4

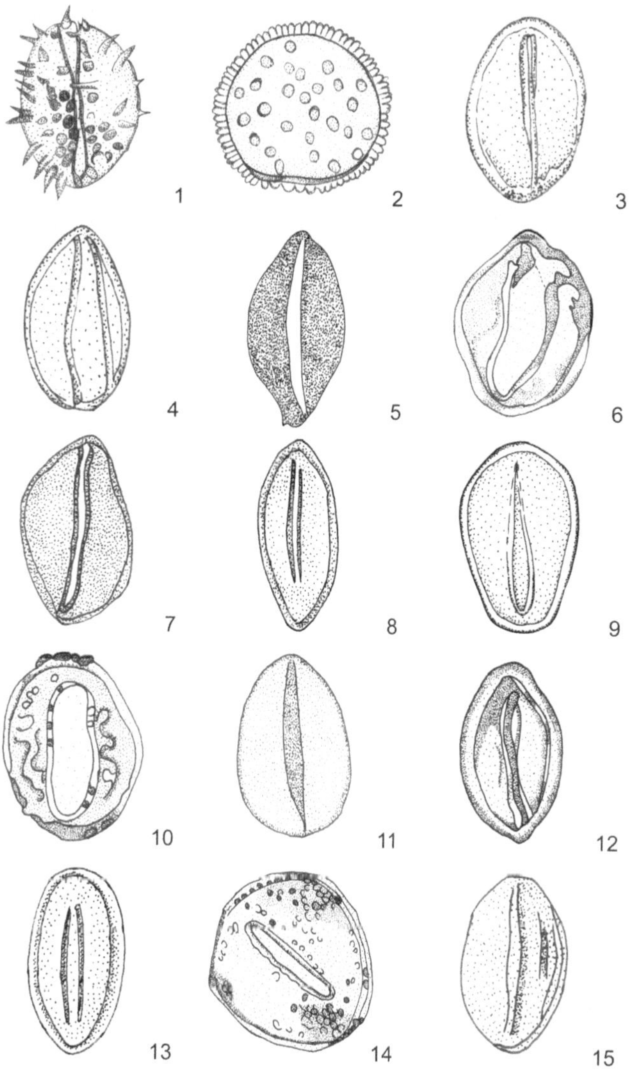

(1) *Neocouperipollis rarispinosus* (Sah and Dutta 1966) Singh 1990. (2) *Thymelaepollis crotonoides* Sah & Kar 1970. (3) *Monocolpopollenites baculatus* (Venkatachala and Rawat 1972) Saxena 2010. (4) *Monocolpopollenites cauveriensis* (Venkatachala and Rawat 1972) Saxena 2010. (5) *Monocolpopollenites communis* (Sah and Dutta 1966) Kumar & Takahashi 1991. (6) *Monocolpopollenites eocenicus* (Biswas 1962) Saxena 2010. (7) *Monocolpopollenites rajpardiensis* (Samant and Phadtare 1997) nom. nov. (8) *Monocolpopollenites indicus* (Ramanujam1966) Saxena 2010. (9) *Monocolpopollenites keralensis* (Rao and Ramanujam 1978) Saxena 2010. (10) *Monocolpopollenites kutchensis* (Venkatachala and Kar 1969) Saxena & Trivedi 2009. (11) *Monocolpopollenites magnus* (Sah and Kar 1970) Saxena 2010. (12) *Monocolpopollenites nadhamunii* (Venkatachala and Kar 1969) Saxena & Trivedi 2009. (13) *Monocolpopollenites neyvelii* (Ramanujam 1966) Saxena 2010. (14) *Monocolpopollenites ovatus* (Sah and Kar 1970) Saxena 2010. (15) *Monocolpopollenites plicatus* Sah and Kar 1970 Saxena 2010

Plate 5

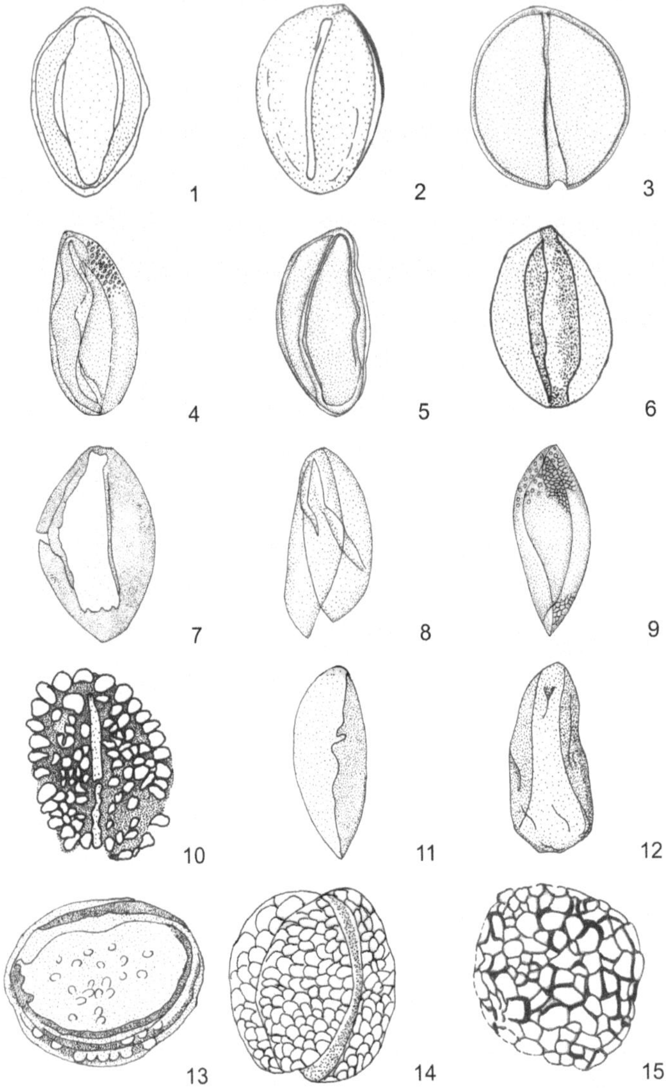

(1) *Monocolpopollenites ratnagiriensis* (Kulkarni et al. 1985) Saxena 2010. (2) *Palmidites aplicatus* Kar & Kumar 1986. (3) *Palmidites excellensus* Kar & Kumar 1986. (4) *Palmidites granulatus* Mehrotra 1983. (5) *Palmidites naviculus* Kar & Saxena 1981. (6) *Palmidites noviculatus* Sarkar & Singh 1988. (7) *Palmidites obtusus* Tripathi & Singh 1985. (8) *Palmidites plicatus* Singh in Sah & Singh 1974. (9) *Palmidites punctatus* Mehrotra 1983. (10) *Paravuripollis mulleri* Rao & Ramanujam 1978. (11) *Pinjoriapollis lanceolatus* Saxena & Singh 1981. (12) *Pinjoriapollis magnus* Saxena & Singh 1981. (13) *Proxapertites assamicus* (Sah and Dutta 1966) Singh 1975. (14) *Proxapertites crassimurus* (Sah and Dutta 1966) Singh 1975. (15) *Proxapertites emendatus* (Sah and Dutta 1966) Kar & Kumar 1986

Plate 6

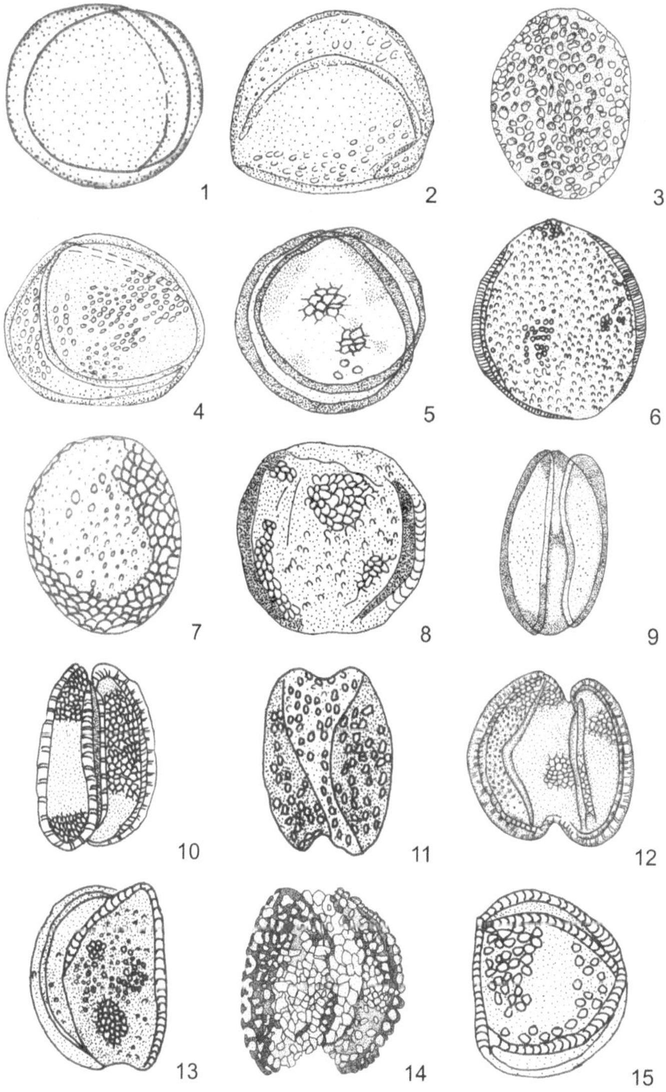

(1) *Proxapertites granulatus* Singh 1977a. (2) *Proxapertites hammenii* Venkatachala & Rawat 1972. (3) *Proxapertites intecta* (Baksi 1962) Singh 1975. (4) *Proxapertites marginatus* (Venkatachala and Kar 1969) Singh 1975. (5) *Proxapertites microreticulatus* Jain et al. 1973. (6) *Proxapertites pramathii* (Biswas 1962) Singh 1975. (7) *Proxapertites reticulatus* (Kar and Saxena 1981) Kar 1985. (8) *Proxapertites rugulatus* Samant & Phadtare 1997. (9) *Proxapertites scabratus* Jain et al. 1973. (10) *Quilonipollenites crassimarginatus* Phadtare & Kulkarni 1984a. (11) *Quilonipollenites minutus* Samant & Phadtare 1997. (12) *Quilonipollenites ornatus* Rao & Ramanujam 1978. (13) *Quilonipollenites reticulatus* Samant & Phadtare 1997. (14) *Quilonipollenites sahnii* Rao & Ramanujam 1978. (15) *Quilonipollenites trianguliformis* Phadtare & Kulkarni 1984a

Plate 7

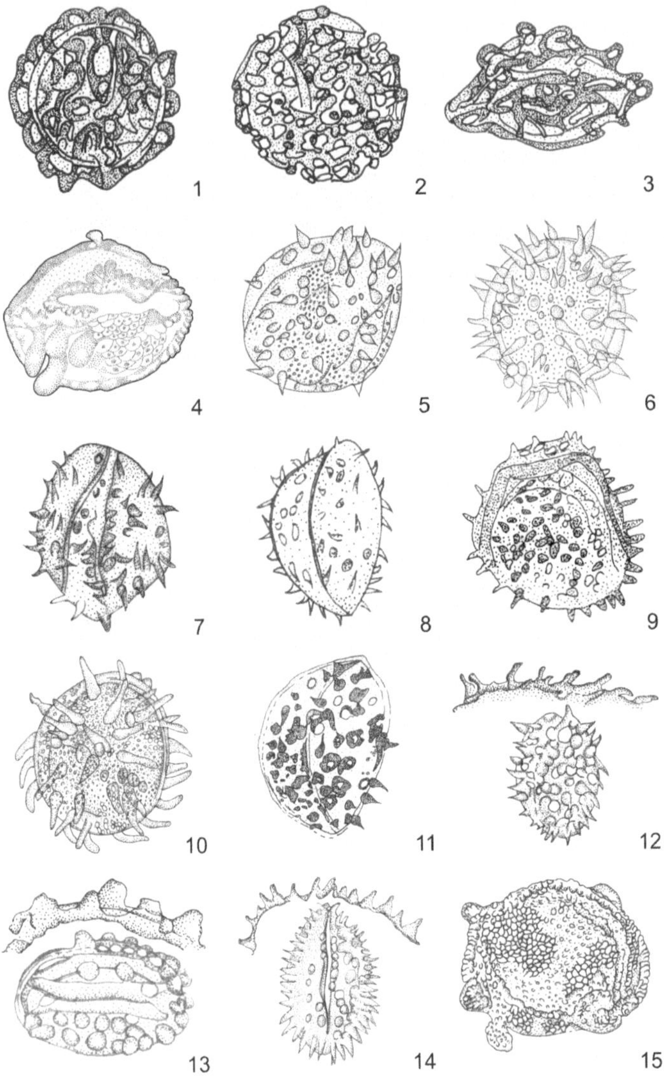

(1) *Retipollenites borassodendrii* Samant & Phadtare 1997. (2) *Retipollenites echinatus* Samant & Phadtare 1997. (3) *Retipollenites monosulcites* Samant & Phadtare 1997. (4) *Retiverrumonosulcites barmerensis* Tripathi 1994. (5) *Spinizonocolpites bulbospinosus* Singh 1990. (6) *Spinizonocolpites indicus* Singh 1990. (7) *Spinizonocolpites quilonensis* Rao & Ramanujam 1978. (8) *Spinizonocolpites spinulosus* Ramanujam 1987. (9) *Spinizonocolpites venkatachalae* Saxena & Khare 2004. (10) *Spinizonocolpites wodehousei* Singh 1990. (11) *Spinomonosulcites achinatus* (Sah and Kar 1970) Singh & Misra 1991c. (12) *Spinomonosulcites magnus* (Dutta and Sah 1970) Singh & Misra 1991c. (13) *Spinomonosulcites spinorobustus* (Kar and Kumar 1986) Singh & Misra 1991c. (14) *Spinomonosulcites varispinosus* Singh & Misra 1991c. (15) *Tuberculozonisulcites retibaculatus* Kar & Sharma 2001

Plate 8

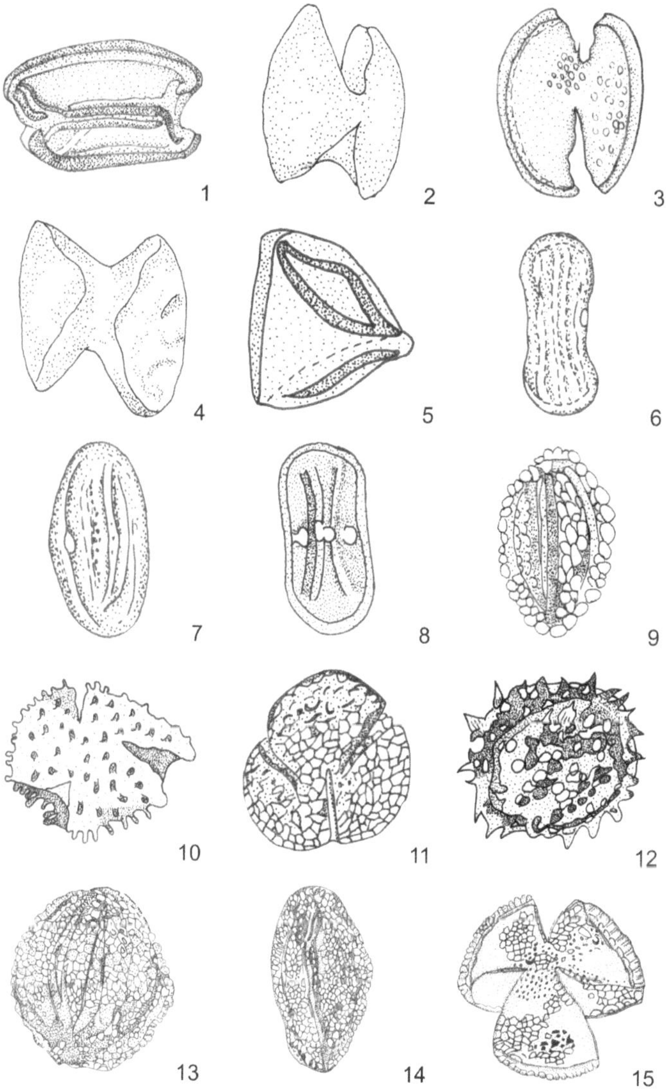

(1) *Dicolpopollis edavensis* Rao & Ramanujam 1978. (2) *Dicolpopollis fragilis* Salujha et al. 1972. (3) *Dicolpopollis microreticulatus* Rao & Ramanujam 1978. (4) *Dicolpopollis proprius* Salujha et al. 1972. (5) *Dicolpopollis triangularis* Kulkarni et al. 1985. (6) *Umbelliferoipollenites constrictus* Venkatachala & Kar 1969. (7) *Umbelliferoipollenites ovatus* Venkatachala & Kar 1969. (8) *Umbelliferoipollenites typicus* Samant & Phadtare 1997. (9) *Albertipollenites aquifoliaceaeformis* (Biswas 1962) Mandal & Rao 2001. (10) *Albertipollenites baculatus* (Jain et al. 1973) Mandal & Rao 2001. (11) *Albertipollenites crassireticulatus* (Dutta & Sah 1970) Mandal & Rao 2001. (12) *Malvacearumpollis rudis* Kar 1979. (13) *Albertipollenites karii* Mandal & Rao 2001. (14) *Albertipollenites kumarii* Mandal & Rao 2001. (15) *Albertipollenites kutchensis* Mandal & Rao 2001

Plate 9

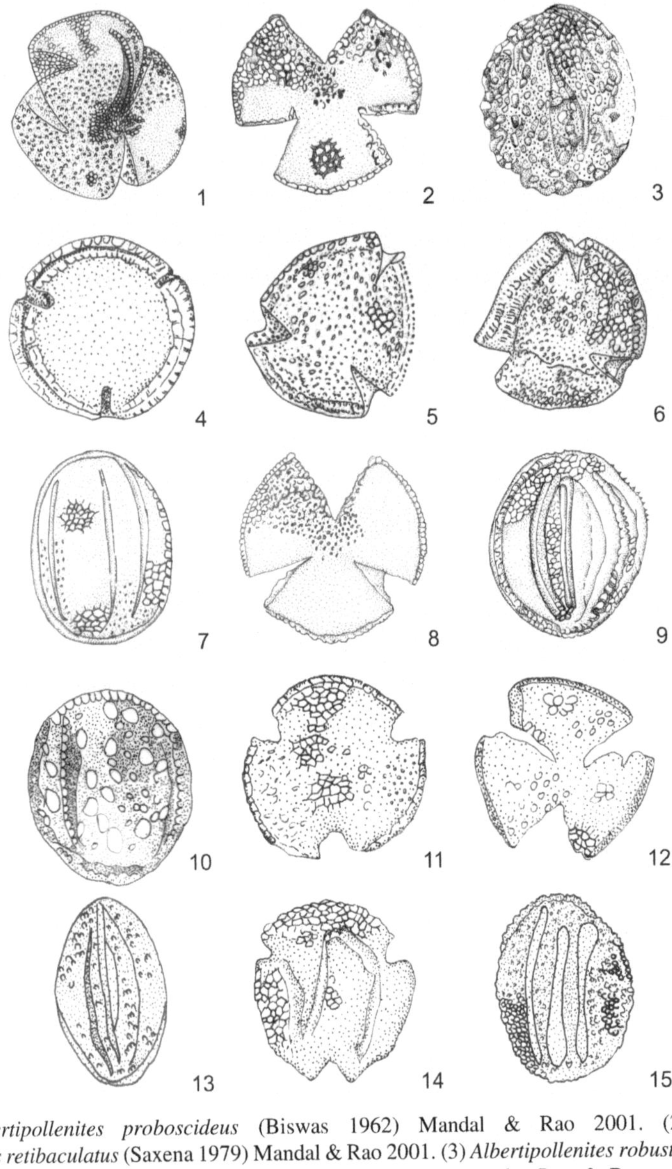

(1) *Albertipollenites proboscideus* (Biswas 1962) Mandal & Rao 2001. (2) *Albertipollenites retibaculatus* (Saxena 1979) Mandal & Rao 2001. (3) *Albertipollenites robustus* (Sah and Kar 1970) Mandal & Rao 2001. (4) *Bacubrevitricolpites rotundus* Rao & Ramanujam 1982. (5) *Bithnokiapollis retipilatus* Kar & Sharma 2001. (6) *Bithnokiapollis striatus* Kar & Sharma 2001. (7) *Cruciferoipollenites elongatus* Navale & Misra 1979. (8) *Dipterocarpuspollenites retipilatus* (Kar and Jain 1981) Kar 1992c. (9) *Foveotricolpites alveolatus* Mandal & Rao 2001. (10) *Cryptopolyporites indicus* Saxena & Khare 2004. (11) *Foveotricolpites piercei* Rao & Ramanujam 1982. (12) *Foveotricolpites prolatus* Rao & Ramanujam 1982. (13) *Foveotricolpites sijuensis* (Baksi 1962) Mandal & Rao 2001. (14) *Foveotricolpites simplex* (Rao and Ramanujam 1982) Mandal & Rao 2001. (15) *Foveotricolpites tamilensis* Saxena & Khare 2004

Plate 10

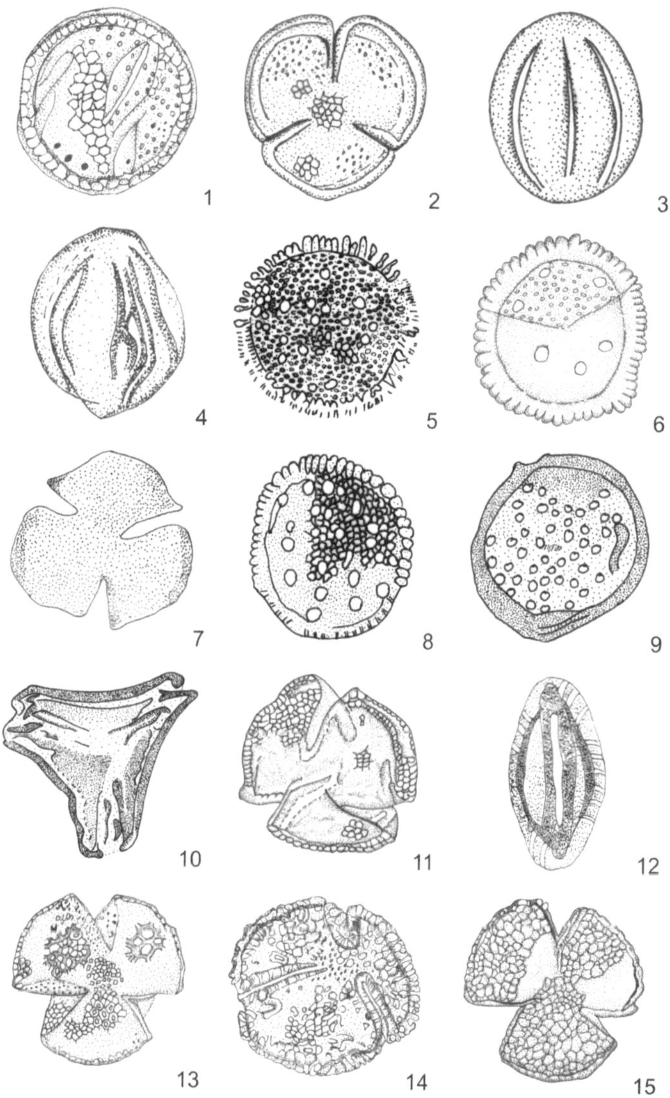

(1) *Icacinoipollenites spinulatus* Navale & Misra 1979. (2) *Intrareticulites brevis* (Sah and Kar 1970) Kar 1985. (3) *Ladakhipollenites constatus* (Dutta and Sah 1970) Mandal & Rao 2001. (4) *Ladakhipollenites elongatus* Tripathi & Singh 1985. (5) *Clavaperiporites homoclavatus* Samant & Phadtare 1997. (6) *Clavaperiporites jacobii* Ramanujam 1966. (7) *Ladakhipollenites levis* (Sah and Dutta1966) Mathur & Jain 1980. (8) *Clavaperiporites ramanujamii* Samant & Phadtare 1997. (9) *Cryptopolyporites cryptus* Venkatachala & Kar 1969. (10) *Loranthipites elegans* Rao & Ramanujam 1982. (11) *Plumbaginacipites neyvelii* Navale & Misra 1979. (12) *Polygonaceaepites zonoides* Baksi 1962. (13) *Rousea matanomadhensis* (Saxena 1979) Mandal & Rao 2001. (14) *Rousea meghalayensis* Mandal & Rao 2001. (15) *Rousea saxenae* Mandal & Rao 2001

Plate 11

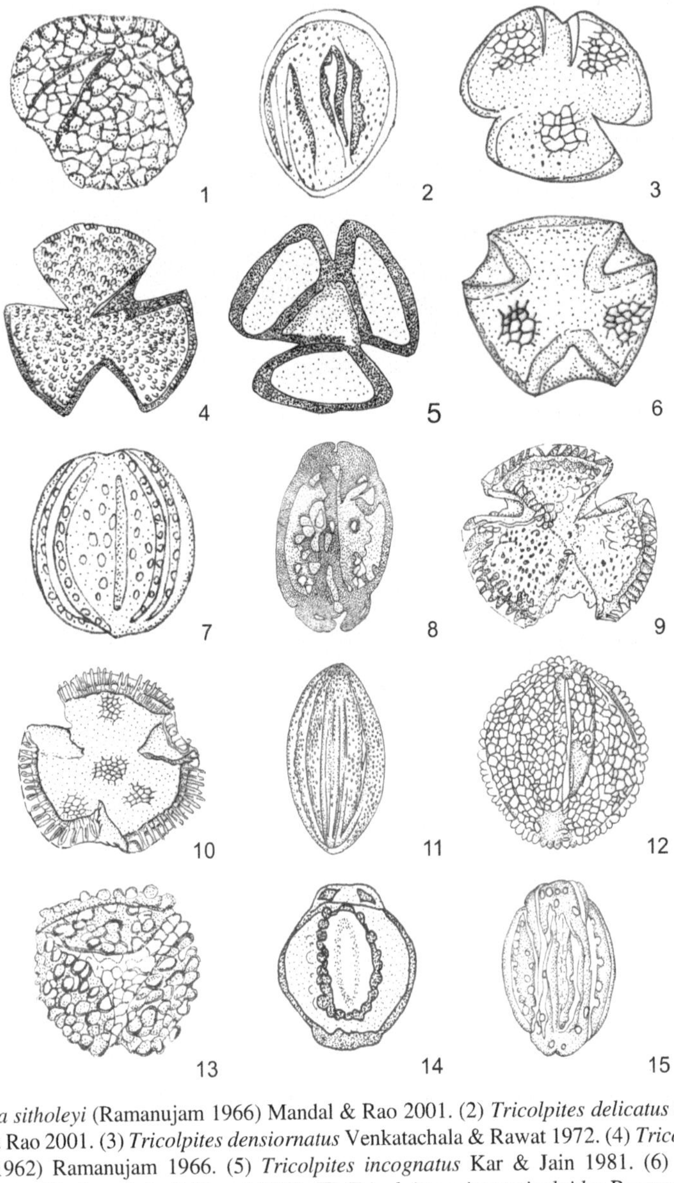

(1) *Rousea sitholeyi* (Ramanujam 1966) Mandal & Rao 2001. (2) *Tricolpites delicatus* (Kar 1979) Mandal & Rao 2001. (3) *Tricolpites densiornatus* Venkatachala & Rawat 1972. (4) *Tricolpites foxii* (Biswas 1962) Ramanujam 1966. (5) *Tricolpites incognatus* Kar & Jain 1981. (6) *Tricolpites margocolpites* Venkatachala & Rawat 1972. (7) *Tricolpites microreticuloides* Ramanujam 1966. (8) *Marginipollis kutchensis* (Venkatachala and Kar 1968) Kar 1978. (9) *Warkallipollenites erdtmanii* Ramanujam & Rao in Thanikaimoni et al. 1984. (10) *Warkallipollenites thanikaimonii* Ramanujam 1987. (11) *Barringtoniapollenites retibaculatus* Kar & Sharma 2001. (12) *Barringtoniapollenites retipilatus* Kar & Sharma 2001. (13) *Clavasyncolpites gracilis* Rao & Ramanujam 1982. (14) *Marginipollis psilatus* Ramanujam et al. 1988. (15) *Marginipollis quilonensis* Rao & Ramanujam 1982

Plate 12

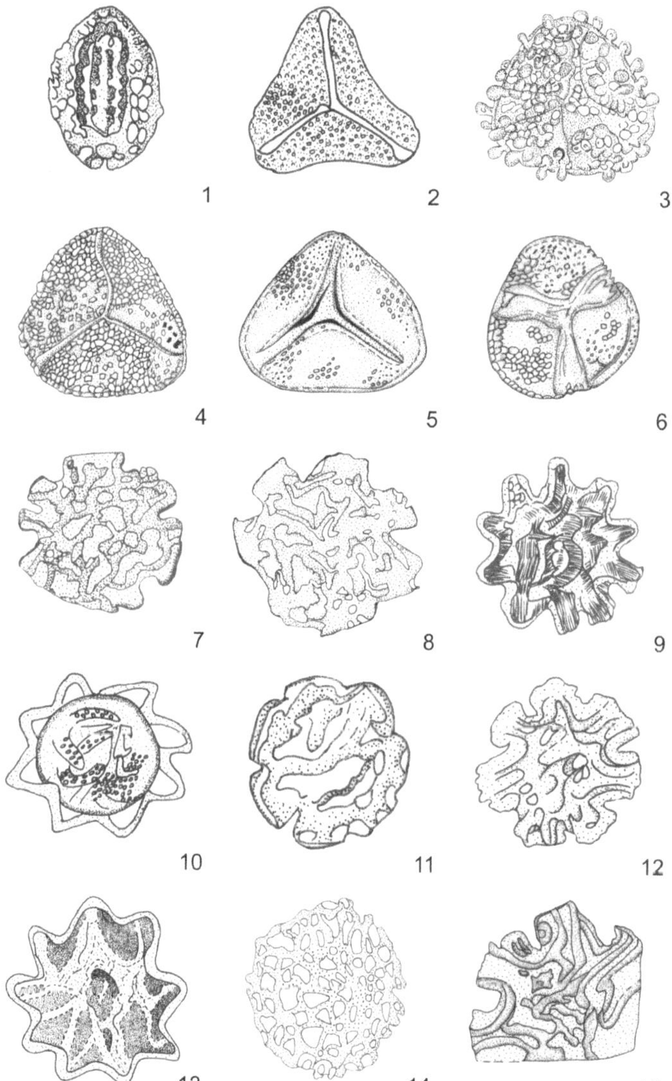

(1) *Marginipollis verrucatus* Ramanujam et al. 1988. (2) *Neotrichotomosulcites foveolatus* Samant & Phadtare 1997. (3) *Pilatrisyncolpites triangulatus* Kar et al. 1994. (4) *Retitrisyncolpites reimannii* Mandal et al. 1994. (5) *Retitrisyncolpites tha-ungii* Mandal et al. 1994. (6) *Trisyncolpites ramanujamii* Kar 1979. (7) *Ctenolophonidites erdtmanii* Ramanujam & Rao 1973. (8) *Ctenolophonidites keralensis* Ramanujam & Rao 1973. (9) *Ctenolophonidites magnus* Saxena & Khare 2004. (10) *Ctenolophonidites neyveliensis* Ambwani et al. 1981. (11) *Ctenolophonidites palaeoparvifolius* Kar & Jain 1981. (12 & 15) *Ctenolophonidites saadii* Ramanujam & Rao 1973. (13) *Ctenolophonidites stellatus* Navale & Misra 1979. (14) *Ocimumpollenites indicus* Kar 1996

Plate 13

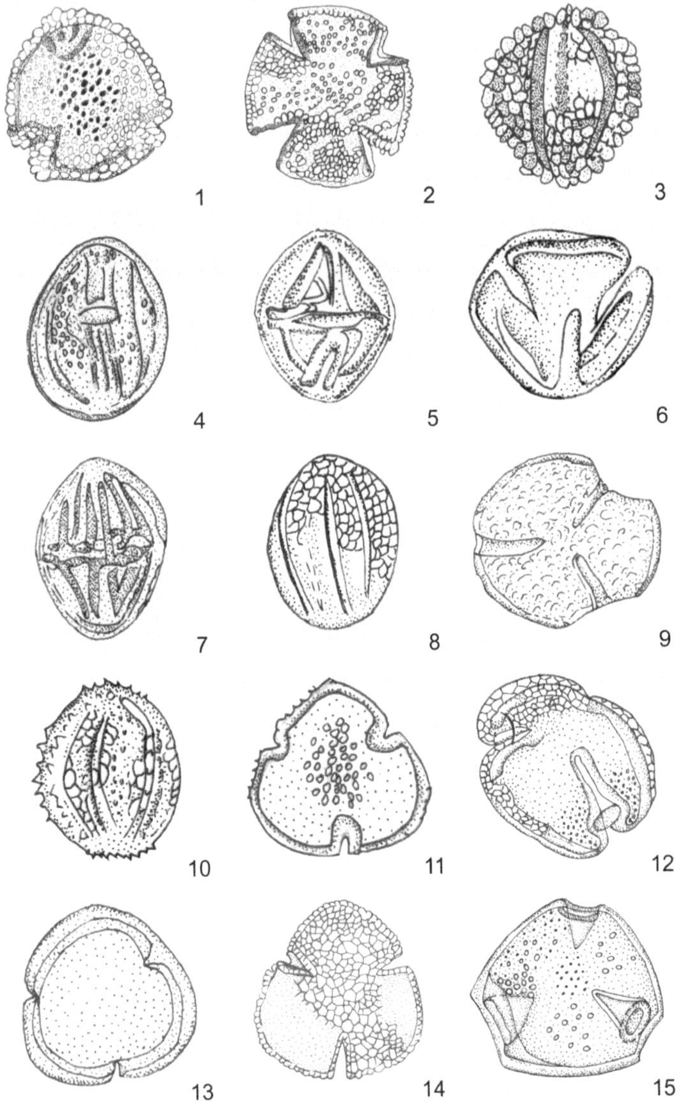

(1) *Alangiopollis gemmatus* Navale & Misra 1979. (2) *Alangiopollis arcotense* Navale & Misra 1979. (3) *Araliaceoipollenites baculatus* Jain et al. 1973. (4) *Araliaceoipollenites descretus* Venkatachala & Rawat 1973. (5) *Araliaceoipollenites potoniaei* Ramanujam 1966. (6) *Araliaceoipollenites psilatus* Dutta & Sah 1970. (7) *Araliaceoipollenites quilonensis* Rao & Ramanujam 1982. (8) *Araliaceoipollenites reticulatus* Dutta & Sah 1970. (9) *Araliaceoipollenites rugulatus* Venkatachala & Rawat 1972. (10) *Asperitricolporites pilosus* Venkatachala & Rawat 1972. (11) *Bombacacidites assamicus* Sah & Dutta 1968. (12) *Angulocolporites microreticulatus* Kar 1985. (13) *Bombacacidites minutus* Rao & Ramanujam 1982. (14) *Bombacacidites triangulatus* Kar 1985. (15) *Calophyllumpollenites rotundus* Sah & Kar 1974

Plate 14

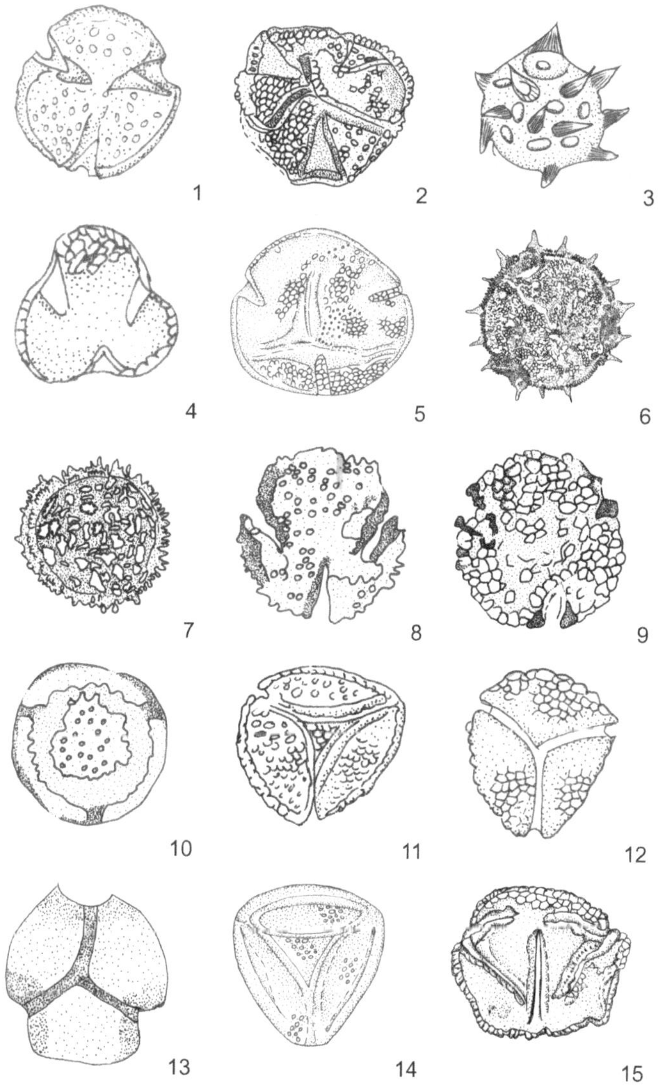

(1) *Caprifoliipites descretus* Venkatachala & Rawat 1972. (2) *Kielmeyerapollenites eocenicus* Sah & Kar 1974. (3) *Malvacearumpollis paucibaculatus* Venkatachala & Rawat 1973. (4) *Caprifoliipites wodehousei* Ramanujam 1966. (5) *Chiranthodendronpollenites bikanerensis* Kar & Sharma 2001. (6) *Compositoipollenites alleppeyensis* Rao 1995. (7) *Polygonacidites frequens* Sah & Dutta 1968. (8) *Coneopollis decorus* Venkatachala & Rawat 1973. (9) *Coneopollis reticulatus* (Venkatachala and Rawat 1973) Venkatachala & Rawat in Thanikaimoni et al. 1984. (10) *Coramandalipollis pauciornatus* (Venkatachala and Rawat 1973) Venkatachala & Rawat in Thanikaimoni et al. 1984. (11) *Cupanieidites cooksonae* Ramanujam 1987. (12) *Cupanieidites flaccidiformis* Venkatachala & Rawat 1972. (13) *Cupanieidites granulatus* Jain et al. 1973. (14) *Cupanieidites punctatus* Rao & Ramanujam 1982. (15) *Dakshinipollenites tripakshii* Navale & Misra 1979

Plate 15

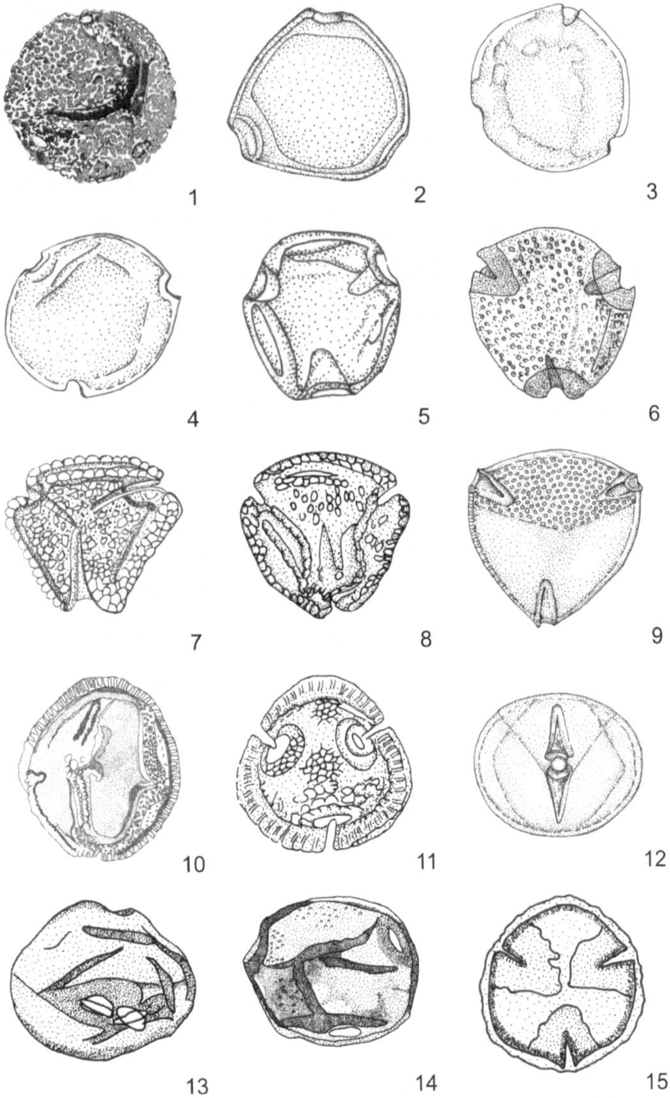

(1) *Densiverrupollenites eocenicus* Tripathi & Singh 1984a. (2) *Dermatobrevicolporites dermatus* (Sah and Kar 1970) Kar 1985. (3) *Dermatobrevicolporites exaltus* Kar 1985. (4) *Dermatobrevicolporites globatus* (Kar and Saxena 1981) Kar 1985. (5) *Dermatobrevicolporites triangulus* (Saxena 1979) Kar 1985. (6) *Favitricolporites complex* Sah & Dutta 1968. (7) *Hippocrateaceaedites constrictus* Sah & Kar 1974. (8) *Hippocrateaceaedites quilonensis* Rao & Ramanujam 1982. (9 & 12) *Hippocrateaceaedites vancampoae* Ramanujam 1966. (10) *Intectocolporites baculatus* Mandal & Guleria 2006. (11) *Lacrimapollis brownlowioides* Ramanujam 1987. (13) *Lakiapollis assamicus* Tripathi & Singh 1985. (14) *Lakiapollis ovatus* Venkatachala & Kar 1969. (15) *Lacrimapollis pilosus* Venkatachala & Rawat 1973

Plate 16

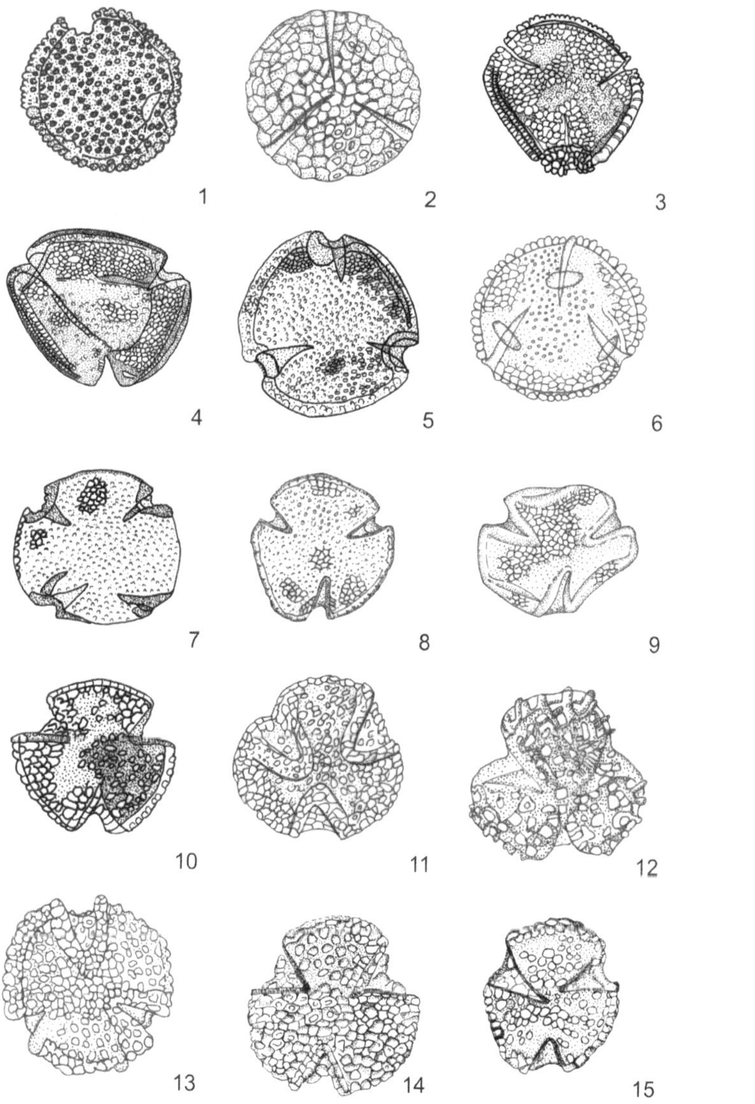

(1) *Lanagiopollis cambayensis* Kumar 1996. (2) *Lanagiopollis meghalayaensis* Tripathi et al. 2000. (3) *Lanagiopollis reitsmae* Phadtare & Thakur in Saxena 2011. (4) *Lanagiopollis rugulatus* Phadtare & Thakur in Saxena 2011. (5) *Lanagiopollis ruguloreticulatus* Phadtare & Thakur in Saxena 2011. (6) *Lanagiopollis subglobosus* Tripathi et al. 2000. (7) *Lanagiopollis tetracolporatus* Kumar 1996. (8) *Margocolporites complexum* Ramanujam 1966. (9) *Margocolporites dubius* Ramanujam 1966. (10) *Margocolporites eocenicus* Samant & Phadtare 1997. (11) *Margocolporites miocenicus* Venkatachala & Rawat 1972. (12) *Margocolporites oligobrochatus* Ramanujam 1966. (13) *Margocolporites prominatus* Mathur & Chopra 1987. (14) *Margocolporites sahnii* Ramanujam 1966. (15) *Margocolporites sitholeyi* Ramanujam 1966

Plate 17

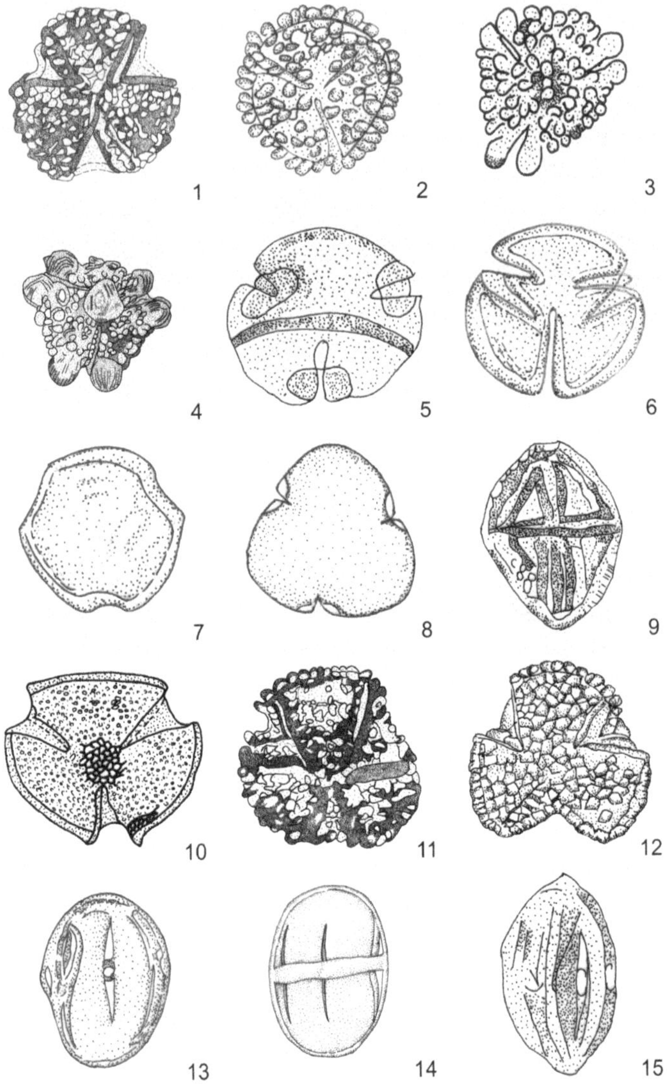

(1) *Margocolporites tsukadae* Ramanujam 1966. (2) *Meyeripollis baksii* Rao & Ramanujam 1982. (3) *Meyeripollis laudabilis* Salujha et al. 1972. (4) *Meyeripollis naharkotensis* Baksi & Venkatachala 1970. (5) *Nyssapollenites barooahii* Sah & Dutta 1968. (6) *Nyssapollenites incertus* Dutta & Sah 1970. (7) *Nyssapollenites kutchensis* Venkatachala & Kar 1969. (8) *Palaeocoprosmadites arcotense* Ramanujam 1966. (9) *Paleoaraliaceaepites indica* Biswas 1962. (10) *Paleoaraliaceaepites indistinctus* Samant & Phadtare 1997. (11) *Paleocaesalpiniaceaepites eocenica* Biswas 1962. (12) *Paleocaesalpiniaceaepites ratnamii* Ramanujam 1966. (13) *Paleosantalaceaepites claricolpus* Venkatachala & Rawat 1972. (14) *Paleosantalaceaepites dinoflagellatus* Biswas 1962 ex Dutta & Sah 1970. (15) *Paleosantalaceaepites ellipticus* Sah & Kar 1970

Plate 18

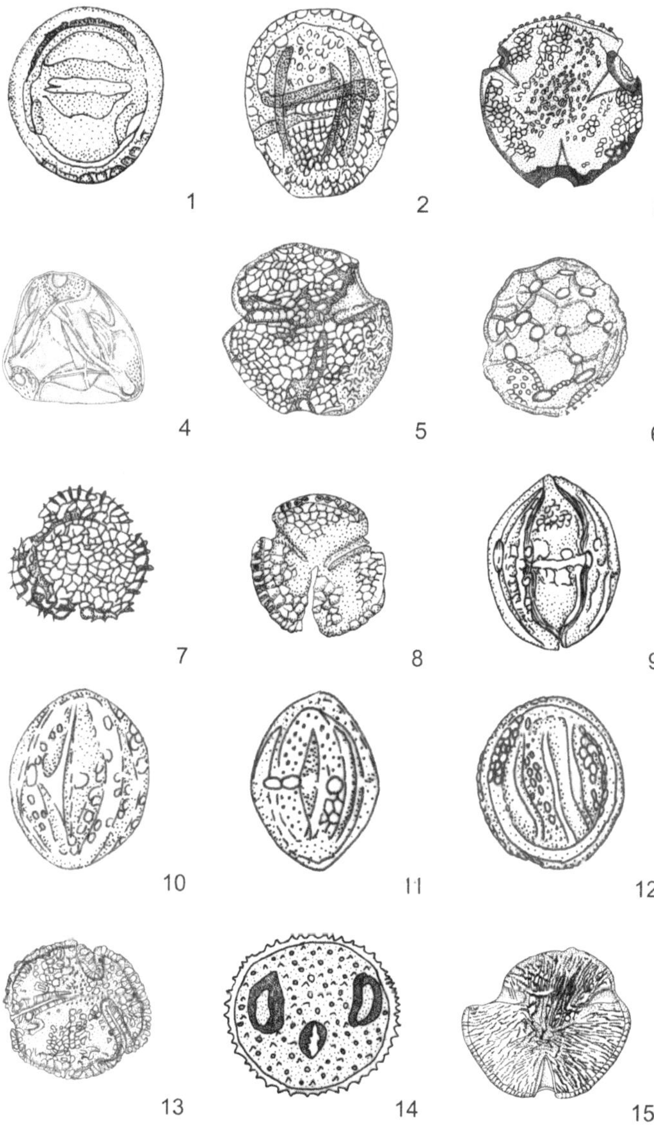

(1) *Paleosantalaceaepites miocenicus* (Ramanujam 1966) Ramanujam et al. 1985. (2) *Paleosantalaceaepites reticulatus* Samant & Phadtare 1997. (3) *Pellicieroipollis langenheimii* Sah & Kar 1970. (4) *Ratariacolporites plicatus* Kar 1985. (5) *Retitrescolpites assamicus* Dutta & Sah 1970. (6) *Polyporina excavatus* Venkatachala & Rawat 1973. (7) *Retitrescolpites minor* Dutta & Sah 1970. (8) *Retitrescolpites singularis* Rao & Ramanujam 1982. (9) *Rhoipites anacardioides* Ramanujam 1987. (10) *Rhoipites baculiferus* Venkatachala & Rawat 1972. (11) *Rhoipites conatus* Venkatachala & Rawat 1972. (12) *Rhoipites kutchensis* Venkatachala & Kar 1969. (13) *Sastriipollenites trilobatus* Venkatachala & Kar 1969. (14) *Spinulotricolporites spinosus* Saxena & Misra 1987. (15) *Striacolporites cephalus* Sah & Kar 1970

Plate 19

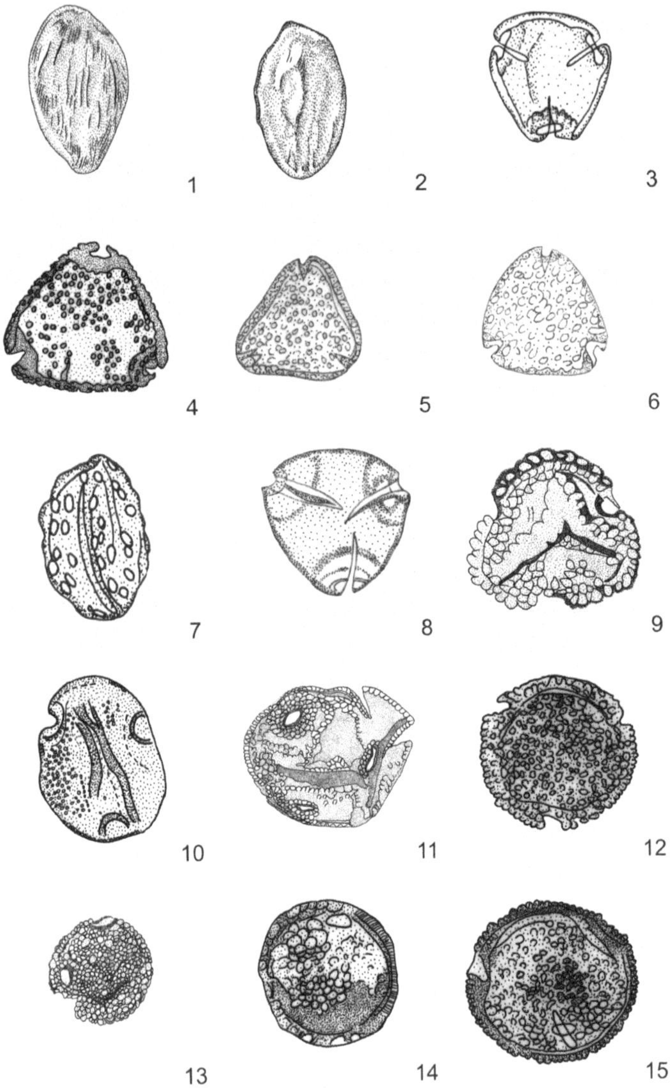

(1) *Striacolporites ovatus* Sah & Kar 1970. (2) *Striacolporites striatus* Sah & Kar 1970. (3) *Symplocoipollenites baksii* Ramanujam 1987. (4) *Symplocoipollenites bhavnagarensis* Samant 2000. (5) *Symplocoipollenites gracilis* Venkatachala & Rawat 1972. (6) *Symplocoipollenites paravurensis* Saxena 1992b. (7) *Polyporina excellens* Sah & Dutta 1968. (8) *Talisiipites wodehousei* Dutta & Sah 1970. (9) *Tinalipollenites duttae* Mandal & Kumar 2000. (10) *Tricolporocolumellites eocenicus* Samant & Phadtare 1997. (11) *Tricolporocolumellites pilatus* Kar 1985. (12) *Tricolporopilites differentialis* Singh & Misra 1991a. (13) *Tricolporopilites robustus* (Kar and Saxena 1981) Kar 1985. (14) *Tricolporopilites tectatus* Singh & Misra 1991a. (15) *Tricolporopilites uniformis* Singh & Misra 1991a

Plate 20

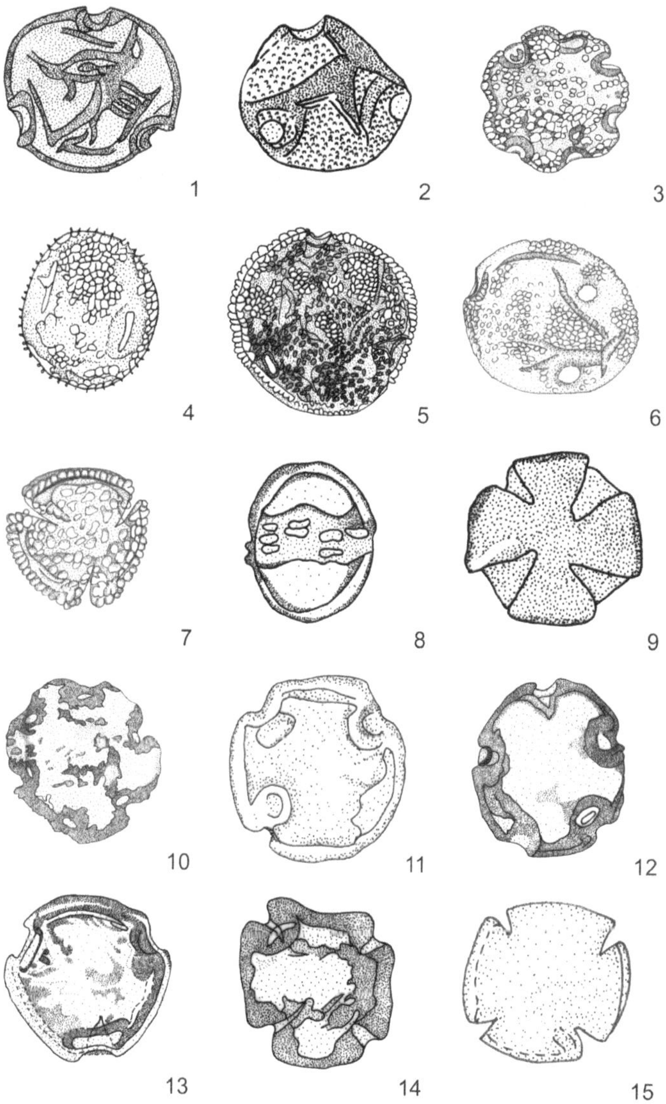

(1) *Tricolporopollis alleppeyensis* Rao & Rajendran 1996. (2) *Tricolporopollis cannanorensis* Rao & Rajendran 1996. (3) *Pseudonothofagidites cerebrus* Venkatachala & Kar 1969. (4) *Tricolporopollis globosa* Dutta & Sah 1970. (5) *Tricolporopollis matanomadhensis* (Venkatachala and Kar 1969) Tripathi & Singh 1985. (6) *Tricolporopollis ruber* (Dutta and Sah 1970) Tripathi & Singh 1985. (7) *Verrucolporites verrucus* Sah & Kar 1970. (8) *Iugopollis tetraporites* Venkatachala & Rawat 1972. (9) *Meliapollis iratus* (Sah and Kar 1974) Navale & Misra 1979. (10) *Pseudonothofagidites kutchensis* Venkatachala & Kar 1969. (11) *Meliapollis melioides* (Ramanujam 1966) Sah & Kar 1970. (12) *Meliapollis navalei* Sah & Kar 1970. (13) *Meliapollis pachydermis* Navale & Misra 1979. (14) *Meliapollis quadrangularis* (Ramanujam 1966) Sah & Kar 1970. (15) *Meliapollis quilonensis* Rao & Ramanujam 1982

Plate 21

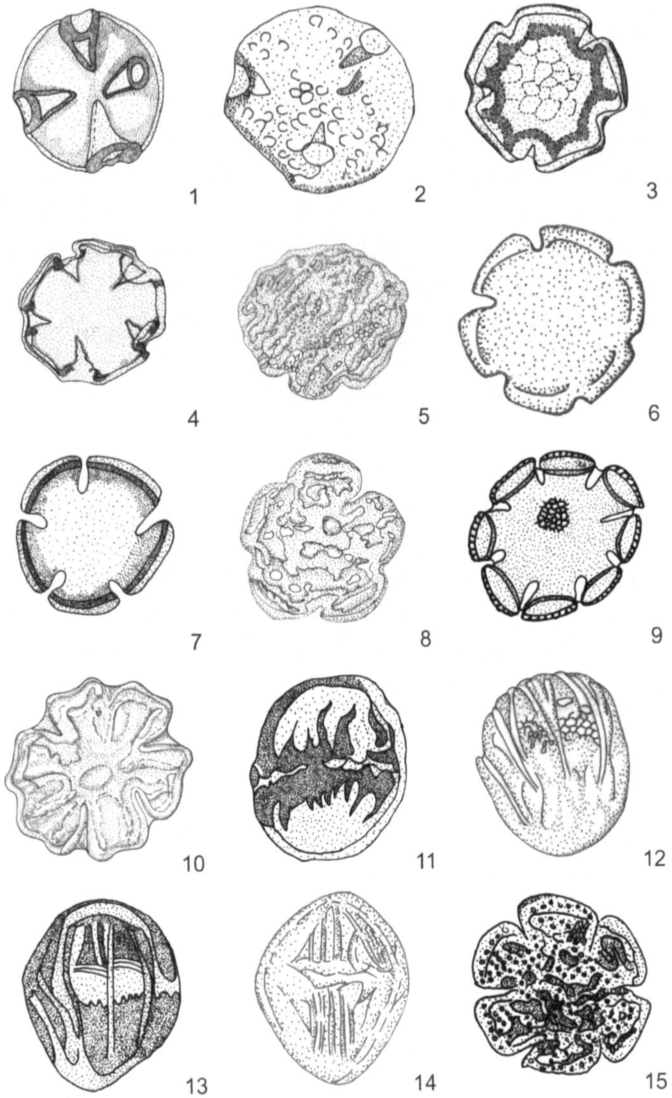

(1) *Meliapollis ramanujamii* Sah & Kar 1970. (2) *Meliapollis raoi* Sah & Kar 1970. (3) *Meliapollis simplex* Navale & Misra 1979. (4) *Meliapollis tamilii* Navale & Misra 1979. (5) *Padappakkarapollis venkatachalae* Rao & Ramanujam 1982. (6) *Polybrevicolporites antiquum* (Ramanujam 1966) Venkatachala & Kar 1969. (7) *Polybrevicolporites cephalus* Venkatachala & Kar 1969. (8) *Polybrevicolporites karii* Rao & Ramanujam 1982. (9) *Polybrevicolporites reticulata* Samant 2000. (10) *Polycolporites indicus* Mehrotra 1983. (11) *Polygalacidites clarus* Sah & Dutta 1966. (12) *Polygalacidites gujratensis* Kar 1985. (13) *Polygalacidites magnus* Samant & Phadtare 1997. (14) *Polygalacidites rhomboidus* Kar & Bhattacharya 1992. (15) *Prectenolophonidites primitiva* Samant & Phadtare 1997

Plate 22

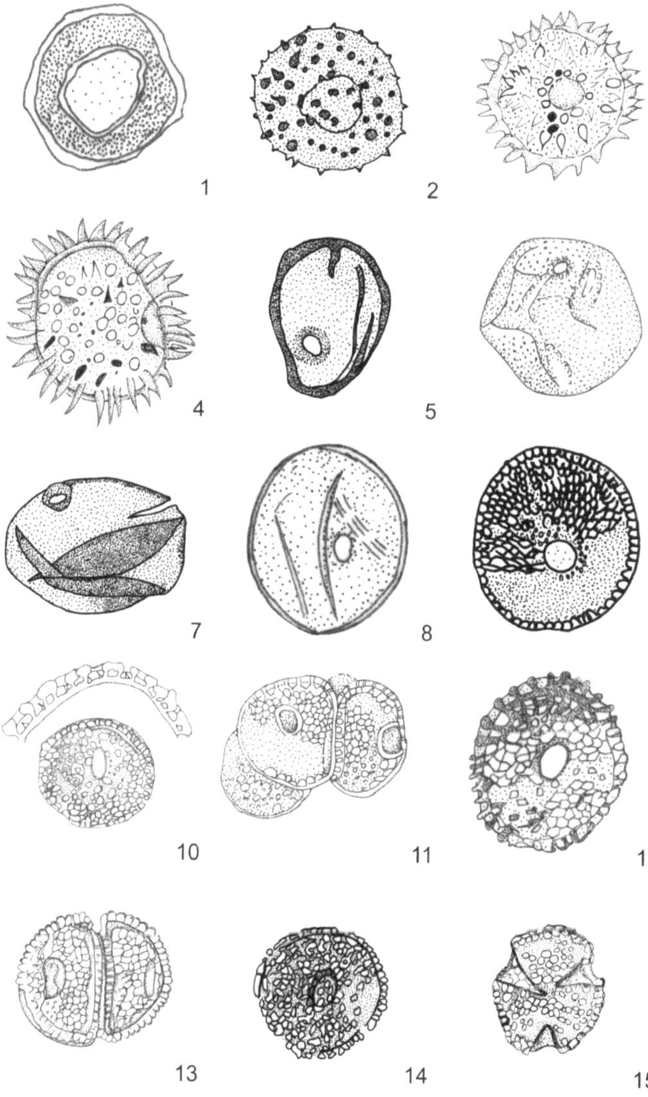

(1) *Dracaenoipollis circularis* Sah & Kar 1970. (2) *Echimonoporopollis circularis* Samant & Tapaswi 2001. (3) *Echimonoporopollis grandiporus* Saxena et al. 1991. (4) *Echimonoporopollis neyveliensis* Saxena et al. 1991. (5) *Graminidites assamicus* Sah & Dutta 1968. (6) *Graminidites granulatus* Kar 1985. (7) *Graminidites maximus* Tripathi & Singh 1985. (8) *Graminidites pliocenicus* Singh & Saxena 1981. (9) *Jacobipollenites arthugalensis* Rao 1990. (10) *Jacobipollenites crassimurus* Singh & Misra 1991b. (11) *Jacobipollenites distinctus* Singh & Misra 1991b. (12) *Jacobipollenites magnificus* Ramanujam 1966. (13) *Jacobipollenites magniporus* Singh & Misra 1991b. (14) *Jacobipollenites ramanujamii* Samant 2000. (15) *Nymphaeacidites clarus* Dutta & Sah 1970

Plate 23

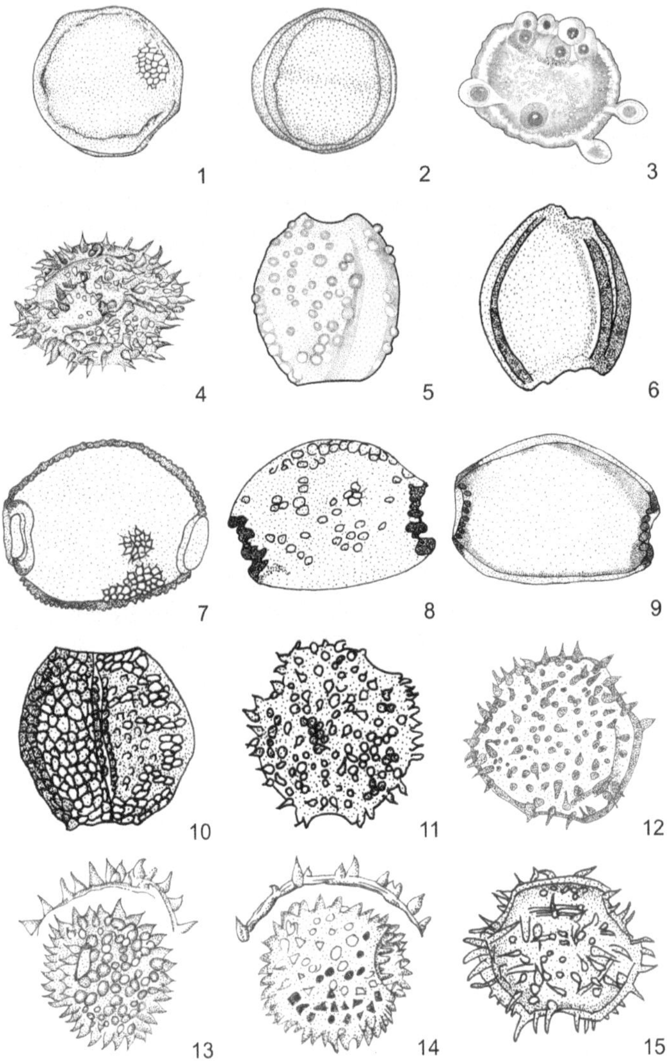

(1) *Nymphaeacidites giganteus* Mehrotra 1983. (2) *Nymphaeacidites sphericus* Dutta & Sah 1970. (3) *Clavadiporopollenites raneriensis* Ambwani & Singh 1996. (4) *Echidiporites indicus* Mandal 1997. (5) *Piladiporocolpites caratinii* Kar 1995b. (6) *Psiladiporocolpites pachyexinus* Kar 1995b. (7) *Psilodiporites bengalensis* (Varma and Rawat 1963) Venkatachala & Rawat 1972. (8) *Psilodiporites erdtmanii* Venkatachala & Rawat 1972. (9) *Psilodiporites hammenii* Varma & Rawat 1963. (10) *Retidiporocolpites excellensus* Kar 1995b. (11) *Acanthotricolpites brevispinosus* Saxena & Khare 2004. (12) *Acanthotricolpites bulbospinosus* (Kar 1985) Singh & Misra 1991c. (13) *Acanthotricolpites complexus* Singh & Misra 1991c. (14) *Acanthotricolpites intermedius* Singh & Misra 1991c. (15) *Acanthotricolpites karii* Saxena & Khare 2004

Plate 24

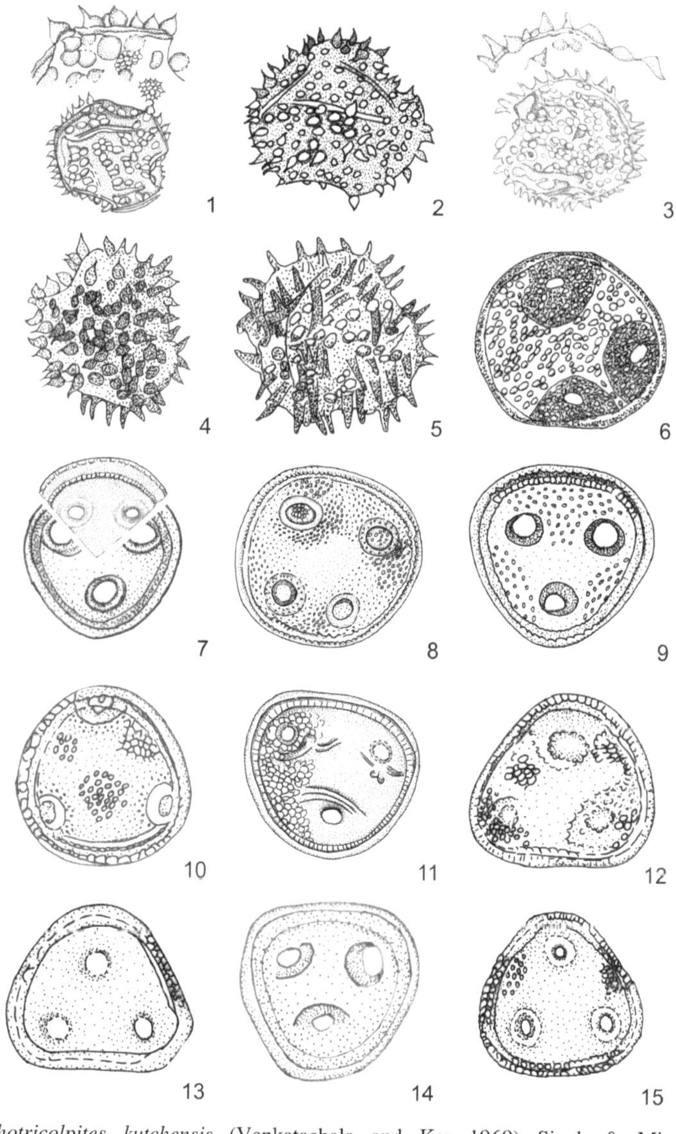

(1) *Acanthotricolpites kutchensis* (Venkatachala and Kar 1969) Singh & Misra 1991c. (2) *Acanthotricolpites microreticulatus* Saxena & Khare 2004. (3) *Acanthotricolpites multitypicus* Singh & Misra 1991c. (4) *Acanthotricolpites robustus* Saxena & Khare 2004. (5) *Acanthotricolpites tiruchirapalliensis* Saxena & Khare 2004. (6) *Dorreenipites bengalensis* Mathur & Chopra 1987. (7) *Dorreenipites cooksoniae* (Ramanujam 1966) Navale & Misra 1979. (8) *Dorreenipites distinctus* Navale & Misra 1979. (9) *Dorreenipites erdtmanii* (Ramanujam 1966) Navale & Misra 1979. (10) *Dorreenipites kutchensis* (Venkatachala and Kar 1969) Navale & Misra 1979. (11) *Dorreenipites medius* Navale & Misra 1979. (12) *Dorreenipites neyveliensis* Misra et al. 1996. (13) *Dorreenipites psilatus* Misra et al. 1996. (14) *Dorreenipites sellingii* (Ramanujam 1966) Navale & Misra 1979. (15) *Dorreenipites undulatus* Misra et al. 1996

Plate 25

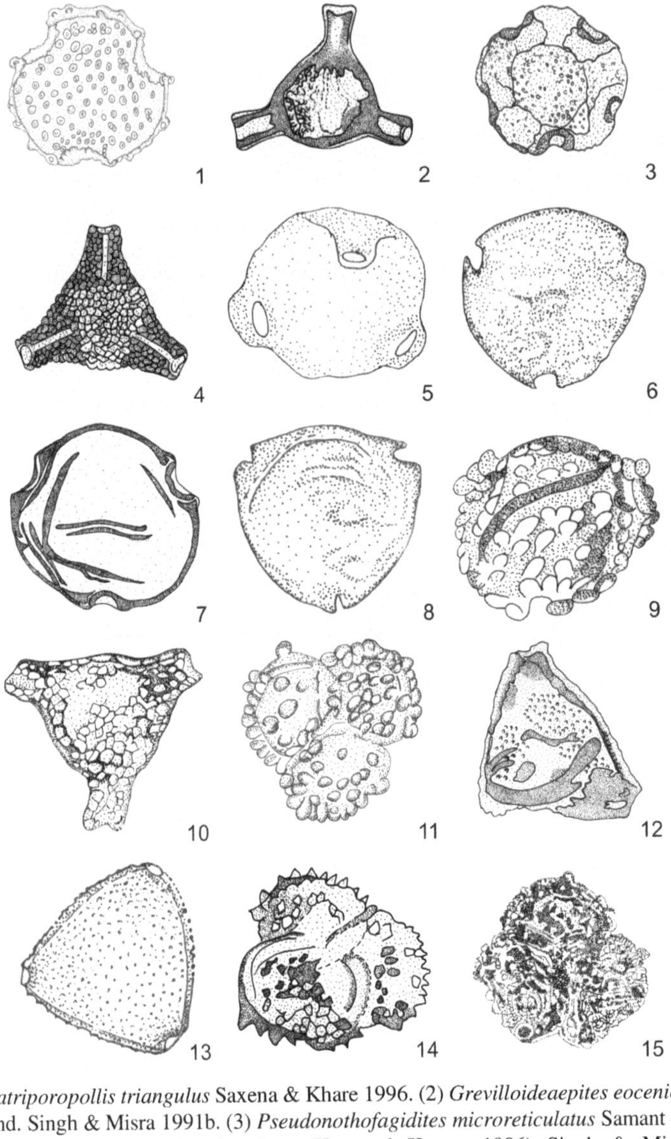

(1) *Gemmatriporopollis triangulus* Saxena & Khare 1996. (2) *Grevilloideaepites eocenicus* Biswas 1962 emend. Singh & Misra 1991b. (3) *Pseudonothofagidites microreticulatus* Samant & Tapaswi 2001. (4) *Grevilloideaepites pachyexinus* (Kar and Kumar 1986) Singh & Misra 1991b. (5) *Grevilloideaepites trilobatus* Mathur & Chopra 1987. (6) *Myricipites harrisii* (Couper 1953) Dutta & Sah 1970. (7) *Myricipites singhii* Rao 1995. (8) *Myricipites vulgaris* Dutta & Sah 1970. (9) *Ornatriporites elegans* Rao & Ramanujam 1982. (10) *Proteacidites excertus* Dutta & Sah 1970. (11) *Ornatetradites chandae* Rao & Ramanujam 1982. (12) *Proteacidites protrudus* Sah & Kar 1970. (13) *Proteacidites triangulus* Kar & Jain 1981. (14) *Ornatetradites droseroides* Rao & Ramanujam 1982. (15) *Ornatetradites keralaensis* Rao 1995

Plate 26

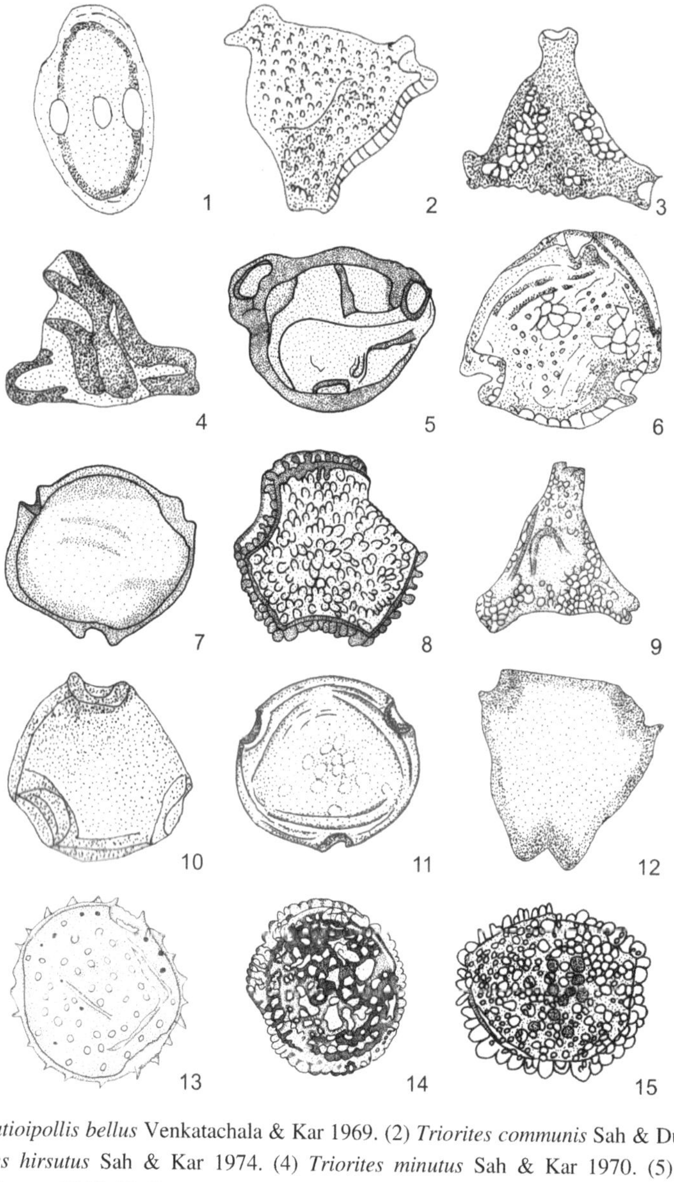

(1) *Sonneratioipollis bellus* Venkatachala & Kar 1969. (2) *Triorites communis* Sah & Dutta 1966. (3) *Triorites hirsutus* Sah & Kar 1974. (4) *Triorites minutus* Sah & Kar 1970. (5) *Triorites protrudus* Samant 2000. (6) *Triorites pseudoreticulatus* Venkatachala & Rawat 1972. (7) *Triorites sahii* Samant 2000. (8) *Triorites triangulus* Sah & Kar 1970. (9) *Triorites tubiferus* Venkatachala & Rawat 1972. (10) *Triporopollenites meenkunnuensis* Rao & Rajendran 1996. (11) *Triporopollenites robustus* Kar & Jain 1981. (12) *Triporopollenites vimalii* Sah & Dutta 1966. (13) *Varispinitriporites ratariaensis* (Kar & Saxena 1981) Kar 1985. (14) *Clavaperiporites clavatus* Navale & Misra 1979. (15) *Clavaperiporites heteroclavatus* Samant & Phadtare 1997

The manufacturer's authorised representative in the EU is Springer Nature Customer Service Centre GmbH, Europaplatz 3, 69115 Heidelberg, Germany. If you have any concerns regarding our products, please contact ProductSafety@springernature.com

Printed and bound by CPI Group (UK) Ltd, Croydon, CR0 4YY
22/09/2026
02221060-0008